T0297529

RELIABILITY, MAINTAINABILITY AND RISK

Also by the same author

Reliability Engineering, Pitman, 1972
Maintainability Engineering, Pitman, 1973 (with A. H. Babb)
Statistics Workshop, Technis, 1974, 1991
Achieving Quality Software, Chapman & Hall, 1995
Quality Procedures for Hardware and Software, Elsevier, 1990 (with J. S. Edge)
Functional Safety: A Straightforward Guide to IEC 61508, 2nd Edition, Butterworth-Heinemann, 2004, ISBN 0 7506 6269 7 (with K. G. L. Simpson)
The Private Pilot's Little Book of Helicopter Safety, Technis, 2010, ISBN 9780951656297

Reliability, Maintainability and Risk

Practical methods for engineers

Eighth Edition

Dr David J Smith
BSc, PhD, CEng, FIET, FCQI, HonFSaRS MIGEM

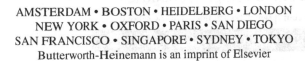

AMSTERDAM • BOSTON • HEIDELBERG • LONDON
NEW YORK • OXFORD • PARIS • SAN DIEGO
SAN FRANCISCO • SINGAPORE • SYDNEY • TOKYO
Butterworth-Heinemann is an imprint of Elsevier

Butterworth-Heinemann
The Boulevard, Langford Lane, Kidlington, Oxford OX5 1GB, UK
225 Wyman Street, Waltham, MA 02451, USA

Eighth edition 2011

First published by Macmillan Education Ltd 1981
Second edition 1985
Third edition 1988
Fourth edition published by Butterworth-Heinemann Ltd 1993
Reprinted 1994, 1996
Fifth edition 1997
Reprinted with revisions 1999
Sixth edition 2001
Reprinted 2002, 2003 (twice)
Seventh edition 2005
Reprinted 2007

British Library Cataloguing in Publication Data
A catalogue record for this book is available from the British Library

Library of Congress Cataloging-in-Publication Data
A catalog record for this book is availabe from the Library of Congress

For information on all Butterworth-Heinemann
publications visit our web site at books.elsevier.com

ISBN 978-0-08-096902-2

Printed and bound in Great Britain

11 12 13 14 15 10 9 8 7 6 5 4 3 2 1

Contents

Preface

After three editions, in 1993, *Reliability, Maintainability in Perspective* became *Reliability, Maintainability and Risk*. The 6th edition, in 2001, included my PhD studies into common cause failure and into the correlation between predicted and achieved field reliability. Once again it is time to update the material as a result of developments in the functional safety area.

The techniques that are explained apply to both reliability and safety engineering and are also applied to optimizing maintenance strategies. The collection of techniques concerned with reliability, availability, maintainability and safety are often referred to as RAMS.

A single defect can easily cost £100 in diagnosis and repair if it is detected early in production, whereas the same defect in the field may well cost £1000 to rectify. If it transpires that the failure is a design fault then the cost of redesign, documentation and retest may well be in tens or even hundreds of thousands of pounds. This book emphasizes the importance of using reliability techniques to discover and remove potential failures early in the design cycle. Compared with such losses, the cost of these activities is easily justified.

It is the combination of reliability and maintainability that dictates the proportion of time that any item is available for use or, for that matter, is operating in a safe state. The key parameters are failure rate and down time, both of which determine the failure costs. As a result, techniques for optimizing maintenance intervals and spares holdings have become popular since they lead to major cost savings.

'RAMS' clauses in contracts, and in invitations to tender, are now commonplace. In defense, telecommunications, oil and gas, and aerospace these requirements have been specified for many years. More recently the transport, medical and consumer industries have followed suit. Furthermore, recent legislation in the liability and safety areas provides further motivation for this type of assessment. Much of the activity in this area is the result of European standards and these are described where relevant.

Software tools have been in use for RAMS assessments for many years and only the simplest of calculations are performed manually. This eighth edition mentions a number of such packages. Not only are computers of use in carrying out reliability analysis but are themselves the subject of concern. The application of programable devices in control equipment, and in particular safety-related equipment, has widened dramatically since the mid-1980s. The reliability/quality of the software and the ways in which it could cause failures and hazards is of considerable interest.

Chapters 17 and 22 cover this area.

Quantifying the predicted RAMS, although important in pinpointing areas for redesign, does not of itself create more reliable, safer or more easily repaired equipment. Too often, the author has to discourage efforts to refine the 'accuracy' of a reliability prediction when an order of magnitude assessment would have been adequate. In any engineering discipline the ability to recognize the degree of accuracy required is of the essence. It happens that RAMS parameters are of wide tolerance and thus judgements must be made on the basis of one- or, at best, two-figure accuracy. Benefit is only obtained from the judgement and subsequent follow-up action, not from refining the calculation.

A feature of the last four editions has been the data ranges in Appendices 3 and 4. These were current for the fourth edition but the full 'up-to-date' database is available in FARADIP. THREE (see last four pages of the book).

DJS

Acknowledgements

Especial thanks to my good friend and colleague Derek Green (who is both a chartered engineer and a barrister) for a thorough overhaul of Chapters 19, 20 and 21 and for valuable updates including a section on Corporate Manslaughter.

I would also particularly like to thank the following friends and colleagues for their help and encouragement in respect of earlier editions:

Ken Simpson and Bill Gulland for their work on repairable systems modelling, the results of which have had a significant effect on Chapter 8 and Appendix 13.

'Sam' Samuel for his very thorough comments and assistance on a number of chapters.

Peter Joyce for his considerable help with earlier editions.

I would also like to thank:

The British Standards Institution for permission to reproduce the lightning map of the UK from BS 6651. The Institution of Gas Engineers and Managers for permission to make use of examples from their guidance document (SR/24, Risk Assessment Techniques).

PART 1

Understanding Reliability Parameters and Costs

CHAPTER 1

The History of Reliability and Safety Technology

Safety/Reliability engineering did not develop as a unified discipline, but grew out of the integration of a number of activities, previously the province of various branches of engineering.

Since no human activity can enjoy zero risk, and no equipment has a zero rate of failure, there has emerged a safety technology for optimizing risk. This attempts to balance the risk of a given activity against its benefits and seeks to assess the need for further risk reduction depending upon the cost.

Similarly, reliability engineering, beginning in the design phase, attempts to select the design compromise that balances the cost of reducing failure rates against the value of the enhanced performance.

The abbreviation RAMS is frequently used for ease of reference to reliability, availability, maintainability and safety-integrity.

1.1 Failure Data

Throughout the history of engineering, reliability improvement (also called reliability growth), arising as a natural consequence of the analysis of failure, has long been a central feature of development. This 'test and correct' principle was practiced long before the development of formal procedures for data collection and analysis for the reason that failure is usually self-evident and thus leads, inevitably, to design modifications.

The design of safety-related systems (for example, railway signaling) has evolved partly in response to the emergence of new technologies but largely as a result of lessons learnt from failures. The application of technology to hazardous areas requires the formal application of this feedback principle in order to maximize the rate of reliability improvement. Nevertheless, as mentioned above, all engineered products will exhibit some degree of reliability growth even without formal improvement programs.

Nineteenth- and early twentieth-century designs were less severely constrained by the cost and schedule pressures of today. Thus, in many cases, high levels of reliability were achieved as a result of over-design. The need for quantified reliability assessment techniques during the design and development phase was not therefore identified.

Reliability, Maintainability and Risk. DOI: 10.1016/B978-0-08-096902-2.00001-5

Therefore, failure rates of engineered components were not required, as they are now, for use in prediction techniques and consequently there was little incentive for the formal collection of failure data.

Another factor is that, until well into the twentieth century, component parts were individually fabricated in a 'craft' environment. Mass production, and the attendant need for component standardization, did not apply and the concept of a valid repeatable component failure rate could not exist. The reliability of each product was highly dependent on the craftsman/manufacturer and less determined by the 'combination' of component reliabilities.

Nevertheless, mass production of standard mechanical parts has been the case for over a hundred years. Under these circumstances defective items can be readily identified, by inspection and test, during the manufacturing process, and it is possible to control reliability by quality-control procedures.

The advent of the electronic age, accelerated by the Second World War, led to the need for more complex mass-produced component parts with a higher degree of variability in the parameters and dimensions involved. The experience of poor field reliability of military equipment throughout the 1940s and 1950s focused attention on the need for more formal methods of reliability engineering. This gave rise to the collection of failure information from both the field and from the interpretation of test data. Failure rate databanks were created in the mid-1960s as a result of work at such organizations as UKAEA (UK Atomic Energy Authority) and RRE (Royal Radar Establishment, UK) and RADC (Rome Air Development Corporation, US).

The manipulation of the data was manual and involved the calculation of rates from the incident data, inventories of component types and the records of elapsed hours. This was stimulated by the advent of reliability prediction modeling techniques that require component failure rates as inputs to the prediction equations.

The availability and low cost of desktop personal computing (PC) facilities, together with versatile and powerful software packages, has permitted the listing and manipulation of incident data with an order of magnitude less effort. Fast automatic sorting of data encourages the analysis of failures into failure modes. This is no small factor in contributing to more effective reliability assessment, since raw failure rates permit only parts count reliability predictions. In order to address specific system failures it is necessary to input specific component failure modes into the fault tree or failure mode analyses.

The requirement for field recording makes data collection labor intensive and this remains a major obstacle to complete and accurate information. Motivating staff to provide field reports with sufficient relevant detail is an ongoing challenge for management. The spread of PC facilities in this area will assist in that interactive software can be used to stimulate the required information input at the same time as other maintenance-logging activities.

With the rapid growth of built-in test and diagnostic features in equipment, a future trend ought to be the emergence of automated fault reporting.

Failure data have been published since the 1960s and each major document is described in Chapter 4.

1.2 Hazardous Failures

In the early 1970s the process industries became aware that, with larger plants involving higher inventories of hazardous material, the practice of learning by mistakes was no longer acceptable. Methods were developed for identifying hazards and for quantifying the consequences of failures. They were evolved largely to assist in the decision-making process when developing or modifying plants. External pressures to identify and quantify risk were to come later.

By the mid-1970s there was already concern over the lack of formal controls for regulating those activities which could lead to incidents having a major impact on the health and safety of the general public. The Flixborough incident in June 1974 resulted in 28 deaths and focused public and media attention on this area of technology. Successive events such as the tragedy at Seveso in Italy in 1976 right through to the Piper Alpha offshore and more recent Paddington rail and Texaco Oil Refinery incidents have kept that interest alive and resulted in guidance and legislation, which are addressed in Chapters 19 and 20.

The techniques for quantifying the predicted frequency of failures were originally applied to assessing plant availability, where the cost of equipment failure was the prime concern. Over the last twenty years these techniques have also been used for hazard assessment. Maximum tolerable risks of fatality have been established according to the nature of the risk and the potential number of fatalities. These are then assessed using reliability techniques. Chapter 10 deals with risk in more detail.

1.3 Reliability and Risk Prediction

System modeling, using failure mode analysis and fault tree analysis methods, has been developed over the last thirty years and now involves numerous software tools which enable predictions to be updated and refined throughout the design cycle. The criticality of the failure rates of specific component parts can be assessed and, by successive computer runs, adjustments to the design configuration (e.g. redundancy) and to the maintenance philosophy (e.g. proof test frequencies) can be made early in the design cycle in order to optimize reliability and availability. The need for failure rate data to support these predictions has therefore increased and Chapter 4 examines the range of data sources and addresses the problem of variability within and between them.

The value and accuracy of reliability prediction, based on the concept of validly repeatable component failure rates, has long been controversial.

First, the extremely wide variability of failure rates of allegedly identical components, under supposedly identical environmental and operating conditions, is now acknowledged. The apparent precision offered by reliability prediction models is thus not compatible with the accuracy of the failure rate parameter. As a result, it can be argued that simple assessments of failure rates and the use of simple models suffice. In any case, more accurate predictions can be both misleading and a waste of money.

The main benefit of reliability prediction of complex systems lies not in the absolute figure predicted but in the ability to repeat the assessment for different repair times, different redundancy arrangements in the design configuration and different values of component failure rate. This has been made feasible by the emergence of PC tools (e.g. fault tree analysis packages) that permit rapid reruns of the prediction. Thus, judgements can be made on the basis of relative predictions with more confidence than can be placed on the absolute values.

Second, the complexity of modern engineering products and systems ensures that system failure is not always attributable to single component part failure. More subtle factors, such as the following, can often dominate the system failure rate:

- failure resulting from software elements
- failure due to human factors or operating documentation
- failure due to environmental factors
- failure whereby redundancy is defeated by factors common to the replicated units
- failure due to ambiguity in the specification
- failure due to timing constraints within the design
- failure due to combinations of component parameter tolerance.

The need to assess the integrity of systems containing substantial elements of software has increased steadily since the 1980s. The concept of validly repeatable 'elements' within the software, which can be mapped to some model of system reliability (i.e. failure rate), is even more controversial than the hardware reliability prediction processes discussed above. The extrapolation of software test failure rates into the field has not yet established itself as a reliable modeling technique. Software metrics that enable failure rate to be predicted from measurable features of the code or design are equally elusive.

Reliability prediction techniques, however, are mostly confined to the mapping of component failures to system failure and do not address these additional factors. Methodologies are currently evolving to model common mode failures, human factor failures and software failures, but there is no evidence that the models that emerge will enjoy any greater precision than the existing reliability predictions based on hardware component failures. In any case the mental discipline involved in setting up a reliability model helps the designer to understand the architecture and can be as valuable as the numerical outcome.

Figure 1.1 illustrates the relationship between a component failure rate based reliability or risk prediction and the eventual field performance. In practice, prediction addresses the component-based 'design reliability', and it is necessary to take account of the additional factors when assessing the integrity of a system.

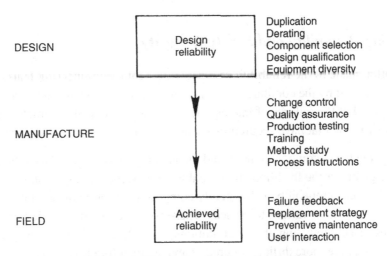

Figure 1.1: 'Design' v. 'achieved' reliability

In fact, Figure 1.1 gives some perspective to the idea of reliability growth. The 'design reliability' is likely to be the figure suggested by a prediction exercise. However, there will be many sources of failure in addition to the simple random hardware failures predicted in this way. Thus the 'achieved reliability' of a new product or system is likely to be an order, or even more, less than the 'design reliability'. Reliability growth is the improvement that takes place as modifications are made as a result of field failure information. A well-established item, perhaps with tens of thousands of field hours, might start to approach the 'design reliability'. Section 12.3 deals with methods of plotting and extrapolating reliability growth.

As a result of the problem, whereby systematic failures cannot necessarily be quantified, it has become generally accepted that it is necessary to consider qualitative defenses against systematic failures as an additional, and separate, activity to the task of predicting the probability of so-called random hardware failures. Thus, two approaches are taken and exist side by side.

1. *Quantitative assessment*: where we predict the frequency of hardware failures and compare them with some target. If the target is not satisfied then the design is adapted (e.g. provision of more redundancy) until the target is met.
2. *Qualitative assessment*: where we attempt to minimize the occurrence of systematic failures (including software related failures) by applying a variety of defenses and design disciplines appropriate to the severity of the target.

The question arises as to how targets can be expressed for the latter (qualitative) approach. The concept is to divide the 'spectrum' of integrity into a number of discrete levels (usually four) and then to lay down requirements for each level. In the safety context these are referred to as SILs and are dealt with in Chapter 22. Clearly, the higher the integrity level then the more stringent the requirements become.

1.4 Achieving Reliability and Safety-Integrity

Reference is often made to the reliability of nineteenth-century engineering feats. Telford and Brunel are remembered by the continued existence of the Menai and Clifton bridges. However, little is remembered of the failures of that age. If we try to identify the characteristics of design and construction that have secured this longevity then three factors emerge:

1. *Complexity*: the fewer component parts and the fewer types of material used then, in general, the greater is the likelihood of a reliable item. Modern equipment, until recently condemned for its unreliability, is frequently composed of thousands of component parts all of which interact within various tolerances. These could be called intrinsic failures, since they arise from a combination of drift conditions rather than the failure of a specific component. They are more difficult to predict and are therefore less likely to be foreseen by the designer. This leads to the qualitative approach involving the rigor of life-cycle techniques mentioned in the previous section. Telford's and Brunel's structures are not complex and are composed of fewer types of material with relatively well-proven modules.

2. *Duplication/replication*: the use of additional, redundant, parts whereby a single failure does not cause the overall system to fail is a method of achieving reliability. It is probably the major design feature that determines the order of reliability that can be obtained. Nevertheless, it adds capital cost, weight, maintenance and power consumption. Furthermore, reliability improvement from redundancy often affects one failure mode at the expense of another type of failure. This is emphasized by an example in the next chapter.

3. *Excess strength*: deliberate design to withstand stresses higher than are anticipated will reduce failure rates. Small increases in strength for a given anticipated stress result in substantial improvements. This applies equally to mechanical and electrical items. Modern commercial pressures lead to the optimization of tolerance and stress margins that just meet the functional requirement. The probability of the tolerance-related failures mentioned above is thus further increased.

The latter two of the above methods are costly and, as will be discussed in Chapter 3, the cost of reliability improvements needs to be paid for by a reduction in failure and operating costs. This argument is not quite so simple for hazardous failures but, nevertheless, there is never an endless budget for improvement and some consideration of cost is inevitable (e.g. cost per life saved).

We can see therefore that reliability and safety are 'built-in' features of a design, be it mechanical, electrical or structural. Maintainability also contributes to the availability of a system, since it is the combination of failure rate and repair/down time that determines unavailability. The design and operating features that influence down time are also taken into account in this book.

Achieving reliability, safety and maintainability results from activities in three main areas.

1. *Design:*
      reduction in complexity
      duplication to provide fault tolerance
      derating of stress factors
      qualification testing and design review
      feedback of failure information to provide reliability growth.
2. *Manufacture:*
      control of materials, methods, changes
      control of work methods and standards.
3. *Field use:*
      adequate operating and maintenance instructions
      feedback of field failure information
      proof testing to reveal dormant failures
      replacement and spares strategies (e.g. early replacement of items with a known wearout characteristic).

It is much more difficult, and expensive, to add reliability/safety after the design stage. The quantified parameters, dealt with in Chapter 2, must be part of the design specification and can no more sensibly be specified retrospectively than power consumption, weight, signal-to-noise ratio, etc.

1.5 The RAMS Cycle

The life-cycle model shown in Figure 1.2 provides a visual link between RAMS activities and a typical design cycle. The top portion shows the specification and feasibility stages of design leading to conceptual engineering and then to detailed design.

RAMS targets should be included in the requirements specification as project or contractual requirements that can include both assessment of the design and demonstration of performance. This is particularly important since, unless called for contractually, RAMS targets may otherwise be perceived as adding to time and budget and there will be little other incentive, within the project, to specify them. Since each different system failure mode will be caused by different parts failures, it is important to realize the need for separate targets for each undesired system failure mode.

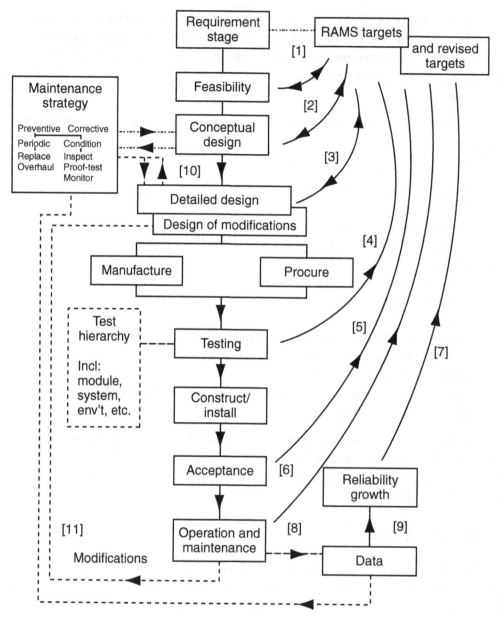

Figure 1.2: RAMS-cycle model

Because one purpose of the feasibility stage is to decide if the proposed design is viable (given the current state of the art) then the RAMS targets can sometimes be modified at that stage, if initial predictions show them to be unrealistic. Subsequent versions of the requirements specification would then contain revised targets, for which revised RAMS predictions will be required.

The feedback loops shown in Figure 1.2 represent RAMS-related activities as follows:

- A review of the system RAMS feasibility calculations against the initial RAMS targets (loop [1]).
- A formal (documented) review of the conceptual design RAMS predictions against the RAMS targets (loop [2]).
- A formal (documented) review, of the detailed design, against the RAMS targets (loop [3]).
- A formal (documented) design review of the RAMS tests, at the end of design and development, against the requirements (loop [4]). This is the first opportunity (usually somewhat limited) for some level of real demonstration of the project/contractual requirements.
- A formal review of the acceptance demonstration, which involves RAMS tests against the requirements (loop [5]). These are frequently carried out before delivery but would preferably be extended into, or even totally conducted in, the field (loop [6]).
- An ongoing review of field RAMS performance against the targets (loops [7,8,9]) including subsequent improvements.

9005 Not every one of the above review loops will be applied to each contract and the extent of review will depend on the size and type of project.

Test, although shown as a single box in this simple RAMS-cycle model, will usually involve a test hierarchy consisting of component, module, subsystem and system tests. These must be described in the project documentation.

The maintenance strategy (i.e. maintenance program) is relevant to RAMS since both preventive and corrective maintenance affect reliability and availability. Repair times influence unavailability as do preventive maintenance parameters. Loop [10] shows that maintenance is considered at the design stage where it will impact on the RAMS predictions. At this point the RAMS predictions can begin to influence the planning of maintenance strategy (e.g. periodic replacements/overhauls, proof-test inspections, auto-test intervals, spares levels, number of repair crews).

For completeness, the RAMS-cycle model also shows the feedback of field data into a reliability growth programme and into the maintenance strategy (loops [8], [9] and [11]). Sometimes the growth program is a contractual requirement and it may involve targets beyond those in the original design specification.

1.6 Contractual and Legal Pressures

As a direct result of the reasons discussed above, it is now common for reliability (including safety) parameters to be specified in invitations to tender and other contractual documents. Failure rates, probabilities of failure on demand, availabilities, and so on, are specified and quantified for both cost- and safety-related failure modes.

This is for two main reasons:

1. *Cost of failure*: failure may lead to huge penalty costs. The halting of industrial processes can involve the loss of millions of pounds per week. Rail and other transport failures can each involve hundreds of thousands of pounds in penalty costs. Therefore system availability is frequently specified as part of the functional requirements.
2. *Legal implications*: there are various legal and implied legal reasons (Chapters 19–21), including fear of litigation, for specifying safety-related parameters (e.g. failure rates, safety integrity levels) in contracts.

There are problems in such contractual relationships arising from:
ambiguity in defining the terms used
hidden statistical risks
inadequate coverage of the requirements
unrealistic requirements
unmeasurable requirements.

These reliability/safety requirements are dealt with in two broad ways:

1. *Demonstration of a black box specification*: a failure rate might be stated and items accepted or rejected after some reliability demonstration test. This is suitable for stating a quantified reliability target for simple component items or equipment where the combination of quantity and failure rate makes the actual demonstration of failure rates realistic.
2. *Ongoing design and project approval*: in this case, design methods, reliability predictions during design, reviews and quality methods, as well as test strategies, are all subject to agreement and audit throughout the project. This approach is applicable to complex systems with long development cycles, and particularly relevant where the required reliability is of such a high order that even zero failures in a foreseeable time frame are insufficient to demonstrate that the requirement has been met. In other words, zero failures in 10 equipment years proves nothing when the required reliability is a mean time between failures of 100 years.

In practice, a combination of these approaches is used and the various pitfalls are covered in the following chapters of this book.

CHAPTER 2

Understanding Terms and Jargon

2.1 Defining Failure and Failure Modes

Before introducing the various reliability parameters it is essential that the word *failure* is fully defined and understood. Unless the failed state of an item is defined, it is impossible to define a meaning for quality or reliability. There is only one definition of failure and that is:

Non-conformance to some defined performance criterion

Refinements that differentiate between terms such as defect, malfunction, failure, fault and reject are sometimes important in contract clauses, and in the classification and analysis of data, but should not be allowed to cloud the issue. These various terms merely include and exclude failures by type, cause, degree or use. For any one specific definition of failure there is no ambiguity in the definition of reliability. Since failure is defined as departure from specification then it follows that revising a definition of failure implies a change to the performance specification. This is best explained by the following example.

Consider Figure 2.1, which shows two valves in physical series in a process line. If the reliability of this 'system' is to be assessed, then one might ask for the failure rate of the individual valves. The response could be, say, 15 failures per million hours (slightly less than one failure per 7 years). One inference would be that the total 'system' reliability is 30 failures per million hours. However, life is not so simple.

If 'loss of supply' from this process line is being considered then the system failure rate is higher than for a single valve, owing to the series nature of the configuration. In fact it is double the failure rate of one valve. Since, however, 'loss of supply' is being specific about the requirement (or specification), a further question arises concerning the

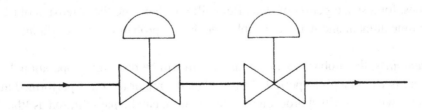

Figure 2.1: Two valves in supply stream

Reliability, Maintainability and Risk. DOI: 10.1016/B978-0-08-096902-2.00002-7

13

15 failures per million hours. Do they all refer to the blocked condition, being the component failure mode that contributes to the system failure mode of interest? This is unlikely because several failure modes are likely to be included in the 15 per million hours and it may well be that the failure rate for modes that cause 'no throughput' is only 7 per million hours.

Suppose, on the other hand, that one is considering loss of control leading to downstream over-pressure rather than 'loss of supply'. The situation changes significantly. First, the fact that there are two valves now enhances rather than reduces the reliability since, for this new definition of system failure, both need to fail. Second, the valve failure mode of interest is the internal leak or fail open mode. This is another, but different, subset of the 15 per million hours – say, 3 per million. A different calculation is now needed for the system reliability and this will be explained in Chapters 7–9. Table 2.1 shows a typical breakdown of the failure rates for various different failure modes of the control valve in the example.

Table 2.1: Control Valve Failure Rates per Million Hours

Fail shut	7
Fail open	3
Leak to atmosphere	2
Slow to move	2
Limit switch fails to operate	1
Total	15

The essential point in all this is that the definition of failure mode totally determines the system reliability and dictates the failure mode data required at the component level. The above example demonstrates this in a simple way, but in the analysis of complex mechanical and electrical equipment, the effect of the defined requirement on the reliability is more subtle.

Given, then, that the word 'failure' is specifically defined, for a given application, quality and reliability and maintainability can now be defined as follows:

Quality: conformance to specification.
Reliability: the probability that an item will perform a required function, under stated conditions, for a stated period of time. Reliability is therefore the extension of quality into the time domain and may be paraphrased as 'the probability of non-failure in a given period'.
Maintainability: the probability that a failed item will be restored to operational effectiveness within a given period of time when the repair action is performed in accordance with prescribed procedures. This, in turn, can be paraphrased as 'the probability of repair in a given time' and is often expressd as a 'percentile down time'.

2.2 Failure Rate and Mean Time Between Failures

Requirements are seldom expressed by specifying targets for reliability or maintainability. There are related parameters such as failure rate, Mean Time Between Failures (MTBF) and Mean Down Time (MDT) that more easily describe them. Figure 2.2 provides a model for the purpose of explaining failure rate.

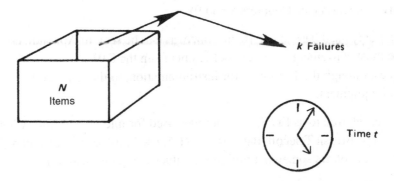

Figure 2.2: Terms useful in understanding failure rate

The symbol for failure rate is λ (lambda). Consider a batch of N items and that at any time, t, a number, k, have failed. The cumulative time, T, will be Nt if it is assumed that each failure is replaced when it occurs whereas in a non-replacement case, T is given by:

$$T = [t_1 + t_2 + t_3 \ldots t_k + (N - k)t]$$

where t_1 is the occurrence of the first failure, etc.

2.2.1 The Observed Failure Rate

This is defined: for a stated period in the life of an item, the ratio of the total number of failures to the total cumulative observed time. If λ is the failure rate of the N items then the observed λ is given by $\hat{\lambda} = k/T$. The ^ (hat) symbol is very important since it indicates that k/T is only an estimate of λ. The true value will be revealed only when all N items have failed. Making inferences about λ from values of k and T is the purpose of Chapters 5 and 6. It should also be noted that the value of $\hat{\lambda}$ is the average over the period in question. The same value might be observed from increasing, constant and decreasing failure rates. This is analogous to the case of a motor car whose speed between two points is calculated as the ratio of distance to time despite the speed having varied during this interval. Failure rate is thus only a meaningful parameter when it is constant.

Failure rate, which has the unit of t^{-1}, is sometimes expressed as a percentage per 1000 hrs and sometimes as a number multiplied by a negative power of ten. Examples, having the same value, are:

8500 per 10^9 hours (8500 FITS known as 'failures in time')

8.5 per 10^6 hours or $\mathbf{8.5 \times 10^{-6}}$ **per hour**

0.85 per cent per 1000 hours

0.074 per year.

Note that these examples are expressed using only two significant figures. It is seldom justified to exceed this level of accuracy, particularly if failure rates are being used to carry out a reliability prediction (see Chapters 8 and 9).

The most commonly used base is per 10^6 hrs since, as can be seen in Appendices 3 and 4, it provides the most convenient range of coefficients from the 0.01 to 0.1 range for microelectronics, through the 1–5 range for instrumentation, to the tens and hundreds for larger pieces of equipment.

The per 10^9 base, referred to as FITS, is sometimes used for microelectronics where all the rates are small. The British Telecom database, HRD5, used this base since it concentrates on microelectronics and offers somewhat optimistic values compared with other sources.

Failure rate can also be expressed in units other than clock time. An example is the emergency shut down valve where the failures per demand are of interest. Another would be a solenoid or relay where the failures per operation provide a realistic measure.

2.2.2 *The Observed Mean Time Between Failures*

This is defined: for a stated period in the life of an item, the mean value of the length of time between consecutive failures, computed as the ratio of the total cumulative observed time to the total number of failures. If $\hat{\theta}$ (theta) is the MTBF of the N items then the observed MTBF is given by $\hat{\theta} = T/k$. Once again the hat indicates a point estimate and the foregoing remarks apply. The use of T/k and k/T to define θ and $\hat{\lambda}$ leads to the inference that $\theta = 1/\lambda$.

This equality must be treated with caution since it is inappropriate to compute failure rate unless it is constant. It will be shown, in any case, that the equality is valid only under those circumstances. See Section 2.3.

2.2.3 *The Observed Mean Time to Fail*

This is defined: for a stated period in the life of an item the ratio of cumulative time to the total number of failures. Again this is T/k. The only difference between MTBF and MTTF is in their usage. MTTF is applied to items that are not repaired, such as bearings and transistors, and MTBF to items which are repaired. It must be remembered that the time between failures excludes the down time. MTBF is therefore mean UP time between failures. In Figure 2.3 it is the average of the values of (t).

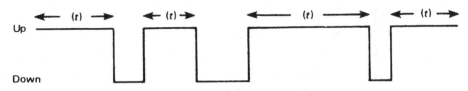

Figure 2.3: Up time and down time

2.2.4 Mean Life

This is defined as the mean of the times to failure but where every item is allowed to fail. This is often confused with MTBF and MTTF. It is important to understand the difference. MTBF and MTTF can be calculated over any period as, for example, confined to the constant failure rate portion of the bathtub curve. Mean life, on the other hand, must include the failure of every item and therefore includes the wearout end of the curve. Only for constant failure rate are MTBF and mean life the same.

To illustrate the difference between MTBF and lifetime compare:

- a match, which has a short life but a high MTBF (few fail, thus a great deal of time is clocked up for a number of strikes)
- a plastic knife, which has a long life (in terms of wearout) but a poor MTBF (they fail frequently).

Again, compare the following:

- the mean life of human beings is approximately 75 years (this combines random and wearout failures)
- our MTBF (early to mid-life) is approximately 2500 years (i.e. a 4×10^{-4} pa risk of fatality).

2.3 Interrelationships of Terms

2.3.1 Reliabilty and Failure Rate

Taking the model in Figure 2.2, and being somewhat more specific, leads us to Figure 2.4.

The number N now takes the form $N_s(t)$ for the number surviving at any time, t. N_0 is the number at time zero. Consider the interval between t and $t + \mathrm{d}t$. The number that will have failed is $\mathrm{d}N_s(t)$ (in other words the change in $N_s(t)$). The time accrued during that interval will have been $N_s(t) \times \mathrm{d}t$ (i.e. the area of the shaded strip). Therefore, from the earlier k/T rule, the instantaneous failure rate, at time t, is:

$$\lambda(t) = -\frac{\mathrm{d}N_s(t)}{N_s(t)\,\mathrm{d}t}$$

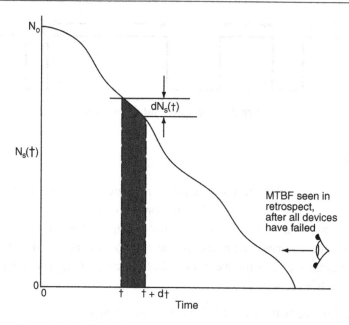

Figure 2.4: Relating instantaneous failure rate to reliability

Multiplying the numerator and denominator by N_0:

$$\lambda(t) = -\frac{dN_s(t)N_0}{N_s(t)\,dt\,N_0}$$

However, from the definition of reliability, $R(t) = N_s(t)/N_0$,

Therefore:

$$-\lambda(t) = \frac{dR(t)}{dt} \cdot \frac{1}{R(t)}$$

Integrating both sides:

$$-\int_0^t \lambda(t)\,dt = \int_1^{R(t)} dR(t)/R(t)$$

A word of explanation concerning the limits of integration is required. $\lambda(t)$ is integrated with respect to time from 0 to t. $1/R(t)$ is However, being integrated with respect to $R(t)$. Now, when $t = 0$, $R(t) = 1$ and at t the reliability $R(t)$ is, by definition, $R(t)$. Integrating then:

$$-\int_0^t \lambda(t)\,dt = \log_e R(t)\big|_1^{R(t)}$$
$$= \log_e R(t) - \log_e 1$$
$$= \log_e R(t)$$

But if $a = e^b$ then $b = \log_e a$, so that:

$$R(t) = \exp\left[-\int_0^t \lambda(t)\, dt\right]$$

If failure rate is now assumed to be constant:

$$R(t) = \exp\left[-\int_0^t \lambda(t)\, dt\right] = \exp - \lambda t|_0^t$$

Therefore $R(t) = e^{-\lambda t}$

Figure 2.5 shows this exponential relationship.

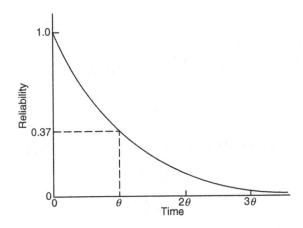

Figure 2.5: Exponential curve

2.3.2 Reliabilty and Failure Rate as an Approximation

In most cases λt is small (that is to say < 0.1) in which case $e^{-\lambda t}$ approaches $(1 - \lambda t)$. For example, if $\lambda = 10^{-5}$ per hour and $t = 10$ hours then $e^{-\lambda t}$ approaches $1 - 10^{-4} = 0.9999$.

The probablity of failure $(1 - R(t))$ therefore approximates to λt. This explains why the following appear to be the same:

- the rate of fatality is 10^{-4} per annum
- the probability of a fatality in a year is 10^{-4} (Derived from 10^{-4} per annum times 1 year.)

However, they are different parameters and thus use different units. One is a rate and the other is dimensionless. The fact that λt is very small $(<< 0.1)$ allows the two statements to appear similar. By contrast consider a group of people who average a rate of one speeding ticket per annum. Clearly:

- the rate is one per annum.

But, the probability is certainly not one. There is of course a statistical distribution. A few individuals may escape conviction, some may indeed receive one ticket and a few might clock up two or more. The probability of receiving a ticket is of course (from the above)

$$(1 - e^{-\lambda t}) = (1 - e^{-1}) = 0.67$$

2.3.3 Reliabilty and MTBF

In order to find the MTBF consider Figure 2.4 again. In each interval, dt, the time accumulated will be $N_s(t)$ dt. At infinity the total will be:

$$\int_0^\infty N_s(t)\, dt$$

Hence the MTBF will be given by:

$$\theta = \int_0^\infty \frac{N_s(t)\, dt}{N} = \int_0^\infty R(t)\, dt$$

$$\theta = \int_0^\infty R(t)\, dt$$

This is the general expression for MTBF and always holds. In the special case of $R(t) = e^{-\lambda t}$ then

$$\theta = \int_0^\infty e^{-\lambda t}\, dt$$

$$\theta = \frac{1}{\lambda}$$

Note that inverting failure rate to obtain MTBF, and vice versa, is valid only for the constant failure rate case.

2.4 The Bathtub Distribution

The much-quoted bathtub curve is an example of the practice of treating more than one failure type (mode) by a single classification. It seeks to describe the variation of failure rate of components during their life. Figure 2.6 shows this generalized relationship as originally assumed to apply to electronic components. The failures exhibited in the first part of the curve, where failure rate is decreasing, are called early failures or infant mortality failures. The middle portion is referred to as the useful life and it is assumed that failures exhibit a constant failure rate, that is to say they occur at random. The latter part of the curve describes the wearout failures and it is assumed that failure rate increases as the wearout mechanisms accelerate.

Figure 2.7, on the other hand, is somewhat more realistic in that it shows the bathtub curve to be the sum of three separate overlapping failure distributions. By labeling sections of the curve as wearout, burn-in and random it can now be seen in a different light. The wearout region implies only that wearout failures predominate, namely that such failures are more likely than the other types. The three distributions are described in Table 2.2.

Table 2.2

Failure Rate	Known As	Notes
Decreasing failure rate	Infant mortality Burn-in Early failures	Usually related to manufacture and QA, e.g. welds, joints, connections, wraps, dirt, impurities, cracks, insulation or coating flaws, incorrect adjustment or positioning. In other words, populations of substandard items owing to microscopic flaws.
Constant failure rate	Random failures Useful life Stress-related failures Stochastic failures	Usually assumed to be stress-related failures. That is, random fluctuations (transients) of stress exceeding the component strength (see Chapter 11). The design reliability referred to in Figure 1.1 is of this type.
Increasing failure rate	Wearout failures	Owing to corrosion, oxidation, breakdown of insulation, atomic migration, friction wear, shrinkage, fatigue, etc.

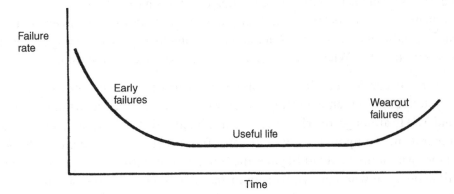

Figure 2.6: Bathtub curve

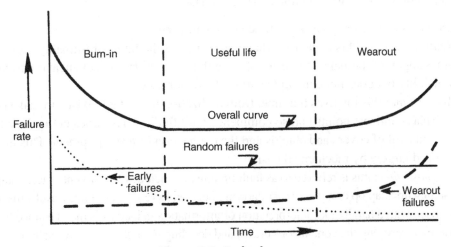

Figure 2.7: Bathtub curve

2.5 *Down Time and Repair Time*

It is now necessary to introduce mean down time and Mean Time to Repair (MDT, MTTR). There is frequently confusion between the two and it is important to understand the difference. Down time, or outage, is the period during which equipment is in the failed state. A formal definition is usually avoided, owing to the difficulties of generalizing about a parameter that may consist of different elements according to the system and its operating conditions. Consider the following examples, which emphasize the problem:

1. A system not in continuous use may develop a fault while it is idle. The fault condition may not become evident until the system is required for operation. Is down time to be measured from the incidence of the fault, from the start of an alarm condition, or from the time when the system would have been required?
2. In some cases it may be economical or essential to leave equipment in a faulty condition until a particular moment or until several similar failures have accrued.
3. Repair may have been completed but it may not be safe to restore the system to its operating condition immediately. Alternatively, owing to a cyclic operating profile it may be necessary to delay. When does down time cease under these circumstances?

It is necessary, as can be seen from the above, to define the down time as required for each system under given operating conditions and maintenance arrangements. MTTR and MDT, although overlapping, are not identical. Down time may commence before repair as in example (1) above. Repair often involves an element of checkout or alignment, which may extend beyond the outage. The definition and use of these terms will depend on whether availability or the maintenance resources are being considered.

The significance of these terms is not always the same, depending upon whether a system, a replicated unit or a replaceable module is being considered.

Figure 2.8 shows the elements of down time and repair time:

(a) *Realization time*: this is the time that elapses before the fault (presumably dormant) becomes apparent (perhaps as a result of a regular proof test). This element contributes to unavailability but does not constitute part of the repair time.

(b) *Access time*: this involves the time, from realization that a fault exists, to make contact with displays and test points and so commence fault finding. This does not include travel but the removal of covers and shields and the connection of test equipment. This is determined largely by mechanical design.

(c) *Diagnosis time*: this is referred to as fault finding and includes adjustment of test equipment (e.g. setting up a laptop or a generator), carrying out checks (e.g. examining waveforms for comparison with a handbook), interpretation of information gained (this may be aided by algorithms), verifying the conclusions drawn and deciding upon the corrective action.

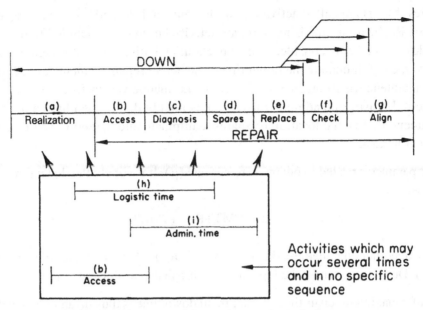

Figure 2.8: Elements of down time and repair time

(d) *Spare part procurement*: part procurement can be from the 'tool box', by cannibalization or by taking a redundant identical assembly from some other part of the system. The time taken to move parts from a depot or store to the system is not included, being part of the logistic time.

(e) *Replacement time*: this involves removal of the faulty LRA (Least Replaceable Assembly) followed by connection and wiring, as appropriate, of a replacement. The LRA is the replaceable item beyond which fault diagnosis does not continue. Replacement time is largely dependent on the choice of LRA and on mechanical design features such as the choice of connectors.

(f) *Checkout time*: this involves verifying that the fault condition no longer exists and that the system is operational. It may be possible to restore the system to operation before completing the checkout, in which case, although a repair activity, it does not all constitute down time.

(g) *Alignment time*: as a result of inserting a new module into the system, adjustments may be required. As in the case of checkout, some or all of the alignment may fall outside the down time.

(h) *Logistic time*: this is the time consumed waiting for spares, test gear, additional tools and manpower to be transported to the system.

(i) *Administrative time*: this is a function of the system user's organization. Typical activities involve failure reporting (where this affects down time), allocation of repair tasks, manpower changeover due to demarcation arrangements, official breaks, disputes, etc.

Activities (b)–(g) are called active repair elements and (h) and (i) passive repair activities. Realization time is not a repair activity but may be included in the MTTR where down time is the consideration. Checkout and alignment, although utilizing manpower, can fall outside the down time. The active repair elements are determined by design, maintenance arrangements, environment, manpower, instructions, tools and test equipment. Logistic and administrative time is mainly determined by the maintenance environment, that is, the location of spares, equipment and manpower and the procedure for allocating tasks.

Another parameter related to outage is repair rate (μ). It is simply the down time expressed as a rate, therefore:

$$\mu = 1/\text{MTTR or} + /\text{MDT}$$

It is very important to remember that down time is a variable (and hence a distributed quantity). Down times tend to be logNormally distributed as illustrated in Figure 2.9.

To talk of a maximum repair time or maximum down time is thus meaningless and dangerous. Given sufficient repairs a particular down time will eventually exceed those already observed. Only mean and percentile times can therefore be specified as a design parameter. This should be borne in mind when studying the contract clauses in Chapter 23.

2.6 Availability, Unavailability and Probability of Failure on Demand

Availability is a useful parameter that describes the proportion of time for which an item is not failed. More usefully unavailability (1 – availability) describes the proportion of time for which an item is failed and can be used to calculated outage costs by multiplying it by the cost of outage per unit time.

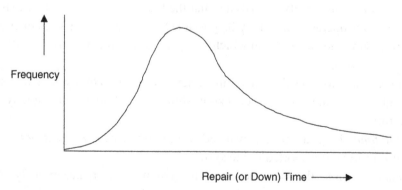

Figure 2.9: LogNormally distributed times

It is determined by both the reliability and the maintainability of the item. Returning to Figure 2.3, it is the ratio of down time to the total time. The total time is made up of the (t) values plus the down time. Unavailability is, therefore:

$$Un = (\text{Down time})/(\text{Total time})$$

$$= (\text{Down time})/(\text{Up time} + \text{Down time})$$

$$= MDT/(MTBF + MDT)$$

Multiplying top and bottom by failure rate we have:

$$= \lambda MDT/(1 + \lambda MDT)$$

Since λMDT is usually small (< 0.1) then:

$$\mathbf{Un \cong \lambda MDT}$$

Again, beware of approximations. When $\lambda MDT > 0.1$ the full expression must be used.

If an item is 'unavailable' for a proportion of the time then, assuming randomness (i.e. constant failure rate) the probability of it being failed at any moment is the same. Thus unavailability is the same thing as the Probability of Failure on Demand (PFD). In reliability work it is usually referred to as unavailability because the word describes the situation (i.e. plant not producing). In safety-related work, where a safety function is inhibited, the term probability of failure on demand (PFD) is more usual.

In the case of unrevealed failures the down time is equal to half the proof-test interval, T (plus the actual MTTR). This can be illustrated by thinking about an annual proof-test interval for the motor car. Consider the unrevealed failure of the air bag which occurs, at random, during the year. If we collect data for enough failures some will have occurred early in the year, some late in the year, and some at other times. The average of the times will be the middle, which is $T/2$. This is developed further in Chapter 8.

Thus the unavailability becomes $\lambda\, MDT = \lambda\, T/2$.

2.7 Hazard and Risk-Related Terms

The terms dealt with in this chapter are equally applicable to hazardous failures. *Hazard* is usually used to describe a situation with the potential for injury or fatality whereas *failure* is the actual event, hazardous or otherwise. The term *major hazard* is different only in degree and refers to certain large-scale potential incidents. These are dealt with in Chapters 10, 21 and 22.

Risk is a term that actually covers two parameters. The first is the probability (or rate) of a particular event. The second is the scale of consequence (perhaps expressed in terms of fatalities). This is dealt with in Chapter 10. Terms such as *societal* and *individual* risk differentiate between failures that cause either multiple or single fatalities.

2.8 Choosing the Appropriate Parameter

It is clear that there are several parameters available for describing the reliability and maintainability characteristics of an item. In any particular instance there is likely to be one parameter more appropriate than the others. Although there are no hard-and-fast rules the following guidelines may be of some assistance:

> *Failure rate*: applicable to most component parts. Useful at the system level, whenever constant failure rate applies, because it is then easy to compute unavailability from $\lambda \times$ MDT. Remember, however, that failure rate is meaningless if it is not constant. The failure distribution should then be described by other means which will be explained in Chapter 6.
> *MTBF and MTTF*: often used to describe equipment or system reliability. However, (asuming constant failure rate) it is the reciprocal of failure rate. Since it is failure rate that we convert into PFD then, more often than not, MTBFs only have to be inverted. Also, as we saw in Section 2.2.4, MTBF is often confused with mean life. For that reason MTBF is not the most useful of parameters and it is better to express in terms of failure rate.
> *Unavailability/PFD*: very useful where the cost of lost revenue, owing to outage, is of interest. Combines reliability and maintainability. Ideal for describing process plants. Unavailability calculates the probability of failure on demand (PFD), commonly needed as a target for safety-related systems.
> *Reliability/Unreliability*: used where the probability of failure is of interest as, for example, in aircraft landings where safety is the prime consideration.
> *Maintainability*: Seldom used as such.
> *Mean time to repair*: often expressed in percentile terms such as the 95 percentile repair time shall be one hour. This means that only 5% of the repair actions shall exceed one hour. Maximum MTTRs are meaningless.
> *Mean down time*: used where the outage affects system reliability or availability. Often expressed in percentile terms. Maximum MDTs are meaningless.
> *Mean life*: beware of the confusion between MTTF and mean life. Whereas the mean life describes the average life of an item taking into account wearout, the MTTF is the average time between failures. The difference is clear if one considers the simple example of the match.

Consider the following example using Figure 2.1.

Fail closed: the consequence is loss of throughput. Thus, the interest is in 'how much product is lost as a proportion of time' or 'the probability of there being no throughput at any particular moment'. Both of these are described by the unavailability/PFD, which is a function of failure rate and down time combined. Failure rate is of somewhat lesser interest because alone, it does not describe how long.

Fail open: *the consequence is excess pressure of the output, leading to potential release and injury. Thus, the interest is in 'how often will I be at risk?'. This is described by the failure rate. The unavailability/PFD are irrelevant in this scenario because once it has occurred the duration is irrelevant.*

There are numerous sources of definitions in standards such as:

> BS 4778: Part 3.2
> BS 4200: Part 1
> IEC Publication 271
> US MIL STD 721B
> UK Defence Standard 00-5 (Part 1)
> Nomenclature for Hazard and Risk in the Process Industries (I Chem E)
> IEC 61508 (Part 4).

It is, however, not always desirable to use standard sources of definitions because this avoids specifying the terms that are needed in a specification or contract. It is all too easy to 'define' the terms by calling up one of the aforementioned standards. It is far more important that terms are fully understood before they are used and if this is achieved by defining them for specific situations, then so much the better. The danger in specifying that all terms shall be defined by a given published standard is that each person assumes that he or she knows the meaning of each term and these are not read or discussed until a dispute arises. The most important area involving definition of terms is that of contractual involvement where mutual agreement as to the meaning of terms is essential. Chapter 19 will emphasize the dangers of ambiguity.

■ Exercises

λ = (a) 1×10^{-6} per hr (b) 100×10^{-6} per hr.

For each of (a) and (b):
1. Calculate the MTBFs in years.
2. Calculate the reliability for one year ($R_{(1yr)}$).
3. If the MDT is 10 hrs, calculate the unavailability.
4. If the MTTR is 1 hr, the failures are dormant, and the inspection interval is 6 months, calculate the unavailability.
5. What is the effect of doubling the MTTR?
6. What is the effect of doubling the inspection interval?

CHAPTER 3

A Cost-Effective Approach to Quality, Reliability and Safety

3.1 Reliability and Optimum Cost

In Section 3.3 manufacturers' quality costs are discussed. However, the costs associated with acquiring, operating and maintaining equipment are equally relevant. The total costs incurred over the period of ownership of equipment are often referred to as *life-cycle costs*. These can be separated into:

Acquisition cost: capital cost plus cost of installation, transport, etc.
Ownership cost: cost of preventive and corrective maintenance and of modifications.
Operating cost: cost of materials and energy.
Administration cost: cost of data acquisition and analysis.

They will be influenced by:

Reliability : determines frequency of repair
    determines spares requirements (see Chapter 16.4)
    determines loss of revenue (together with maintainability).
Maintainability: affects training, test equipment, down time and manpower (see Chapters 14 and 15).
Safety factors: affects operating efficiency, maintainability and liability costs.

Life-cycle costs will clearly be reduced by improving reliability, maintainability and safety but will be increased by the activities needed to achieve them. Therefore, we need to find an optimum set of parameters which minimizes the total cost. This concept is illustrated in Figures 3.1 and 3.2. Each curve represents cost against availability. Figure 3.1 shows the general relationship between availability and cost. The manufacturer's pre-delivery costs, those of design, procurement and manufacture, increase with availability. On the other hand, the manufacturer's after-delivery costs, those of warranty, redesign, and loss of reputation, decrease as availability improves. The total cost is shown by a curve indicating some value of availability at which minimum cost is incurred. Price will be related to this cost. Taking, then, the price/availability curve and plotting it again in Figure 3.2, the user's costs involve the addition of another curve representing losses and expense, owing to failure, borne by the user. The result is a curve also showing an optimum availability that incurs minimum cost. These diagrams serve to illustrate the idea that cost is minimized by finding reliability and maintainability enhancements whose savings exceed the initial expenditure.

Reliability, Maintainability and Risk. DOI: 10.1016/B978-0-08-096902-2.00003-9

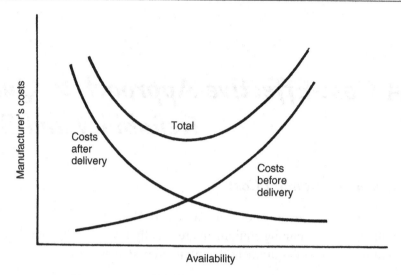

Figure 3.1: Availability and cost – manufacturer

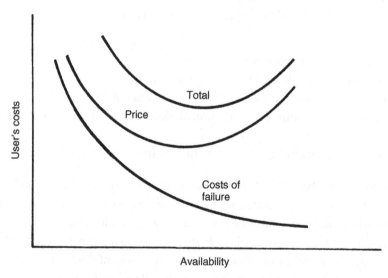

Figure 3.2: Availability and cost – user

A typical example is as follows:

- A duplicated process control system has a spurious shutdown failure rate of 1 per annum.
- Triplication reduces this failure rate to 0.8 per annum.
- The mean down time, in the event of a spurious failure, is 24 hours.
- The total cost of design and procurement for the additional unit is £60 000.

- The cost of spares, preventive maintenance, weight and power arising from the additional unit is £1000 per annum.
- The continuous process throughput, governed by the control system, is £5 million per annum.
- The potential saving is therefore $(1 - 0.8) \times 1/365 \times £5$ million per annum $= £2740$ per annum, which is equivalent to a capital investment of approximately £55 000 (assuming a 5% return on investment).
- The cost of the modification is £60 000 plus £1000 per annum, which is equivalent to a capital investment of £60 000 + £20 000 = £80 000.
- On that basis the proposed modification is therefore NOT justified.

There will be other factors influencing the decision such as safety, weight, space available, etc. From the reliability cost point of view, however, we saw that the expenditure is not justified.

The cost of carrying out RAMS-cycle predictions will usually be small compared with the potential safety or life-cycle cost savings as shown in the following examples.

A cost justification may be requested for carrying out these RAMS prediction activities; in which case the costs of the following activities should be estimated, for comparison with the predicted savings. RAMS prediction costs (i.e. resources) will depend upon the complexity of the equipment. The following two budgetary examples, expressing RAMS prediction costs as a percentage of the total development and procurement costs, are given in Table 3.1:

Example (A) A simple safety subsystem consisting of a duplicated 'shut down' or 'fire detection' system with up to 100 inputs and outputs, including power supplies, annunciation and operator interfaces.

Example (B) A single stream plant process (e.g. chain of gas compression, chain of H_2S removal reactors and vessels) and associated pumps and valves (up to 20) and the associated instrumentation (up to 50 pressure, flow and temperature transmitters).

Life-cycle costs (for both safety and unavailability) can be orders greater than the above typical project costs. Thus, even relatively small enhancements in MTBF/availability will easily lead to costs far in excess of the example expenditures quoted above.

The cost of carrying out RAMS prediction activities is of the order of 5% of total project cost. Although definitive records are not readily available it is credible that the assessment process, with its associated comparison of alternatives and proposed modifications, will lead to savings that exceed this outlay. In the above examples, credible results of the RAMS studies might be:

(A) ESD system:
The unavailability might typically be improved from 0.001 to 0.0005 as a result of the RAM study leading to some reduced proof test intervals and, perhaps, some duplicated instrumentation.

Table 3.1: RAMS prediction costs

	Man-days for (A)	Man-days for (B)
Figure 1.2 loop [1]: Feasibility RAMS prediction. This will consist of a simple block diagram prediction with the vessels or electronic controllers treated as units.	4	6
Figure 1.2 loop [2]: Conceptual design prediction. Similar to [1] but with more precise input/output quantities.	10	13
Figure 1.2 loop [3]: Detailed design prediction. Includes Failure Modes, Effects and Criticality Analysis Module (FMECA) at circuit level for 75% of the units, attention to common cause, human error and proof-test intervals.	6	18
Figure 1.2 loop [4]: RAMS testing. This refers to preparing subsystem and system test plans and analysis of test data rather than the actual test effort.	2	10
Figure 1.2 loop [5]: Acceptance testing. This refers to preparing test plans and analysis of test data rather than the actual test effort.	2	6
Figure 1.2 loop [6]: First year, reliability growth reviews. This is a form of design review using field data.	1	2
Figure 1.2 loop [7]: Subsequent reliability growth, data analysis.	2	3
Figure 1.2 loop [9]: First year, field data analysis. Not including effort for field data recording but analysis of field returns.	2	8
Figure 1.2 loop [10]: RCM planning. This includes identification of major components, establishing RAMS data for them, calculation of optimum discard, spares and proof-test intervals.	3	8
Overall totals	32	74
Cost @ £500/man-day	£16K	£37K
Typical project cost (design and procure)	£250K	£800K
RAMS cost as % of total project cost	6.4%	4.6%

Spurious shutdown, resulting from failure of the ESD, might typically be £500 000 per day for a small gas production platform. Thus, the £8000 expenditure on RAM saves:

$$£500\,000 \times (0.001 - 0.0005) \times 365 = \textbf{£91 000 per annum}$$

(B) H$_2$S system:
The availability might typically be improved from 0.95 to 0.98 as a result of the RAM study leading to measures similar to those mentioned in (A) above. Loss of throughput, resulting from failure, might typically cost £5000 per day. Thus, the £18 500 expenditure on RAM saves:

$$£5000 \times (0.98 - 0.95) \times 365 = \textbf{£55 000 per annum}$$

Non RAMS-specialist engineers should receive training in RAMS techniques in order that they acquire sufficient competence to understand the benefits of those activities. The IET/BCS competency guidelines document 1999 offers a framework for assessing such competencies.

3.2 Costs and Safety

3.2.1 The Need for Optimization

Once the probability of a hazardous event has been assessed, attention will inevitably be drawn to the cost of the various measures that can be taken to reduce the risk. If the risk to life is so high that it must be reduced as a matter of priority, or if the measures involved are a legal requirement, then the economics are of little or no concern – the equipment or plant must be made safe or closed down.

If, however, the risk to life is perceived to be sufficiently low then further reduction in risk, at a given cost, can be examined to see if such expenditure can be justified. At this point the concept of ALARP (As Low As Reasonably Practicable) arises. A risk is said to be ALARP if the cost of further risk reduction is disproportionate to the benefit. This is determined by comparing the cost per life saved, from any proposed risk reduction, with some agreed criterion. In this way risk reduction expenditure is focused onto areas of greatest benefit by ranking proposed expenditures so as to apply funds to the most effective area of risk improvement. Any technique that appears to put a price on human life is, however, potentially distasteful and thus attempts to use it are often resisted. It should not, in any case, be used as the sole criterion for deciding upon expenditure. The topic is dealt with fully in Section 10.2 of Chapter 10.

3.2.2 Costs and Savings Involved with Safety Engineering

Although costs vary considerably, according to the scale and complexity of a system or project, the following typical resources have been seen in meeting various aspects of safety-integrity.

> Typical safety-integrity targeting with random hardware failures predictions and the demonstration of ALARP (Chapter 10.2) – 2 to 6 man-days.
> Assessing safe failure fraction (described in Chapter 22) – 1 to 5 man-days.
> Bringing an ISO 9001 management system up to IEC 61508 functional safety capability – 5 man-days for the purpose of a product demonstration, 20 to 50 man-days for the purpose of accredited certification.

As far as savings are concerned:

> There is an intangible but definite benefit due to enhanced credibility in the market place. Additional sales vis-à-vis those who have not demonstrated integrity are likely.
> Major savings are claimed due to reduced maintenance for those systems where reduced integrity target can be shown to be adequate (i.e. ALARP). This also has the side effect of focusing resources onto systems requiring higher-integrity targets.

Conversely, there are penalty costs associated with hazardous failures:

> The manufacturer and the user will incur far higher costs of retrospective redesign if design changes are needed in order to meet the maximum tolerable risk.
> The user could face enormous legal costs in the event of a major incident that infringes the H&SW Act (Chapter 20), especially if appropriate functional safety standards had not been addressed when it was reasonably practicable to do so.

3.3 The Cost of Quality

The practice of identifying quality costs is by no means new, although it is usually only very large organizations that identify and analyze this highly significant proportion of their turnover. Attempts to set budgets for the various elements of quality costs are even rarer. This is unfortunate, since the contribution of any activity to a business is measured ultimately in financial terms, and quality, reliability and maintainability are no exception. If the costs of failure and repair were more fully identified and compared with the costs of improvement then great benefit could be obtained from the exercise. The pursuit of quality and reliability for their own sake is no justification for the investment of labor, plant and materials and greater recognition of costs would lead to the better allocation of resources.

Quality cost analysis involves extracting various items from the accounts and grouping them under three headings:

> *Prevention costs:* costs of preventing failures.
> *Appraisal costs:* costs related to measurement.
> *Failure costs:* costs incurred as a result of scrap, rework, failure, etc.

Each of these categories can be broken down into identifiable items and Table 3.2 shows a typical breakdown of quality costs for a six-month period in a manufacturing organization. The totals are expressed as a percentage of sales, this being the usual ratio. It is understood by those who collect these costs that they are usually under-recorded and that the failure costs obtained can be as little as a quarter of the true value. The ratios shown in Table 3.2 are typical of a manufacturing and assembly operation involving light machining, assembly, wiring and functional test of electrical equipment. The items are as follows:

Prevention Costs

> *Design review:* review of new and modified designs prior to the release of drawings.
> *Quality and reliability training:* training of QA staff. Q, R and Functional-safety training of other staff.
> *Vendor quality planning:* evaluation of vendors' abilities to meet requirements.
> *Audits:* audits of systems, products, processes and procedures.

Table 3.2: Quality costs: 1 January 2010 to 30 June 2010 (Sales £2 Million)

	£'000	% of Sales
Prevention Costs		
Design review	0.5	
Quality and reliability training	2.	
Vendor quality planning	2.1	
Audits	2.4	
Installation prevention activities	3.8	
Product qualification	3.5	
Quality engineering	3.8	
	18.1	0.91
Appraisal Costs		
Test and inspection	45.3	
Maintenance and calibration	2.	
Test equipment depreciation	10.1	
Line quality engineering	3.6	
Installation testing	5.	
	66.0	3.3
Failure Costs		
Design changes	18.	
Vendor rejects	1.5	
Rework	20.	
Scrap and material renovation	6.3	
Warranty	10.3	
Commissioning failures	5.	
Fault finding in test	26.	
	87.1	4.36
Total quality cost	171.2	**8.57**

Installation prevention activities: any of these activities applied to installations and the commissioning activity.

Product qualification: comprehensive testing of a product against all its specifications prior to the release of final drawings to production. Some argue that this is an appraisal cost. Since it is prior to the main manufacturing cycle the author includes it in prevention since it always attracts savings far in excess of the costs incurred.

Quality engineering: preparation of quality plans, workmanship standards, inspection procedures.

Appraisal Costs

Test and inspection: all line inspection and test activities but excluding rework and waiting time. If the inspectors or test engineers are direct employees then the costs should be suitably loaded to reflect overheads. It will be necessary to obtain, from the cost accountant, a suitable overhead rate that allows for the fact that the QA overheads are already reported elsewhere in the quality cost report.

Maintenance and calibration: the cost of labor and subcontract charges for the calibration, overhaul, upkeep and repair of test and inspection equipment.

Test equipment depreciation: include all test and measuring instruments.

Line quality engineering: that portion of quality engineering which is related to answering test and inspection queries.

Installation testing: test during installation and commissioning.

Failure Costs

Design changes: all costs associated with engineering changes due to defect feedback.

Vendor rejects: rework or disposal costs of defective purchased items where this is not recoverable from the vendor.

Rework: loaded cost of rework in production and, if applicable, test.

Scrap and material renovation: cost of scrap less any reclaim value. Cost of rework of any items not covered above.

Warranty: labor and parts as applicable. Cost of inspection and investigations to be included.

Commissioning failures: rework and spares resulting from defects found and corrected during installation.

Fault finding in test: where test personnel carry out diagnosis over and above simple module replacement then this should be separated out from test and included in this item. In the case of diagnosis being carried out by separate repair operators then that should be included.

It can be seen from the above list that reliability and maintainability are directly related to these items.

UK industry turnover is of the order of £250 billion. The total quality cost for a business is likely to fall between 4% and 15%, the average being somewhere in the region of 8%. Failure costs are usually approximately 50% of the total – higher if insufficient is being spent on prevention. It is likely then that about £10 billion was wasted in defects and failures. A 10% improvement in failure costs would release into the economy approximately

£1 billion.

Prevention costs are likely to be approximately 1% of the total and therefore of the order of £2.5 billion.

In order to introduce a quality cost system it is necessary to:

Convince top management: initially a draft quality cost report similar to Table 3.1 should be prepared. The accounting system may not be 'geared-up' for the automatic collection and grouping of the items but this can be carried out on a one-off basis. The object of the exercise is to draw attention to the magnitude of quality costs and to show that prevention costs are small by comparison with the total.

Collect and analyze quality costs: the data should be drawn from the existing accounting system and no major change should be made. In the case of change notes and scrapped items, the effort required to analyze every one may be prohibitive. In this case the total may be estimated from a representative sample. It should be remembered, when analyzing change notes, that some may involve a cost saving as well as an expenditure. It is the algebraic total that is required.

Quality cost improvements: the third stage is to set budget values for each of the quality cost headings. Cost-improvement targets are then set to bring the larger items down to an acceptable level. This entails making plans to eliminate the major causes of failure. Those remedies that are likely to generate the greatest reduction in failure cost for the smallest outlay should be chosen first.

Things to remember about quality costs are:

* They are not a target for individuals but for the company.
* They do not provide a comparison between departments because quality costs are rarely incurred where they are caused.
* They are not an absolute financial measure but provide a benchmark against which to make comparisons.
* Consistency in their presentation is the main consideration.

PART 2

Interpreting Failure Rates

CHAPTER 4

Realistic Failure Rates and Prediction Confidence

4.1 Data Accuracy

There are several sources of failure rate data compiled by defense, telecommunications, process industries, oil and gas and other organizations. Some are published in the form of data handbooks such as:

> US MILITARY HANDBOOK 217 (Electronics)
> CNET (French PTT) Data
> HRD (Electronics, British Telecom)
> RADC Non-Electronic Parts Handbook NPRD
> OREDA (Offshore data)

Some are databanks that are accessible by virtue of membership or fee such as:

> FARADIP.THREE (Data ranges and modes) – updated annually
> SRD (Systems Reliability Department of UKAEA) databank
> Technis databank [the author] (Tonbridge)

Some are in-house data collections that are not generally available. These occur in:

> large industrial manufacturers
> public utilities.

Data collection activities were at their peak in the 1980s but, sadly, they declined during the 1990s and the majority of published sources have not been updated since that time.

Failure data are usually, unless otherwise specified, taken to refer to random failures (i.e. constant failure rates). It is important to read, carefully, any covering notes since, for a given temperature and environment, a stated component, despite the same description, may exhibit a wide range of failure rates because:

1. Some failure-rate data include items replaced during preventive maintenance whereas others do not. These items should, ideally, be excluded from the data but, in practice, it is not always possible to identify them. This can affect rates by an order of magnitude.

2. Failure rates are affected by the tolerance of a design and this will cause a variation in the values. Because definitions of failure vary, a given parametric drift may be included in one database as a failure, but ignored in another.

3. Although nominal environmental and quality assurance levels are described in some databases, the range of parameters covered by these broad descriptions is large. They represent, therefore, another source of variability.

4. Component parts often are only described by reference to their broad type (e.g. signal transformer). Data are therefore combined for a range of similar devices rather than being separately grouped, thus widening the range of values. Furthermore, different failure modes are often mixed together in the data.

5. The degree of data screening will affect the relative numbers of intrinsic and induced failures in the quoted failure rate. An example would be not including a systematic failure whose re-occurrence is designed out.

6. Reliability growth occurs because field experience is used to enhance reliability as a result of modifications. This will influence the failure rate data.

7. Trial and error replacement is sometimes used as a means of diagnosis and this can artificially inflate failure rate data.

8. Some data record undiagnosed incidents and 'no fault found' visits. If these are included in the statistics as faults, then failure rates can be inflated. Quoted failure rates are therefore influenced by the way they are interpreted by an analyst.

Failure rate values can span one or two orders of magnitude as a result of different combinations of these factors. Prediction calculations are explained in Chapters 8 and 9 but it will be seen (Section 4.4) that the relevance of failure rate data is more important than refinements in the model used for the calculation. The data sources described in Section 4.2 can at least be subdivided into 'site/company specific', 'industry specific' and 'generic' and research, described in Section 4.4, confirms that the more specific the data source the greater the confidence in the prediction.

Data are presented in one of two forms:

1. *Tables*: lists of failure rates such as those in Appendices 3 and 4, with or without multiplying factors, for such parameters as quality and environment. Sometimes failure rates are tabulated, for a given component type, against ambient temperature and the ratio of applied to rated stress (power or voltage).

2. *Regression Models*: obtained by regression analysis of the data. These are presented in the form of equations that provide a failure rate as a result of inserting the device parameters into the appropriate expression. Because of the large number of variables involved in describing microelectronic devices, data are often expressed in the form of models. These regression equations (WHICH GIVE A TOTALLY MISLEADING IMPRESSION OF PRECISION) involve some or all of the following:

    *   complexity (number of gates, bits, equivalent number of transistors)
    *   number of pins

- junction temperature (see Arrhenius, Section 11.2)
- package (ceramic and plastic packages)
- technology (CMOS, NMOS, bipolar, etc.)
- type (memory, random LSI, analogue, etc.)
- voltage or power loading
- quality level (affected by screening and burn-in)
- environment
- length of time in manufacture.

Although empirical relationships have been established relating certain device failure rates to specific stresses, such as voltage and temperature, no precise formula exists which links specific environments to failure rates. The permutation of different values of environmental factors, such as those listed in Chapter 12, is immense. General adjustment (multiplying) factors have been evolved and these are often used to scale up basic failure rates to particular environmental conditions.

Because failure rate is, probably, the least precise engineering parameter, it is important to bear in mind the limitations of a reliability prediction. The research described in Section 4.4 makes it possible to express predictions using confidence intervals. The resulting MTBF, availability (or whatever) should not be taken as an absolute parameter but rather as a general guide to the design reliability. Within the prediction, however, the relative percentages of contribution to the total failure rate are of a better accuracy and provide a valuable tool in design analysis.

Because of the differences between data sources, comparisons of reliability should always involve the same data source in each prediction.

For a reliability assessment to be meaningful, it must address a specific system failure mode. To predict that a safety (shutdown) system will fail at a rate of, say, once per annum is, on its own, saying very little. It might be that 90% of the failures lead to a spurious shutdown and 10% to a failure to respond. If, on the other hand, the ratios were to be reversed then the picture would be quite different.

The failure rates, mean times between failures or availabilities must therefore be assessed for defined failure types (modes). In order to achieve this, the appropriate component level failure modes must be applied to the prediction models that are described in Chapters 8 and 9. Component failure mode data are sparse but a few of the sources do contain some information. The following sections indicate where this is the case.

4.2 Sources of Data

Sources of failure rate and failure mode data can be classified as:

1. *Site/company specific*: failure-rate data that have been collected from similar equipment being used on very similar sites (e.g. two or more gas compression sites where

environment, operating methods, maintenance strategy and equipment are largely the same). Another example would be the use of failure rate data from a flow corrector used throughout a specific distribution network. These data might be applied to the RAMS prediction for a new design of circuitry for the same application.

2. *Industry specific*: an example would be the use of the OREDA offshore failure rate data book for a RAMS prediction of a proposed offshore process package.
3. *Generic*: a generic data source combines a large number of applications and sources.

As will be emphasized in Chapters 7–9, predictions require failure rates for specific modes of failure (e.g. open circuit, signal high, valve closes). Some, but unfortunately only a few, data sources contain specific failure mode percentages. Mean time to repair data are even more sparse although the OREDA database is very informative in this respect.

The following are the more widely quoted sources.

4.2.1 Electronic Failure Rates

4.2.1.1 US Military Handbook 217 (generic, no failure modes)

This is one of the better known data sources and was from RADC (Rome Air Data Center in the USA). Opinions are sharply divided as to its value due to the unjustified precision implied by virtue of its regression model nature of its microelectronics sections. It covers:

> microelectronics
> discrete semiconductors
> tubes (thermionic)
> lasers
> resistors and capacitors
> inductors
> connections and connectors
> meters
> crystals
> lamps, fuses and other miscellaneous items.

The Microelectronics sections present the information as a number of regression models. For example, the Monolithic Bipolar and MOS Linear Device model is given as:

Part operating failure rate model (λ_p):

$$\lambda_p = \pi_Q \left(C_1 \pi_t \pi_V + C_2 \pi_E \right) \pi_L \text{ Failures}/10^6 \text{ hours}$$

> where

π_Q is a multiplier for quality,
π_t is a multiplier for junction temperature,

π_V is a multiplier for applied voltage stress,

π_E is an application multiplier for environment,

π_L is a multiplier for the amount of time the device has been in production,

C_1 is based on the equivalent transistor count in the device,

C_2 is related to the packaging.

There are two reservations about this approach. First, it is not possible to establish the original application of the items from which the data are derived and it is not clear what mix of field and test data pertains. Second, a regression model both interpolates and extrapolates the results of raw data. There are similar models for other microelectronic devices and for discrete semiconductors. Passive components are described using tables of failure rates and the use of multipliers to take account of quality and environment.

The trend in successive issues of MIL-217 was towards lower failure rates, particularly in the case of microelectronics. This is also seen in other databanks and may reflect the steady increase in manufacturing quality and screening techniques over the last twenty-five years. On the other hand, it may be due to reassessing the earlier data. MIL-217 is available (as MILSTRESS) on disk from ITEM software. Between 1965 and 1991, it moved from Issue A to Issue F (amended 1992). It seems unlikely that it will be updated again.

4.2.1.2 HRD5 Handbook of Reliability Data for Electronic Components used in Telecommunications Systems *(industry specific, no failure modes)*

This document was produced, from field data, by British Telecom's Laboratories at Martlesham Heath and offers failure rate lists for integrated circuits, discrete semiconductors, capacitors, resistors, electromechanical and wound components, optoelectronics, surge protection, switches, visual devices and a miscellaneous section (e.g. microwave).

The failure rates obtained from this document are generally optimistic compared with the other sources, often by as much as an order of magnitude. This is due to an extensive 'screening' of the data whereby failures that can be attributed to a specific cause are eliminated from the data once remedial action has been introduced into the manufacturing process. Considerable effort is also directed towards eliminating maintenance-induced failures from the data.

Between 1977 and 1994 it moved from Issue 1 to Issue 5 but it seems unlikely that it will be updated again.

4.2.1.3 Recueil de Donnés de Fiabilité du CNET *(industry specific, no failure modes)*

This document is produced by the *Centre National d'Etudes des Telecommunications* (CNET), now known as France Telecom R&D. It was first issued in 1981 and has been subject to subsequent revisions. It has a similar structure to US MIL-217 in that it consists of regression models for the prediction of component failure rates as well as generic tables. The models involve a simple regression equation with graphs and tables that enable each parameter to be specified. The model is also stated as a parametric equation in terms of voltage, temperature, etc. The French PTT use the CNET data as their standard.

4.2.1.4 Bellcore (Reliability Prediction Procedure for Electronic Equipment) TR-NWT–000332 Issue 5 1995 (industry specific, no failure modes)

Bellcore is the research center for the Bell telephone companies in the USA. Bellcore data are electronic failure rate data for telecommunications.

4.2.1.5 Electronic Data NOT Available for Purchase

A number of companies maintain failure rate databanks including Nippon Telephone Corporation (Japan), Ericsson (Sweden) and Thomson CSF (France) but these data are not generally available outside the organizations.

4.2.2 Other General Data Collections

4.2.2.1 Non-Electronic Parts Reliability Data Book – NPRD (generic, some failure modes)

This document is also produced by RADC and was first published as NPRD 1 in 1978 and as NPRD 5 in 1995. It contains many hundreds of pages of failure rate information for a wide range of electromechanical, mechanical hydraulic and pneumatic parts. Failure rates are listed for a number of environmental applications. Unlike MIL-217, these are field data. It provides failure rate data against each component type and there are one or more entries per component type depending on the number of environmental applications for which a rate is available.

Each piece of data is given with the number of failures and hours (or operations/cycles). Thus there are frequently multiple entries for a given component type. Details for the breakdown of failure modes are given. NPRD 5 is available on disk.

4.2.2.2 OREDA – Offshore Reliability Data (1984/92/95/97/2002) (industry specific, detailed failure modes, mean times to repair)

This data book was prepared and published in 1984 and subsequently updated by a consortium of: BP Petroleum Development Ltd Norway, Elf Aquitaine Norge A/S, Norsk Agip A/S, A/S Norske Shell, Norsk Hydro a.s, Statoil, Saga Petroleum a.s and Total Oil Marine plc.

OREDA is managed by a steering committee made up from the participating companies. It is a collection of offshore failure rate and failure mode data with an emphasis on safety-related equipment. It covers components and equipment from:

  fire and gas detection systems
  process alarm systems
  fire fighting systems
  emergency shut-down systems

pressure relieving systems

general alarm and communication systems.

4.2.2.3 TECHNIS (the author) (industry and generic, many failure modes, some repair times)

For over twenty-five years, the author has collected a wide range of failure rate and mode data as well as recording the published data mentioned here. This is available to clients on a report basis. An examination of these data has revealed a 40% improvement in failure rates between the 1980s and the 1990s.

4.2.2.4 UKAEA (industry and generic, many failure modes)

This databank is maintained by the Systems Reliability Department (SRD) of UKAEA at Warrington, Cheshire, who have collected the data as a result of many years of consultancy. It is available on disk to members who pay an annual subscription.

4.2.2.5 Sources of Nuclear Generation Data (industry specific)

In the UK, UKAEA, above, has some nuclear data, as has NNC (National Nuclear Corporation) although this may not be openly available.

In the USA Appendix III of the WASH 1400 study provided much of the data frequently referred to and includes failure rate ranges, event probabilities, human error rates and some common cause information. The IEEE standard IEEE500 also contains failure rates and restoration times. In addition there is NUCLARR (Nuclear Computerized Library for Assessing Reactor Reliability), which is a PC-based package developed for the Nuclear Regulatory Commission and contains component failure rates and some human error data. Another US source is the NUREG publication. Some of the EPRI data are related to nuclear plants.

In France, *Électricité de France* (EDF) provides the EIReDA mechanical and electrical failure rate database, which is available for sale.

In Sweden the TBook provides data on components in Nordic nuclear power plants.

4.2.2.6 US Sources of Power Generation Data (industry specific)

The EPRI (Electric Power Research Institute) of GE Co., New York, data scheme is largely gas turbine generation failure data in the USA.

There is also the GADS (Generating Availability Data System) operated by NERC (North American Electric Reliability Council). They produce annual statistical summaries based on experience from power stations in the USA and Canada.

4.2.2.7 SINTEF (industry specific)

SINTEF (at Trondheim) is part of the Norwegian Institute of Technology and, amongst many activities, collects failure rate data as, for example, data sheets on fire and gas detection equipment.

4.2.2.8 Data NOT Available for Purchase

Many companies (e.g. Siemens) and for that matter firms of RAMS consultants (e.g. RM Consultants Ltd) maintain failure-rate data but only for use by that organization.

4.2.3 Some Older Sources

A number of sources have been much used and are still frequently referred to. They are, however, somewhat dated but are listed here for completeness.

> Reliability Prediction Manual for Guided Weapon Systems (UK MOD) – DX99/013–100
> Reliability Prediction Manual for Military Avionics (UK MOD) – RSRE250
> UK Military Standard 00–41
> Electronic Reliability Data – INSPEC/NCSR (1981)
> Green and Bourne, *Reliability Technology*, Wiley 1972 (book)
> Frank Lees, *Loss Prevention in the Process Industries*, Butterworth-Heinemann (book).

4.3 Data Ranges

For some components there is fairly close agreement between different sources whereas in other cases there is a wide range of failure rate values, the reasons for which were summarized in Section 4.1.

The FARADIP.THREE database was created to show the ranges of failure rate for most component types. This database, CURRENTLY version 6.5 in 2010 (but updated annually), is a summary of Technis data together with most of the other databases and shows, for each component, the range of failure rate values that is to be found from them. Where a value in the range tends to predominate then this is indicated. Failure mode percentages are also included. It is available as a software package (with FMEA facilities) from the author at 26 Orchard Drive, Tonbridge, Kent TN10 4LG, UK technis.djs@virgin.net and includes:

Microelectronics:
    logic and linear
    memory.
Discrete:
    diodes and transistors
    optoelectronics
    lamps and displays
    crystals and piezo devices
    tubes.
Passive:
    capacitors
    resistors

inductive
microwave.

Instruments and analyzers:

analyzers
fire and gas detection
meters
flow instruments
pressure instruments
level instruments
temperature instruments.

Connection:

connections and connectors
switches and breakers
pCBs cables and leads.

Electromechanical:

relays and solenoids
rotating machinery (fans, motors, engines).

Power:

cells and chargers
supplies and transformers.

Mechanical:

pumps
valves and parts
bearings
miscellaneous.

Pneumatics (including leakages sources).
Hydraulics.
Computers, data processing and communications.
Alarms, fire protection, arresters and fuses.

The ranges are presented in three ways:

1. *A single value*: where the various references are in good agreement.
2. *Two values indicating a range*. It is not uncommon for the range to be an order of magnitude wide. The user, as does the author, must apply engineering judgement in choosing a value. This involves consideration of the size, application and type of device in question. Where two values occupy the first and third columns, then an even spread of failure rates is indicated. Where the middle and one other column are occupied then a spread with predominance to the value in the middle column is indicated.
3. *Three values indicating a range*. This implies that there is a fair amount of data available but that it spans more than an order of magnitude in range. Where the data tend to

predominate in one area of the range then this is indicated in the middle column. The most likely explanation of the range widths is the fact that some data refer only to catastrophic failures, whereas other data include degraded performance and minor defects revealed during preventive maintenance. This should be taken into account when choosing a failure rate from the tables.

As far as possible, the data given are for a normal ground-fixed environment and for items procured to a good standard of quality assurance as might be anticipated from a reputable manufacturer operating to ISO 9001. The variation that might be expected due to other environments and quality arrangements is dealt with by means of multiplying factors.

SAMPLE FARADIP SCREEN (not current version) – Fire and Gas Detection

	Failure Rates, per Million Hours		
Gas pellister (fail 0.003)	5.00	10	30
Detector smoke ionization	1.00	6.00	40
Detector ultraviolet	5.00	8.00	20
Detector infrared (fail 0.003)	2.00	7.00	50
Detector rate of rise	1.00	4.00	12
Detector temperature	0.10	2.00	–
Firewire/rod + psu	25	–	–
Detector flame failure	1.00	10	200
Detector gas IR (fail 0.003)	1.50	5.00	80
Failure modes (proportion):			
Rate of rise	Spurious 0.6	Fail 0.4	
Temp, firewire/rod	Spurious 0.5	Fail 0.5	
Gas pellister	Spurious 0.3	Fail 0.7	
Infrared	Spurious 0.5	Fail 0.5	
Smoke (ionize) and UV	Spurious 0.6	Fail 0.4	

4.3.1 Using the Ranges

The average range ratio for the entire FARADIP.THREE database is 7:1. In all cases, site-specific failure rate data or even that acquired from identical (or similar) equipment, and being used under the same operating conditions and environment, should be used in place of any published data.

Such data should, nevertheless, be compared with the appropriate range. In the event that it falls outside the range, there is a case for closer examination of the way in which the data were collected or in which the accumulated component hours were estimated.

Where the ranges contain a single value it can be used without need for judgement unless the specific circumstances of the assessment indicate a reason for a more optimistic or pessimistic failure rate estimate. Two or three values with predominating center column: in the absence of any specific reason to favor the extreme values the predominating value is the most credible choice.

Where there are wide ranges with ratios > 10:1 the use of the geometric mean is justified for the following reasons. The use of the simple arithmetic mean is not satisfactory for selecting a representative number when the two estimates are so widely spaced, since it favors the higher figure. The following example compares the arithmetic and geometric means where:

1. the arithmetic mean of n values of λ_i is given by

$$\sum_{i}^{n} \lambda_i / n$$

and

2. the geometric mean by:

$$\left(\prod_{i}^{n} \lambda_i \right)^{1/n}$$

Consider two estimates of failure rate, 0.1 and 1.0 (per million hours). The arithmetic mean (0.55) is five times the lower value and only a half of the upper value, thereby favoring the 1.0 failure rate. Where the range is an order or more, the larger value has significantly more bias on the arithmetic mean than the smaller.

The geometric mean (0.316) is, on the other hand, related to both values by a multiple of three and the excursion is thus the same. The geometric mean is, of course, derived from the arithmetic mean of the logarithms and therefore provides an average of the orders of magnitude involved. It is thus a more desirable parameter for describing the range.

In order to express the ranges as a single failure rate it is thus proposed to utilize the geometric mean. Appendix 3 shows microelectronic data in three columns giving the minima, maxima and geometric means. They can be interpreted as follows:

1. In general the lower figure in the range, used in a prediction, is likely to yield an assessment of the credible design objective reliability. That is the reliability that might reasonably be targeted after some field experience and a realistic reliability growth program. The initial (field trial or prototype) reliability might well be an order of magnitude less than this figure.
2. The center column figure indicates a failure rate that is more frequently indicated by the various sources. It is therefore a matter of judgement, depending on the type of prediction being carried out, as to whether it should be used in place of the lower figure.
3. The higher figure will probably include a high proportion of maintenance revealed defects and failures. The fact that data collection schemes vary in the degree of screening of maintenance revealed defects explains the wide ranges of quoted values.

4.4 Confidence Limits of Prediction

The ratio of predicted failure rate (or system unavailability) to field failure rate (or system unavailability) was calculated for each of 44 examples and the results (part of the author's PhD study) were classified in three categories:

1. *Predictions using site-/company-specific data*: these are predictions based on failure rate data which have been collected from similar equipment being used on very similar sites (e.g. two or more sites where environment, operating methods, maintenance strategy and equipment are largely the same).
2. *Predictions using industry-specific data*: an example would be the use of the OREDA offshore failure rate data book for a RAMS prediction of a proposed offshore gas compression package.
3. *Predictions using generic data*: these are predictions for which neither of the above two categories of data are available. Generic data sources (listed above) are used. FARADIP. THREE is also a generic data source in that it combines a large number of sources.

The results are:
1. For a prediction using site-/company-specific data (illustrated in Figure 4.1)

One can be this confident	*That the eventual field failure rate will be BETTER than:*
95%	$3\frac{1}{2}$ times the predicted
90%	$2\frac{1}{2}$ times the predicted
60%	$1\frac{1}{2}$ times the predicted
One can be this confident	*That the eventual field failure rate will be in the range:*
90%	$3\frac{1}{2}$:1 to 2/7:1

2. For a prediction using industry-specific data (illustrated in Figure 4.2)

One can be this confident	*That the eventual field failure rate will be BETTER than:*
95%	5 times the predicted
90%	4 times the predicted
60%	$2\frac{1}{2}$ times the predicted
One can be this confident	*That the eventual field failure rate will be in the range:*
90%	5:1 to 1/5:1

3. For a prediction using generic data (illustrated in Figure 4.3)

One can be this confident	*That the eventual field failure rate will be BETTER than:*
95%	8 times the predicted
90%	6 times the predicted
60%	3 times the predicted
One can be this confident	*That the eventual field failure rate will be in the range:*
90%	8:1 to 1/8:1

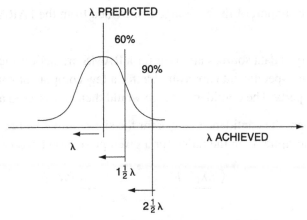

Figure 4.1: Using site/company data

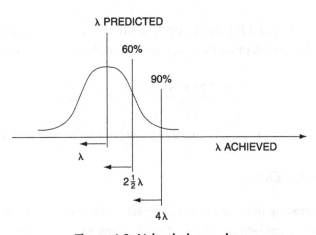

Figure 4.2: Using industry data

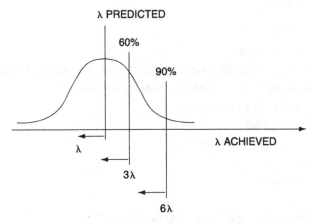

Figure 4.3: Using generic data

Additional evidence in support of the 8:1 range is provided from the FARADIP databank, which suggests 7:1.

It often occurs that mixed data sources are used for a RAMS prediction such that, for example, site-/company-specific data are available for a few component parts but generic data are used for the other parts. The confidence range would then be assessed as follows:

If *Range$_s$* and *Range$_g$* are the confidence ranges for the site-/company-specific and generic data expressed as a multiplier then the range for a given prediction becomes

$$\frac{(\Sigma\lambda_s \times Range_s) + (\Sigma\lambda_g \times Range_g)}{\Sigma\lambda_s \times \Sigma\lambda_g}$$

where $\Sigma\lambda_s$ and $\Sigma\lambda_g$ are the total failure rates of the site-/company-specific and generic items respectively.

For example, using the $3\frac{1}{2}$:1 and 8:1 ranges (90% confidence) given above, if $\Sigma\lambda_s = 20$ per million hrs (pmh) and $\Sigma\lambda_g = 100$ pmh, the range for the prediction (at 90% confidence) would be:

$$\frac{(20 \times 3.5) + (100 \times 8)}{120} = 7.25:1$$

At the end of Chapter 9 these ranges are used to compare predictions with targets.

4.5 Manufacturers' Data

There is a rapidly increasing trend to quote failure rates offered by equipment manufacturers. Extreme care should be exercised in the use of such failure rate data. Only users can claim to record all failures. There are numerous reasons why these failure rates can be highly optimistic. Reasons include:

- items in store before use
- items still in the supply chain
- failed item tolerated due to replacement causing process disruption and the ability to continue in degraded mode due to information redundancy
- item replaced by user without returning
    - disillusioned by supplier
    - not worth the cost (low-value item)
    - no warranty incentive
    - feedback not encouraged
    - user fixes it
- transient fault subsequently appears as 'no fault found'
- mismatch between perceived calendar versus operating hours for the item (standby items etc.)

- failure discounted due to inappropriate environment despite the fact that real-life failure rates include these
- vested interest in optimism
- the data were actually only a reliability prediction.

The author's studies indicate that manufacturers' data can be up to an order of magnitude optimistic (on average 5:1).

4.6 Overall Conclusions

The use of stress-related regression models implies an unjustified precision in estimating the failure rate parameter.

Site-/company-specific data should be used in preference to industry-specific data, which, in turn, should be used in preference to generic data.

Predictions should be expressed in confidence limit terms using the above information.

The FARADIP.THREE software package provides maximum and minimum rates together with failure modes.

In practice, failure rate is a system level effect. It is closely related to but not entirely explained by component failure. A significant proportion of failures encountered with modern electronic systems are not the direct result of parts failures but of more complex interactions within the system. The reason for this lack of precise mapping arises from such effects as human factors, software, environmental interference, interrelated component drift and circuit design tolerance.

The primary benefit to be derived from reliability engineering is the reliability growth that arises from continuing analysis and follow-up as well as corrective actions following failure analysis. Reliability prediction, based on the manipulation of failure-rate data, involves so many potential parameters that a valid repeatable model for failure rate estimation is not possible. Thus, failure rate is the least accurate of engineering parameters and prediction from past data should be carried out either:

- as an indicator of the approximate level of reliability of which the design is capable, given reliability growth in the field
- to provide relative comparisons in order to make engineering decisions concerning optimum redundancy
- as a contractual requirement
- in response to safety-integrity requirements.

It should not be regarded as an accurate indicator of future field reliability.

CHAPTER 5

Interpreting Data and Demonstrating Reliability

5.1 The Four Cases

From the following table it can be seen that there are four cases to be considered when interpreting k failures and T hours. First, there may be reason to assume constant failure rate, which includes two cases. If k is large (say, more than five) then the sampling inaccuracy in such a wide-tolerance parameter may be ignored. Chapter 4 has emphasized the wide ranges that apply and thus, for large values of k the formulae:

$$\lambda = k/T \text{ and } \theta = T/k$$

can be used. When k is small (even zero), the need arises to make some statistical interpretation of the data and that is the purpose of this chapter. The table also shows the second case where constant failure rate cannot be assumed. Again there may be few or many failures to interpret. Chapter 6 deals with this problem where the concept of a failure rate parameter is not relevant to describe the failure distribution.

	Constant Failure Rate	Variable Failure Rate
Many failures	Use $\lambda = k/T$	Chapter 6 (use probability plotting)
Few failures	Chapter 5 (statistical interpretation)	(Inadequate data so assume constant failure)

5.2 Inference and Confidence Levels

In Section 2.2 the concept of a point estimate of failure rate ($\hat{\lambda}$) or MTBF ($\hat{\theta}$) was introduced. Figure 2.2 showed N items having k failures in T cumulative hours. The observed failure rate ($\hat{\lambda}$) of that sample measurement was k/T. If the test were to be repeated, and another value of k/T obtained, it would not be exactly the same as the first and, indeed, a number of tests would yield a number of values of estimates of failure rates. Since these estimates are the result of sampling they are called point estimates and have the symbols $\hat{\lambda}$ and $\hat{\theta}$. It is the true failure rate or MTBF of the batch that is of interest and the only way to obtain this is to allow everything to fail and then to evaluate k/T or T/k. This is why the theoretical expression for MTBF in Section 2.3.3 has the integration limits of zero and infinity:

$$\text{MTBF} = \int_0^\infty \frac{N_s(t)}{N}\,\mathrm{d}t$$

Reliability, Maintainability and Risk. DOI: 10.1016/B978-0-08-096902-2.00005-2

In other words, all devices must fail if the true failure rate or MTBF is to be determined. Such a test will, of course, yield accurate data but, alas, no items left to use. In practice, we are forced to truncate tests after a given number of hours or failures. One is called a time-truncated test and the other a failure-truncated test. The problem is that an assessed failure rate or MTBF is required when only sample data are available. In many cases, where there is high reliability and hence few failures, the time required to accumulate several failures would be unrealistic.

The process of making a statement about a population of items based on the evidence of a sample is known as statistical inference. It involves, therefore, the additional concept of confidence level.

One way of illustrating the idea of confidence is to use the example in Figure 5.1, which shows a distribution of heights of a group of people in histogram form. Superimposed onto the histogram is a curve of the normal distribution. The practice in statistical inference is to select a mathematical distribution that closely fits the data. Statements based on the distribution are then assumed to apply to the data. In the figure there is a good fit between the normal curve, which has a mean of 5'10" and a standard deviation (measure of spread) of 1", and the heights of the group in question. Consider a person drawn, at random, from the group. It is permissible to state, from a knowledge of the normal distribution, that the person will be 5'10" tall or more providing that it is stated that the prediction is made with 50% confidence. This really means that we anticipate being correct 50% of the time if we continue to take samples. On this basis, an indefinite number of statements can be made, providing that an appropriate confidence level accompanies each value. For example:

  5'11" or more at 15.9% confidence
  6'0" or more at 2.3% confidence
  6'1" or more at 0.1% confidence
  OR between 5'9" and 5'11" at 68.2% confidence.

The inferred range of measurement and the confidence level can, hence, be traded off against each other. Thus, the lower the choice of height then the greater is the confidence of not being proved wrong by an unlucky random sample.

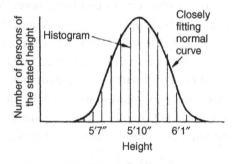

Figure 5.1: Distribution of heights

For random failures the situation is slightly different, in that we are not sampling an analog measurement (like height) but the occurrence of a discrete event with an underlying frequency. Figure 5.2 illustrates how a stream of events (failures) may lead to three different results according to the random positioning of the sample.

Indeed sample (3) 'sees' no failures. This emphasizes the fact that a zero-failures sample does not necessarily imply a zero failure rate. There will be a different estimate of failure rate for each choice of probability (i.e. confidence) that the sample size in question would, at random, 'see' no failures. This leads to a method for inferring an underlying failure rate from sample data.

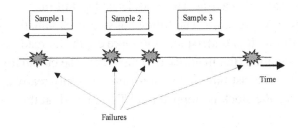

Figure 5.2: Random failures being sampled

5.3 The Chi-Square Test

Returning to the point estimates of failure rate and MTBF it is therefore possible to infer a value, together with a confidence level, if we assume constant failure rate (i.e. random failures). It can be shown that the expression

$$\frac{2k\hat{\theta}}{\theta} \text{ (random failures assumed)}$$

follows a x^2 distribution with $2k$ degrees of freedom, where the test is truncated at the kth failure. We know already that

$$\hat{\theta} = \frac{T}{k} = \frac{\text{Accumulated test hours}}{\text{Number of failures}}$$

Therefore

$$\frac{2k\hat{\theta}}{\theta} = \frac{2kT}{k\theta} = \frac{2T}{\theta}$$

so therefore **$2T/\theta$ is χ^2** distributed.

This leads us to

$$\lambda = \chi^2/2T \text{ and } \theta = 2T/\chi^2$$

If a value of χ^2 can be fixed for a particular test then failure rate or MTBF can be stated to lie between specified limits at some confidence value. In practice, the lower limit of failure rate is usually set at zero and one speaks of a failure rate of some value or less. Similarly one would have an MTBF or greater. This method is known as the single-sided confidence limit. Sometimes the double-sided limit method is used and this is described in Section 5.4. It is, however, more usual to use a single-sided approach.

In order to determine a value of χ^2 it is necessary to specify two parameters. The first is the number of degrees of freedom (usually two times one more than the number of failures) and the second is the confidence level. The tables of χ^2 at the end of this book (Appendix 2) have columns and rows labeled α and n. The confidence level of the χ^2 distribution is α and n is the number of degrees of freedom. The following simple explanation is sufficient to use the method. Readers who wish to understand it in more depth can read Section 5.4. Time-truncated refers to the situation (the majority of data) where the cumulative time, T, did not cease at the occurrence of the last failure. Failure-truncated (only really applicable to formal tests) refers to data where the clock is stopped (and T computed) at the occurrence of the last failure.

The following list of steps summarizes the use of the χ^2 tables for interpreting the results of reliability tests:

SINGLE-SIDED INTERPRETATION
FAILURE RATE LESS THAN; MTBF GREATER THAN

1. Observe T (accumulated test hours) and k (number of failures).
2. Select a confidence level and let $\alpha = (1 - confidence\ level)$.
3. Let n = 2(k + 1) for time-truncated data; and n = 2k for failure-truncated data.
4. Note the value of χ^2 from the tables at the end of this book (Appendix 2).
5. Let failure rate, at the given confidence level, be $\chi^2/2T$ (or MTBF $2T/\chi^2$).

DOUBLE-SIDED INTERPRETATION
FAILURE RATE and MTBF BETWEEN TWO LIMITS

For double-sided limits use the above procedure twice at

Lower Limit λ, Upper Limit MTBF

$n = 2k$ and use $1 - [\alpha/2]$

Upper Limit λ, Lower Limit MTBF

$n = 2k$ (failure-truncated); $n = 2(k + 1)$ (time-truncated) and use $\alpha/2$

It should be noted that, for constant failure rate, 100 components under test for 20 hrs yield the same number of accumulated test hours as 10 components for 200 hrs. Other methods of converting test data into statements of failure rate or MTBF are available but the χ^2 distribution method is the most flexible and easy to apply. Failure rates and MTBFs are usually inferred using 60% and 90% confidence levels.

Now look at the following examples.

■ Example 1

In a replacement test (i.e. each failed device is replaced immediately) 100 devices are tested for 1000 hrs during which three failures occur. We shall now calculate the failure rate of the batch at 90% and 60% confidence levels.

1. Since this is a replacement test T is obtained from the number under test multiplied by the linear test time. Therefore T = 100 000 hrs and k = 3.
2. Let $n = 2(k + 1)$ = 8 degrees of freedom. For 90% confidence $\alpha = (1 − 0.9)$ = 0.1 and for 60% confidence $\alpha = 1 − 0.6$ = 0.4.
3. Read off χ^2 values of 13.4 and 8.35 respectively (see Appendix 2).
4. $\lambda_{90\%}$ = 13.4/100 000 = 1.34 10^{-4} = 135 per million hours.
$\lambda_{60\%}$ = 8.35/100 000 = 8.35 10^{-5} = 83.5 per million hours.

■

■ Example 2

In a test 100 devices are tested for 1000 hrs during which there are no failures. We shall now calculate the failure rate of the batch at 90% and 60% confidence levels.
T is obtained from the number under test multiplied by the linear test time. Therefore T = 100 000 hrs and k = 0.

1. Let $n = 2(k + 1)$ = 2 degrees of freedom. For 90% confidence $\alpha = (1 − 0.9)$ = 0.1 and for 60% confidence $\alpha = 1 − 0.6$ = 0.4.
2. Read off χ^2 values of 4.61 and 1.83 respectively (see Appendix 2).
3. $\lambda_{90\%}$ = 4.61/100 000 = 4.61 10^{-5} = 46.1 per million hours.
$\lambda_{60\%}$ = 1.83/100 000 = 1.83 10^{-5} = 18.3 per million hours.

■

Note that in Example 2 we have been able to infer an 'underlying' failure rate despite zero failures in the sample. A word of caution is needed here. Imagine that a group of (say 10) car drivers at a meeting depart and re-convene in one year. They all survive the year (i.e. zero deaths). Using the above method, we can infer, at 90% confidence, that the fatality rate on the roads is 4.61/[2 × 10 years] = 2.3 10^{-1} pa. In fact it is nearer 5 10^{-5} pa, which is four orders of magnitude less pessimistic. What is wrong with the experiment? Clearly the sample is too small. If 80 000 people were involved in the experiment and none died then the answer becomes 4.61/[2 × 80 000 years] = 2.9 10^{-5} pa. In this second case there is sufficient accumulated time (with no deaths) to prove a more optimistic result. Thus, zero-failures inferences must be treated with care and only results that are OPTIMISTIC, when compared with some benchmark, should be used.

The examples, so far, have involved time-truncated data since this is the more usual situation. For a failure-truncated test, the 'plus one' is not added to the number of failures for the upper limit of failure rate (lower limit of MTBF). The added one failure in time-truncated data takes account of the possibility that, had the test continued for a few more seconds, a failure might have occurred.

5.4 Understanding the Method in More Detail

For those who wish to understand the method in a little more detail then Figure 5.3 shows a distribution for the χ^2 statistic. The area of the shaded portion is the probability of χ^2 exceeding that particular value at random.

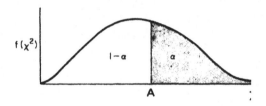

Figure 5.3: Single-sided confidence limits

As we have already seen, the confidence level of the χ^2 distribution is α and the number of degrees of freedom is n. The limits of failure rate are required between some value (shown as A in Figure 5.3) and zero. For the case of MTBF it would be between A and infinity. Since $\theta = 2T/\chi^2$ the value of χ^2 corresponding to infinite MTBF is zero. The limits are therefore zero and A. In Figure 5.3 and in Appendix 2, since α is the area to the right of A then $1 - \alpha$ must be the confidence level.

If the confidence limit is to be at 60%, the lower single-sided limit would be that value for which the failure rate is less (and the MTBF exceeds), by chance, 6 times out of 10. Since the degrees of freedom can be obtained from *2k or 2(k + 1)* and $\alpha = (1 - 0.6) = 0.4$, then a value of χ^2 can be obtained from the tables.

From $2T/\chi^2$ it is now possible to state a value of failure rate or MTBF at 60% confidence. In other words, such a value, or better, would be observed 60% of the time. It is written $\theta_{60\%}$; or, alternatively, $\lambda_{60\%}$, which is equal to $\chi^2/2T$.

5.5 Double-Sided Confidence Limits

So far, single-sided statements of failure rate and MTBF have been made. Sometimes it is required to state that the parameter lies between two confidence limits. Once again $\alpha = (1 - \text{confidence level})$ and is split equally on either side of the limits as shown in Figure 5.4.

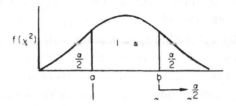

Figure 5.4: Double-sided confidence limits

The two values of χ^2 are found by using the tables twice, first at $n = 2k$ and at $1 - \alpha/2$ (this gives the lower limit of χ^2) and second at $n = 2k$ ($2k + 2$ for time-truncated) and at $\alpha/2$ (this gives the upper limit of χ^2). Once again, the upper limit of χ^2 corresponds with the lower limit of MTBF and vice versa. Figure 5.4 shows how $\alpha/2$ and $1 - \alpha/2$ are used. The probabilities of χ^2 exceeding the limits are the areas to the right of each limit and the tables are given accordingly.

Each of the two values of χ^2 can be used to obtain the limits of MTBF from the expression $\theta = 2T/\chi^2$. Assume that the upper and lower limits of MTBF for an 80% confidence band are required. In other words, limits of MTBF are required such that 80% of the time it will fall within them. $T = 100\,000\,\text{hrs}$ and $k = 3$. The two values of χ^2 are obtained:

$$n = 6,\ \alpha = 0.9,\ \chi^2 = 2.2$$

$$n = 8,\ \alpha = 0.1,\ \chi^2 = 13.4$$

This yields the two values of MTBF $14\,925\,\text{hrs}$ and $90\,909\,\text{hrs}$, in the usual manner from the expression $\theta = 2T/\chi^2$.

Hence the MTBF lies between $14\,925$ and $90\,909\,\text{hrs}$ with a confidence of 80%.

5.6 Reliability Demonstration

Imagine that, as a manufacturer, you have evaluated the failure rate of your instruments at some confidence level using the technique outlined above, and that you have sold them to me on the basis of such a test. I may well return, after some time, and say that the number of failures experienced in a given number of hours indicates a higher failure rate, at the same confidence, than did your earlier test. You could then suggest that I wait for another month, by which time there is a chance that the number of failures and the number of test hours will have swung the calculation back in your favor. Since this is hardly a suitable way of conducting business, it is necessary for consumer and producer to agree on a mutually acceptable test for accepting or rejecting batches of items. Once the test has been passed there is to be no question of later rejection on discovering that the batch passed on the strength of an optimistic sample. On the other hand, there is no redress if the batch is rejected, although otherwise acceptable, on the basis of a pessimistic sample. The risk that the batch, although within specification, will fail owing to a pessimistic sample being drawn is known as the producer's risk and has the symbol α (*not to be confused with the α used in the previous sections of this chapter*). The risk that a 'bad' batch will be accepted owing to an optimistic sample is known as the consumer's risk, β. The test consists of accumulating a given number of test hours and then accepting or rejecting the batch on the basis of whether or not a certain number of failures have been observed.

Imagine such a test where the sample has to accumulate T test hours with no failures in order to pass. If the failure rate, λ, is assumed to be constant then the probability of observing no failures in T test hours is the reliablity, which is $e^{-\lambda T}$ (being the zero failures curve of the Poisson distribution). Such a zero failures test is represented in Figure 5.5, which is a graph of the probability of observing no failures (in other words, of passing the test) against the anticipated number of failures given by λT. This type of test is known as a fixed-time demonstration test and it can be seen from the graph that, as the failure rate increases, the probability of passing the test falls.

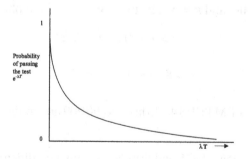

Figure 5.5: Zero failures test

The problem with this type of testing is known as the degree of *discrimination*, which depends on the statistical risks involved and is highlighted by the following example.

Assume, for the sake of argument, that the acceptable proportion of bad eggs (analogous to failure rate) is 10^{-4} (1 in 10 000). If the reader were to purchase six eggs each week then he or she would be carrying out a demonstration test having a zero-failures criterion. That is, with no bad eggs all is well, but if there is just one defective egg then a complaint will ensue. On the surface, this appears to be a valid test that carries a very high probability of being passed if the proportion of bad eggs is as stated.

Consider, however, the situation where the proportion increases dramatically to 10^{-3}, in other words by 10 times. What of the test? The next purchase of six eggs is still very unlikely to reveal a bad one. This test is therefore a poor discriminator and the example displays, albeit lightheartedly, the problem of demonstrating a very high reliability (low failure rate). In many cases a statistical demonstration can be totally unrealistic for the reasons described above.

A component has an acceptable failure rate of 1×10^{-6}/hrs (approx 1 in 100 yr). Fifty are tested for 2000 hrs (approx. 11 cumulative years of test) with a zero-failures pass criterion. λT is therefore

$$1 \times 10 - 6 \times 100\ 000 = 0.1 \text{ and the probability of passing is e} - 0.1 = 90\%$$

Suppose that an alternative test is made from a batch whose failure rate is four times that of the first batch (i.e. 4×10^{-6}/hr). Now the probability of passing the test is $e^{-\lambda T} = e^{-0.4} = 67\%$. Whereas the acceptable batch is 90% sure of acceptance (i.e. $\alpha = 10\%$) the 'bad' batch is only 33% sure of rejection (i.e. $\beta = 67\%$). In other words, although the test is satisfactory for passing batches of the required failure rate, it is a poor discriminator whose acceptance probability does not fall sufficiently quickly as the failure rate increases. This is shown in Figure 5.7.

A test is required that not only passes acceptable batches (a sensible producer's risk, α, would be between 5% and 15%) but rejects batches with a significantly higher failure rate. Four times the failure rate should reduce the acceptance probability to the order of 10% or less. The only way that this can be achieved is to increase the test time so that the acceptance criterion is much higher than zero failures (in other words, buy many more eggs!).

In general, the criterion for passing the test is n or fewer failures and the probability of passing the test is:

$$P_{0-n} = \sum_{i=0}^{n} \frac{\lambda^i T^i e^{-\lambda T}}{i!}$$

This expression yields the family of curves shown in Figure 5.6, which includes the special case $(n = 0)$ shown in Figure 5.5. These curves are known as Operating Characteristics (OC) curves, each one representing a test plan.

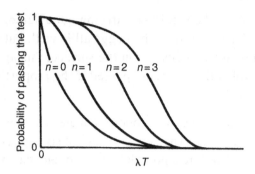

Figure 5.6: Family of OC curves

Each of these curves represents a valid test plan, and to demonstrate a given failure rate there is a choice of 0, 1, 2, 3,..., n failure criterion tests with corresponding values of T. The higher the number of failures, the greater is the number of test hours required. Figure 5.7, as did the above examples, shows the improvement in discrimination as n increases. Note that n is replaced by the symbol c, which is the usual convention.

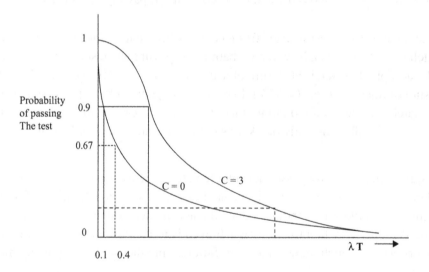

Figure 5.7: OC curves showing discrimination

Consider the $c = 0$ plan and note that a change from λ_0 to $3\lambda_0$ produces little decrease in the acceptance probability and hence a poor consumer's risk. If the consumer's risk were to be 10% the actual failure rate would be a long way to the right on the horizontal axis and would be many times λ_0. This ratio is known as the reliability design index or discrimination ratio. Looking now at the $c = 3$ curve, both producer and consumer risks are reasonable for a 4:1 change in failure rate. In the extreme case of 100% failures both risks would of course reduce to zero. Figure 5.8, being a set of cumulative poisson curves, is effectively the same as

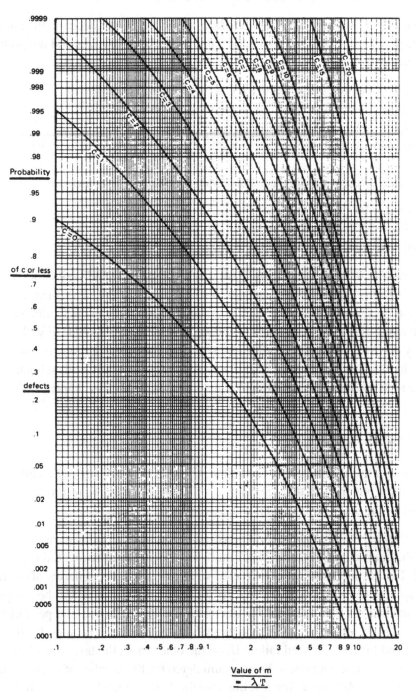

Figure 5.8: Cumulative poisson curves

Figure 5.7 but looks different in shape due to being on a logarithmic scale. It enables the test plans and risks to be evaluated as in the following example.

A failure rate of 3×10^{-4}/hr is to be demonstrated using 10 items. Calculate the number of test hours required if the test is to be passed with four or fewer failures and the probability of rejecting acceptable items (α) is to be 10%:

1. Probability of passing test $= 1 - 0.1 = 0.9$.
2. Using Figure 5.8 the corresponding value for $c = 4$ at 0.9 is 2.45.
3. $\lambda T = 3 \times 10^{-4} \times T = 2.45$. Therefore $T = 8170$ hrs.
4. Since there are 10 items the test must last 817 hrs with no more than four failures.

If the failure rate is four times the acceptable value, calculate the consumer's risk, β:

1. $3\lambda T = 4 \times 3 \times 10^{-4} \times 8170 = 9.8$.
2. Using Figure 5.8 for $m = 9.85$ and $c = 4$: $P_{0-4} = 0.03$.
3. The consumer's risk is therefore 3%.

Readers might care to repeat this example for a zero failures test and verify for themselves that, although T is as little as 333 hrs, β rises quickly to nearly 70%. The difficulty with high-reliability testing can now be appreciated. For example, equipment that should have a one-year MTBF requires at least three years of testing to demonstrate its MTBF with acceptable risks. If only one item is available for test then the duration of the demonstration would be three years. In practice, far larger MTBFs are aimed for, particularly with submarine cable and satellite systems, and demonstration testing, as described in this chapter, is therefore not appropriate.

5.7 Sequential Testing

The above type of test is known as a fixed-time demonstration. Owing to the difficulties of discrimination, any method that results in a saving of accumulated test hours without changing any of the other parameters is to be welcomed.

Experience shows that the sequential demonstration test tends to achieve results slightly faster than the equivalent fixed-time test. Figure 5.9 shows how a sequential reliability test is operated. Two parallel lines are constructed so as to mark the boundaries of the three areas – accept, reject and continue testing. As test hours are accumulated the test proceeds along the x-axis and as failures occur the line is moved vertically one unit per failure. Should the test line cross the upper boundary, too many failures have been accrued for the hours accumulated and the test has been failed. If, on the other hand, the test crosses the lower boundary, sufficient test hours have been accumulated for the number of failures and the test has been passed. As long as the test line remains between the boundaries the test must continue.

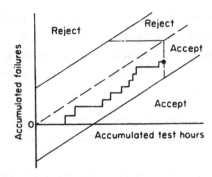

Figure 5.9: Truncated sequential demonstration test

If a time limit is set to the testing then a truncating line is drawn as shown to the right of the diagram so that if the line crosses above the mid-point, the test has been failed. If, as shown, it crosses below the mid-point, the test has been passed. If a decision is made by crossing the truncating line rather than one of the boundary lines, then the consumer and producer risks calculated for the test no longer apply and must be recalculated.

As in the fixed-time test, the consumer's risk, producer's risk and the MTBF associated with each are fixed. The ratio of the two MTBFs (or failure rates) is the reliability design index. The lines are constructed from the following equations:

$$y_{upper} = \frac{(1/\theta_1) - (1/\theta_0)}{\log_e(\theta_0/\theta_1)}T + \frac{\log_e A}{\log_e(\theta_0/\theta_1)}: A \approx \frac{1-\beta}{\alpha} \text{ and } B \approx \frac{\beta}{1-\alpha}$$

provided α and β are small (less than 25%).

The equation for y_{lower} is the same with $\log_e B$ substituted for $\log_e A$. If the risks are reduced then the lines move further apart and the test will take longer. If the design index is reduced, bringing the two MTBFs closer together, then the lines will be less steep, making it harder to pass the test.

5.8 Setting Up Demonstration Tests

In order to conduct a *demonstration test* (sometimes called a verification test) the following conditions, in addition to the statistical plans already discussed, must be specified:

1. Values of consumer's risk and acceptable failure rate or MTBF. The manufacturer will then decide on the risk and upon a reliability design index. This has already been examined in this chapter. A failure distribution must be assumed (this chapter has dealt only with random failures). A test plan can then be specified.
2. The sampling procedure must be defined in terms of sample size and from where and how the samples should be drawn.
3. Both environmental and operational test conditions must be fixed. This includes specifying the location of the test and the test personnel.

4. Failure must be defined so that there will be no argument over what constitutes a failure once the test has commenced. Exceptions should also be defined, i.e. failures that are to be disregarded (failures due to faulty test equipment, wrong test procedures, etc.).
5. If a 'burn-in' period is to be allowed, in order that early failures may be disregarded, this too must be specified.

The emphasis in this chapter has been on component testing and demonstration, but if equipment or systems are to be demonstrated, the following conditions must also be specified:

1. Permissible corrective or preventive maintenance during the test (e.g. replacement of parts before wearout, routine care).
2. Relevance of secondary failures (failures due to fluctuations in stress caused by other failures).
3. How test time relates to real time (24-hours operation of a system may only involve 3 hours of operation of a particular unit).
4. Maximum setting-up and adjustment time permitted before the test commences.

US Military Standard 781C – *Reliability Design Qualification and Production Acceptance Tests* – contains both fixed-time and sequential test plans. Alternatively, plans can be easily constructed from the equations and curves given in this chapter.

■ Exercises

1. A replacement test involving 50 devices is run for 100 hrs and then truncated. Calculate the MTBF (single-sided lower limit) at 60% confidence:
   (a) If there are two failures;
   (b) If there are zero failures.
2. The items in Exercise 1 are required to show an MTBF of 5000 hrs at 90% confidence. What would be the duration of the test, with no failures, to demonstrate this?
3. The producer's risk in a particular demonstration test is set at 15%. How many hours must be accumulated, with no failures, to demonstrate an MTBF of 1000 hrs? What is the result if a batch is submitted to the test with an MTBF of 500 hrs? If the test were increased to five failures what would be the effect on T and the consumer's risk, β?

Variable Failure Rates and Probability Plotting

6.1 The Weibull Distribution

The bathtub curve in Figure 2.7 showed that, as well as random failures, there are distributions involving increasing or decreasing failure rate. In these variable failure rate cases there is no point in considering the failure rate since it is continually changing. Only reliability and MTBF are meaningful. In Chapter 2 we saw that:

$$R(t) = \exp\left[-\int_0^t \lambda(t)dt\right]$$

Only the random failures case enabled us to simplify this to $R(t) = e^{-\lambda t}$ and to make use of the failure rate parameter. Since the relationship between failure rate and time can take many forms, and depends on the device/component in question, the above integral is of little direct use. Even if the variation of failure rate with time were known, it might well be of such a complex nature that the integration would prove far from simple.

In practice it is found that the distribution can usually be described by the following three-parameter distribution known as the Weibull distribution, named after Professor Waloddi Weibull:

$$R(t) = \exp\left[-\left(\frac{t-\gamma}{\eta}\right)^\beta\right]$$

In fact, for the majority of cases, a two-parameter model proves sufficient to describe the data. Hence:

$$R(t) = \exp\left[-\left(\frac{t}{\eta}\right)^\beta\right]$$

The constant failure rate case is therefore a special one-parameter case of the Weibull distribution (with $\beta = 1$). However, as we have seen, it is only randomness that can be described by a single parameter (i.e. failure rate).

The three parameters (γ, β, η) do not have physical meanings in the same way as does failure rate. They are parameters that allow us to compute reliability and MTBF. In the special case of $\gamma = 0$ and $\beta = 1$ the expression reduces to the exponential case with η giving the MTBF. In the general case, however, η is not the MTBF and is known as the scale parameter. β is known as

Reliability, Maintainability and Risk. DOI: 10.1016/B978-0-08-096902-2.00006-4

the shape parameter and describes the rate of change of failure rate (increasing or decreasing). γ is known as the location parameter, in other words a displacement of the time origin. $\gamma = 0$ means that the time origin is, in fact, at $t = 0$.

The following explanation shows how data, described by a Weibull function, can be made to fit a straight line. It is not essential to follow the explanation and the reader may, if desired, move to the next block of text.

The Weibull expression can be reduced to a straight-line equation by taking logarithms twice:

If $1 - R(t) = Q(t)$ ... the unreliability (probability of failure in t)

then

$$1 - Q(t) = \exp\left[-\left(\frac{t - \gamma}{\eta}\right)^{\beta}\right]$$

so that

$$\frac{1}{1 - Q(t)} = \exp\left(\frac{t - \gamma}{\eta}\right)^{\beta}$$

Therefore

$$\log\frac{1}{1 - Q(t)} = \left(\frac{t - \gamma}{\eta}\right)^{\beta}$$

and

$$\log\log\frac{1}{1 - Q(t)} = \beta\log(t - \gamma) - \beta\log\eta$$

which is $Y = mX + C$, the equation of a straight line.

If $(t - \gamma)$ is replaced by t' then:

$$Y = \log\log\frac{1}{1 - Q(t)} \quad \text{and} \quad X = \log t' \text{ and the slope } m = \beta.$$

If $Y = 0$

$$\log\log\frac{1}{1 - Q(t)} = 0$$

then

$$\beta\log t' = \beta\log\eta$$

so that

$$t' = \eta$$

This occurs if

$$\text{loglog}\frac{1}{1 - Q(t)} = 0 \quad \text{so that} \quad \log\frac{1}{1 - Q(t)} = 1$$

i.e.

$$\frac{1}{1 - Q(t)} = e \quad \text{and} \quad Q(t) = 0.63$$

If a group of failure times can be modeled by a Weibull function, and it is initially assumed that $\gamma = 0$, then by plotting the times to failure against time on double logarithmic paper (failure percentage on loglog scale and time on log scale), a straight line should be obtained. The three Weibull parameters and hence the expression for reliability may then be obtained from measurements of the slope and intercept.

Figure 6.1 is loglog by log graph paper with suitable scales for cumulative percentage failure and time. Cumulative percentage failure is effectively the unreliability and is estimated by taking each failure in turn from median ranking tables of the appropriate sample size. It should be noted that the sample size, in this case, is the number of failures observed. However, a test yielding 10 failures from 25 items would require the first 10 terms of the median ranking table for sample size 25.

6.2 Using the Weibull Method

6.2.1 Curve Fitting to Interpret Failure Data

Assume that the failure rate is not constant **OR**, alternatively, that we want to determine whether or not it is constant.

Whereas in the case of random failures (dealt with in Chapter 5) it was only necessary to know the total time T applying to the k failures, it is now essential to know the individual times to failure of the items. Without this information it would not be possible to fit the data to a distribution.

The Weibull technique assumes, initially, that the distribution of failures, whilst not random, is at least able to be modeled by a simple two-parameter distribution. It assumes that:

$$R(t) = \exp - (t/\eta)^{\beta}$$

The technique is to carry out a curve-fitting (probability-modeling) exercise first to establish that the data will fit this assumption and second to estimate the values of the two parameters.

Traditionally this was done by 'pencil and paper' curve-fitting methods, which are described here. Later in this chapter a software tool for performing this task is also described.

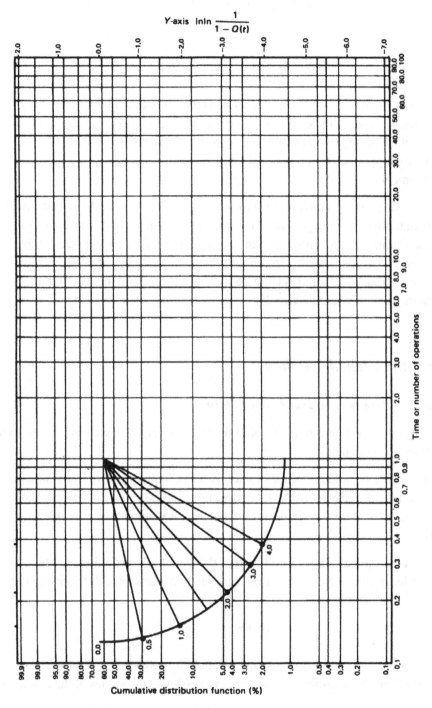

Figure 6.1: Graph paper for Weibull plot

If $\beta = 1$ then the failures are random and a constant failure rate can be assumed where failure rate $= 1/\eta$.

If $\beta > 1$ then the failure rate is increasing.

If $\beta < 1$ then the failure rate is decreasing.

In some cases, where the two-parameter distribution is inadequate to model the data, the three-parameter version can be used. In that case:

$$R(t) = \exp - [(t - \gamma)/\eta]^{\beta}$$

γ can be estimated by successive iteration until a fit to the two-parameter distribution is obtained. This will be described in Section 6.3.

6.2.2 Manual Plotting

Ten devices were put on test and permitted to fail without replacement. The time at which each device failed was noted and from the test information we are required to determine:

1. if there is a Weibull distribution that fits these data;
2. if so, the values of γ, η and β;
3. the probability of items surviving for specified lengths of time;
4. if the failure rate is increasing, decreasing or constant;
5. the MTBF.

The results are shown in Table 6.1 against the median ranks for sample size 10. The 10 points are plotted on Weibull paper as in Figure 6.2 and a straight line is obtained.

Table 6.1

Cumulative failures, Q_t (%) median rank	6.7	16.2	25.9	35.6	45.2	54.8	64.5	74.1	83.8	93.3
Time, t (hours × 100)	1.7	3.5	5.0	6.4	8.0	9.6	11.	13.	18.	22.

The straight line tells us that the Weibull distribution is applicable and the parameters are determined as follows:

   γ: It was shown in Section 6.1 that if the data yield a straight line then $\gamma = 0$.

   β: The slope yields the value of β, which is obtained by taking a line parallel to the data line but through the origin of the construction in Figure 6.2. The value of β is shown by the intersection with the arc. Here $\beta = 1.5$.

   η: We have already shown that $\eta = t$ for $Q(t) = 0.63$, hence η is obtained by taking a horizontal line from the origin of the construction across to the data line and then reading the corresponding value of t.

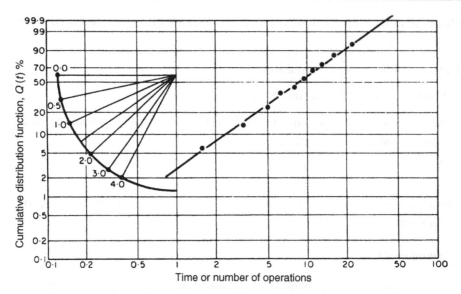

Figure 6.2: Results plotted on Weibull paper

The reliability expression is therefore:

$$R(t) = \exp\left[-\left(\frac{t}{1110}\right)^{1.5}\right]$$

The probability of survival to $t = 1000$ hrs is therefore:

$$R(1000) = e^{-0.855} = 42.5\%$$

The test indicates wearout since β, which is known as the shape parameter, >1.

    For increasing failure rate $\beta > 1$

    for decreasing failure rate $\beta < 1$

    for constant failure rate $\beta = 1$.

It now remains to evaluate the MTBF. This is, of course, the integral from zero to infinity of $R(t)$. Table 6.2 enables us to short-cut this step.

Since $\beta = 1.5$ then MTBF/$\eta = 0.903$ and MTBF $= 0.903 \times 1110 = 1002$ hrs. Since median rank tables have been used, the MTBF and reliability values calculated are at the 50% confidence level. In the example, time was recorded in hours but there is no reason why a more appropriate scale should not be used such as number of operations or cycles. The MTBF would then be quoted as mean number of cycles between failures.

For samples of other than 10 items, a set of median ranking tables is required. Since space does not permit a full set to be included the following approximation is given. For sample size N the rth rank is obtained from Bernard's approximation:

Table 6.2

β	$\dfrac{MTBF}{\eta}$	β	$\dfrac{MTBF}{\eta}$	β	$\dfrac{MTBF}{\eta}$	β	$\dfrac{MTBF}{\eta}$
0.0	∞	1.0	1.000	2.0	0.886	3.0	0.894
0.1	10!	1.1	0.965	2.1	0.886	3.1	0.894
0.2	5!	1.2	0.941	2.2	0.886	3.2	0.896
0.3	9.261	1.3	0.923	2.3	0.886	3.3	0.897
0.4	3.323	1.4	0.911	2.4	0.886	3.4	0.898
0.5	2.000	1.5	0.903	2.5	0.887	3.5	0.900
0.6	1.505	1.6	0.897	2.6	0.888	3.6	0.901
0.7	1.266	1.7	0.892	2.7	0.889	3.7	0.902
0.8	1.133	1.8	0.889	2.8	0.890	3.8	0.904
0.9	1.052	1.9	0.887	2.9	0.892	3.9	0.905
						4.0	0.906

$$\frac{r - 0.3}{N + 0.4}$$

Care must be taken in the choice of the appropriate ranking table. N is the number of items in the test and r the number that failed, in other words, the number of data points. In our example N was 10 not because the number of failures was 10 but because it was the sample size. As it happens, we considered the case where all 10 failed.

Had there been 20 items, of which 10 did not fail, the median ranks from Bernard's formula would have been:

%:   3.4   8.3   13   18   23   28   33   38   43   48

Although this method allows for the ranking of the failures it does not take account of the actual hours contributed by the censored items. In the next section, the maximum likelihood technique is introduced partly for this purpose.

6.2.3 Using the COMPARE Computer Tool

The COMPARE software package provides a tool for probability plotting whereby Weibull parameters are found that best fit the data being analyzed.

Repair times and censored data are entered and estimates of the Weibull parameters, as well as a graphical plot, are provided. The term 'censored data' refers to items that have not failed but, nevertheless, whose operating time needs to be taken account of. There are four types of censoring:

1. Items that continue after the last failure (the most usual type of censored data).
2. Items removed (for some reason other than failure) before the test finishes.
3. Items that are added after the beginning of the test and whose operating hours need to be included.

4. Failed items that, having been restored to 'as new' condition, then clock up further oper-
    ating time.

In the latter case it is important to be satisfied that the refurbishment really is 'as new'.
If so the additional hours count from the refurbishment and can be treated as if from an
extra item.

In practice it may happen that there is a time to failure for a particular failure mode. The item
might be repaired 'as new' and continue until it fails again. IMPORTANT – if the second
failure is the same mode then the time to failure is counted from the refurbishment. If the
second failure is a different mode then the time to failure is the whole operating time from the
commencement of the test.

It MUST be remembered, however, that any computerized algorithm will allocate parameters
to any data for a given distribution. It is, therefore, important to be aware of the limitations of
probability plotting.

Two methods of estimating the Weibull parameters from a set of times to failure are LEAST
SQUARES and MAXIMUM LIKELIHOOD.

The least squares method is used as an initial calculation and involves calculating the
hypothetical line for which the sum of the squares of the distances of the horizontal distances
from the data points to the line is a minimum. The Weibull parameters, BETA and ETA, are
obtained from the line. For the two-parameter Weibull distribution, the least squares estimates
are obtained from:

$$\text{BETA} = (\Sigma(Y_i)^2 - \bar{Y} \ \Sigma Y_i)/(\Sigma(X_iY_i - \bar{X} \ \Sigma Y_i)$$

$$\text{ETA} = \exp{(\bar{X} - \bar{Y}/\text{Beta})}$$

where

$$Y = \log_e \{\log_e[1/(1 - F(t))]\}$$

$$X = \log_e t$$

$$t = \text{time}$$

Because this least squares method involves treating each of the squared distances with
equal importance, it favors the higher values of time. Nevertheless, the least squares
estimates of BETA and ETA may well be adequate if there are very few or, better still, no
censored data. Nevertheless it does not take account of the censored data involving the
times with no failure (the survivors). To deal with this the maximum likelihood estimate is
required.

In COMPARE, the least squares estimates of BETA and ETA are used as the most reasonable estimate for commencing the iterative process of determining maximum likelihood values that give equal weight to each data point by virtue of calculating its probability of causing the estimated parameter. The algorithm generates the Weibull BETA and ETA parameters from which the data are most likely to have come by setting up a likelihood equation, differentiating with respect to BETA and ETA, and setting this equal to zero (in other words the standard calculus method of obtaining a minimum). The process is iterated for alternate BETA and ETA estimates until the values do not significantly change.

The maximum likelihood values are then taken as the best estimates of the Weibull parameters.

A large number of data collection schemes do not readily provide the times to failure of the items in question. For example, if an assembly (such as a valve) is replaced from time to time then its identity and its time to failure and replacement might be obtainable from the data. However, it might well be the diaphragm that is eventually the item of interest. Diaphragms may have been replaced during routine maintenance and the identity of each diaphragm not recorded. Subsequent Weibull analysis of the valve diaphragm would not then be possible. Careful thought has to be given when implementing a data collection scheme as to what subsequent data analysis will take place.

As in the above example of a valve and its diaphragm each of several failure modes will have its own failure distribution for which Weibull analysis may be appropriate. It is very likely, when attempting this type of modeling, that data not fitting the two-parameter distribution actually contain more than one failure mode. Separating out the individual failure modes may permit successful Weibull modeling.

6.2.4 Significance of the Result

The dangers of attempting to construct a Weibull plot with too few data points should be noted. A satisfactory result will not be obtained with fewer than at least six points. Tests yielding zero, one, two and even three failures do not allow a variable failure rate to be observed. In these cases constant failure rate must be assumed and the chi-square test used, which is a valid approach provided that the information extracted is applied only to the same time range as the test.

The comparison between the results obtained from least squares and maximum likelihood estimations (described above) provides an initial feel for how good a fit the data are to the inferred Weibull parameters.

If (in addition to the confidence obtained from the physical plot) the two values of shape parameter, obtained from least squares and maximum likelihood, are in good agreement then there is a further test.

This is provided by way of the Gnedenko test, which tests for constant failure rate. This is an 'F' test which tests the hypothesis that the failure times are at random, i.e. $\beta = 1$. The screen will state whether or not it is valid to reject the assumption that $\beta = 1$. The lower the value of the significance per cent then the more likely it is that the failure rate is significantly different from constant.

Essentially the test compares the MTTF of the failure times as grouped either side of the middle failure time and tests for a significant difference.

If the total number of failure times is n, and the time of the $n/2$th failure is T, the two estimates are:

$$\frac{\sum\limits_{i=1}^{n/2} t_i (n/2 \times T)}{n/2} \quad \text{and} \quad \frac{\sum\limits_{i=n/2+1}^{n} (t_i - T)}{n/2}$$

That is to say we are comparing the MTTF of the 'first half' of the failures and the MTTF of the 'second half'. The ratio should be one if the failure rate is constant. If it is not then the magnitude of the ratio gives an indication of significance. The ratio follows an 'F' distribution and the significance level can therefore be calculated. The two values of MTTF are shown on the screen. If this test were applied to the graphical plot in Section 6.2.2, we would see that, despite a fairly good straight line, the confidence that β is not 1 is only 32%!

It should be remembered that a small number of failure times, despite a high value of β, may not show a significant departure from the 'random' assumption. In practice 10 or more failure times is a minimum desirable data set for Weibull analysis. Nevertheless, engineering judgement should always be used to temper statistical analysis. The latter looks only at numbers and does not take account of known component behaviors.

> *Note*: If a poor fit is obtained from the two-parameter model, and the plot is a simple curve rather than 'S'-shaped or disjointed, then it is possible to attempt a three-parameter model by estimating the value of γ described in section 6.3. The usual approach is to assume that γ takes the value of the first failure time and to proceed, as above, with the two-parameter model to find η and β. Successive values of γ can be attempted, by iteration, until the two-parameter model provides a better fit. It must be remembered, however, that if the reason for a poor fit with the two-parameter model is that only a few failure times are available then the use of the three-parameter model is unlikely to improve the situation.

If the plot is 'S'-shaped, then it is possible that two failure modes are present in the data.

In the author's experience only a limited number of components show a significantly increasing failure rate. This is often due to the phenomenon (known as Drenick's law) whereby a mixture of three or more failure modes will show a random failure distribution irrespective of the BETAs of the individual modes.

6.2.5 Optimum Preventive Replacement

In Chapter 3 (Figure 3.1) the concept of optimum reliability/availability was introduced. Exactly the same picture applies to replacement interval for items with a wearout characteristic (i.e. $\beta > 1$). We may choose to replace an item that wears out at some arbitrary point in its life. The longer we leave the replacement then the greater is the chance of incurring the penalty costs associate with an 'unexpected' failure. On the other hand the more frequently we replace the item the more we spend on replacement items. There will be an optimum point at which to carry out the preventive replacement and this is dealt with in Chapter 16.3.

6.3 More Complex Cases of the Weibull Distribution

Suppose that the data in our example had yielded a curve rather than a straight line. It is still possible that the Weibull distribution applies but with γ greater than zero. The approach is to choose an assumed value for γ, usually the first value of t in the data, and replot the line against t', where $t' = t - \gamma$. The first data point is now not available and the line will be constructed from one fewer point. Should the result be a straight line then the value of γ is as estimated and one proceeds as before to evaluate the other two parameters. Mean time between failure is calculated as before plus the value of γ. If, on the other hand, another curve is generated then a further value of γ is tried until, by successive approximations, the correct value is found. This trial-and-error method of finding γ is not as time-consuming as it might seem. It is seldom necessary to attempt more than four approximations of γ before either generating a straight line or confirming that the Weibull distribution will not fit the data. One possible reason for the Weibull distribution not applying could be the presence of more than one failure mechanism in the data. Two mechanisms are unlikely to follow the same distribution and it is important to confine the analysis to one mechanism at a time.

So far, a single-sided analysis at 50% confidence has been described. It is possible to plot the 90% confidence bands by use of the 5% and 95% rank tables. First, Table 6.3 is constructed and the confidence bands plotted as follows.

Table 6.3

Time, t (hours × 100)	1.7	3.5	5.0	6.4	8.0	9.6	11.	13.	18.	22.
Median rank	6.7	16.2	25.9	35.6	45.2	54.8	64.5	74.1	83.8	93.3
5% rank	0.5	3.7	8.7	15.	22.	30.	39.	49.	61.	74.
95% rank	26.	39.	51.	61.	70.	78.	85.	91.	96.	99.

Consider the point corresponding to the failure at 500 hrs. The two points A and B are marked on the straight line corresponding to 8.7% and 51% respectively. The median rank for this point was 25.9% and vertical lines are drawn from A and B to intersect the horizontal. These

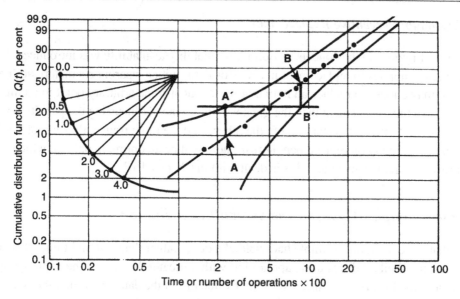

Figure 6.3: Ninety per cent confidence bands

two points lie on the confidence bands. The other points are plotted in the same way and confidence bands are produced as shown in Figure 6.3. Looking at the curves, the limits of $Q(t)$ at 1000 hrs are 30% and 85%. At 90% confidence the reliability for 1000 hrs is therefore between 15% and 70%.

6.4 Continuous Processes

There is a very strict limitation to the use of this Weibull method, which is illustrated by the case of filament lamps. It is well known that these do not fail at random. Indeed, they have a pronounced wearout characteristic with a β of two. However, imagine a brand new building with brand new lamps. Due to the distribution of failures, very few will fail in the first few months, perhaps only a few in the next few months and several towards the end of the year. After several years, however, the lamps in the building will all have been replaced at different times and the number failing in any month will be approximately the same. Thus, a population of items with increasing failure rate appears as a constant failure rate system. This is an example of a continuous process, and Figure 6.4 shows the failure characteristic of a single lamp and the superimposition of successive generations.

If the intervals between failure were observed, ranked and plotted in a Weibull analysis then a β of 1 would be obtained. Weibull analysis must not therefore be used for the times between failure within a continuous process but only for a number of items whose individual times to failure are separately recorded. It is not uncommon for people to attempt the former and obtain a totally false picture of the process.

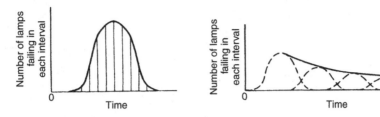

Figure 6.4: Failure distribution of a large population

More suitable models for this case are the reliability growth models (CUSUM and Duane) described in Chapter 12. Another is to apply the Laplace test, which provides a means of indicating if the process failure rate has a trend.

If a system exhibits a number of failures after time zero at times $x_1, x_2, x_3, \ldots, x_i$, then the test statistic for the process is

$$U = \frac{(\Sigma x_i / n) - (x_0 / 2)}{x_0 \sqrt{(1/12n)}}$$

x_0 is the time at which the test is truncated. If $U = 0$ then there is no trend and the failure rate is not changing. If $U < 0$ then the failure rate is decreasing and if $U > 0$ it is increasing.

This test could be applied to the analysis of software failures since they are an example of a continuous repair process.

Exercises

1. Components, as described in the example of Section 6.2, are to be used in a system. It is required that these are preventively replaced such that there is only a 5% probability of their failing beforehand. After how many hours should each item be replaced?
2. A sample of 10 items is allowed to fail and the time for each failure is as follows:

    4, 6, 8, 11, 12, 13, 15, 17, 20, 21 (thousand hours)

    Use the Weibull paper in this chapter to determine the reliability characteristic and the MTBF.

Predicting Reliability and Risk

CHAPTER 7

Basic Reliability Prediction Theory

7.1 Why Predict RAMS?

Reliability prediction (i.e. modeling) is the process of calculating the anticipated system RAMS from assumed component failure rates. It provides a quantitative measure of how close a proposed design comes to meeting the design objectives and allows comparisons to be made between different design proposals. It has already been emphasized that reliability prediction is an imprecise calculation, but it is nevertheless a valuable exercise for the following reasons:

- It provides an early indication of a system's potential to meet the design reliability requirements.
- It enables an assessment of life-cycle costs to be carried out.
- It enables one to establish which components, or areas, in a design contribute to the major portion of the unreliability.
- It enables trade-offs to be made as, for example, between reliability, maintainability and proof-test intervals in achieving a given availability.
- Its use is increasingly called for in invitations to tender, contracts and in safety-integrity standards.

It must be stressed that prediction is a design tool and not a precise measure of reliability. The main value of a prediction is in showing the relative reliabilities of modules so that allocations can be made. Whatever the accuracy of the exercise, if one module is shown to have double the MTBF of another then, when calculating values for modules in order to achieve the desired system MTBF, the values allocated to the modules should be in the same ratio. Prediction also permits a reliability comparison between different design solutions. Again, the comparison is likely to be more accurate than the absolute values. The accuracy of the actual predicted value will depend on:

1. Relevance of the failure rate data and the chosen environmental multiplication factors.
2. Accuracy of the mathematical model.
3. The absence of gross over-stressing in operation.
4. Tolerance of the design to component parametric drift.

Reliability, Maintainability and Risk. DOI: 10.1016/B978-0-08-096902-2.00007-6

The greater the number of different component types involved, the more likely that individual over- and under-estimates will cancel each other out.

It is important to state, at this point, that quantified reliability prediction (the subject of this chapter and of Chapters 9 and 10) generally applies to random hardware failures only. Systematic failures, which were introduced in Chapter 1 (and are dealt with in Chapter 17), are NOT quantified by failure rates because they are not device specific and do not repeat themselves once they have been designed out. They are therefore dealt with quite separately by addressing the 'rigor of life-cycle activities'.

7.2 Probability Theory

The following basic probability rules are sufficient for an understanding of the system modeling involved in reliability prediction.

7.2.1 The Multiplication Rule

If two or more events can occur simultaneously, and their individual probabilities of occurring are known, then the probability of simultaneous events is the product of the individual probabilities provided, of course, that they are INDEPENDENT of each other. The shaded area in Figure 7.1 represents the probability of events A and B occurring simultaneously. Hence the probability of A and B occurring is:

$$Pab = Pa \times Pb$$

Generally

$$Pan = Pa \times Pb, ..., \times Pn$$

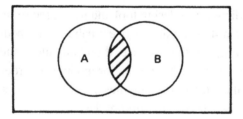

Figure 7.1

7.2.2 The Addition Rule

It is also required to calculate the probability of either event A *or* event B *or both* occurring. This is the area of the two circles in Figure 7.1. This probability is:

$$P(a \text{ or } b) = pa + pb - papb$$

being the sum of Pa and Pb less the area $PaPb$, which is included twice. This becomes:

$$P(a \text{ or } b) = 1 - (1 - pa)(1 - pb)$$

Hence the probability of one or more of n events occurring is:

$$= 1 - (1 - pa)(1 - pb),\ldots,(1 - pn)$$

7.2.3 The Binomial Theorem

The above two rules are combined in the binomial theorem. Consider the following example involving a pack of 52 playing cards. A card is removed at random, its suit noted, and then replaced. A second card is then removed and its suit noted. The possible outcomes are:

two hearts
one heart and one other card
two other cards.

If p is the probability of drawing a heart then, from the multiplication rule, the outcomes of the experiment can be calculated as follows:

probability of two hearts p^2
probability of one heart $2pq$
probability of no hearts q^2.

Similar reasoning for an experiment involving three cards will yield:

probability of three hearts p^3
probability of two hearts $3p^2q$
probability of one heart $3pq^2$
probability of no hearts q^3.

The above probabilities are the terms of the expressions $(p + q)^2$ and $(p + q)^3$. This leads to the general statement that if p is the probability of some random event, and if $q = 1 - p$ then the probabilities of 0, 1, 2, 3,..., outcomes of that event in n trials are given by the terms of the expansion:

$$(p + q)^n \text{ which equals}$$

$$p^n, np^{(n-1)}q, \frac{n(n-1)p^{(n-2)}q^2}{2!}, \ldots, q^n$$

This is known as the binomial expansion.

7.2.4 Bayes Theorem

The marginal probability of an event is its simple probability. Consider a box of seven cubes and three spheres, in which case the marginal probability of drawing a cube is 0.7. To introduce the concept of a conditional probability, assume that four of the cubes are black and three white and that, of the spheres, two are black and one is white, as shown in Figure 7.2.

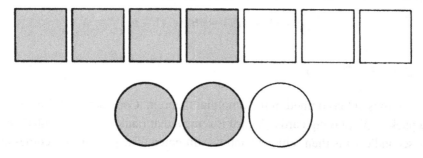

Figure 7.2

The probability of drawing a black article, given that it turns out to be a cube, is a conditional probability of 4/7 and ignores the possibility of drawing a sphere. Similarly the probability of drawing a black article, given that it turns out to be a sphere, is 2/3. On the other hand, the probability of drawing a black sphere is a joint probability. It acknowledges the possibility of drawing cubes and spheres and is therefore 2/10.

Comparing joint and conditional probabilities, the conditional probability of drawing a black article given that it is a sphere is the joint probability of drawing a black sphere (2/10) divided by the probability of drawing any sphere (3/10). The result is therefore:

$$P_{b/s} = \frac{P_{bs}}{P_s}$$

given that: $P_{b/s}$ is the conditional probability of drawing a black article given that it is a sphere; P_s is the simple or marginal probability of drawing a sphere; P_{bs} is the joint probability of drawing an article which is both black and a sphere.

This is known as the Bayes theorem. It follows then that $P_{bs} = P_{b/s} \cdot P_s$ or $P_{s/b} \cdot P_b$. Consider now the probability of drawing a black sphere (P_{bs})and the probability of drawing a white sphere (P_{ws}):

$$P_s = P_{bs} + P_{ws}$$

Therefore

$$P_s = P_{s/b} \cdot P_b + P_{s/w} \cdot P_w$$

and, in general,

$$P_x = P_{x/a} \cdot P_a + P_{x/b} \cdot P_b, \ldots, + P_{x/n} \cdot P_n$$

which is the form applicable to prediction formulae.

7.3 Reliability of Series Systems

Consider the two valves connected in series that were described in Figure 2.1 of Chapter 2.

One of the failure modes discussed was loss of supply, which occurs if *either* valve fails closed. This situation, where any failure causes the system to fail, is known as *series* reliability. This must not be confused with the series configuration of the valves shown in Figure 2.1. It so happens that, for this loss of supply failure mode, the physical series and the reliability series diagrams coincide. When we consider the over-pressure case in the next section it will be seen that, although the valves are still in series, the reliability block diagram changes.

For loss of supply then, the reliability of the system is the probability that Valve A does not fail *and* Valve B does not fail.

From the multiplication rule in Section 7.2.1 then:

$$R_{ab} = R_a \cdot R_b \text{ and, in general,}$$
$$R_{an} = R_a \cdot R_b, \ldots, R_n$$

In the constant failure rate case where:

$$R_a = e^{-\lambda_a t}$$

then

$$R_{an} = \exp\left[-(\lambda_a + \lambda_b, \ldots, \lambda_n)t\right]$$

from which it can be seen that the system is also a constant failure rate unit whose reliability is of the form e^{-Kt}, where K is the sum of the individual failure rates. Provided that the two assumptions of constant failure rate and series modeling apply, then it is valid to speak of a system failure rate computed from the sum of the individual unit or component failure rates.

The practice of adding up the failure rates in a component count type prediction assumes that any single failure causes a system failure. It is therefore a worst-case prediction since, clearly, a failure mode analysis against a specific failure mode will involve only those components which contribute to that top event.

Returning to the example of the two valves, assume that each has a failure rate of 7×10^{-6} per hour for the fail closed mode and consider the reliability for one year. One year has 8760 hours.

From the above:

$$\lambda_{\text{system}} = \lambda_a + \lambda_b = 14 \times 10^{-6} \text{ per hour}$$
$$\lambda t = 8760 \times 14 \times 10^{-6} = 0.1226$$
$$R_{\text{system}} = e^{-\lambda t} = 0.885$$

7.4 Redundancy Rules

7.4.1 General Types of Redundant Configuration

There are a number of ways in which redundancy can be applied. These are summarized in diagrammatic form in Figure 7.3. So far, we have considered only the special case of full active redundancy. The models for the other cases will be described in the following sections. At present, we are considering redundancy without repair and it is assumed that failed redundant units remain failed until the whole system fails. The point concerning variable failure rate applies equally to each of the models.

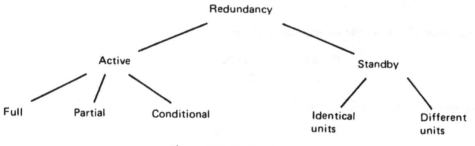

Figure 7.3: Redundancy

7.4.2 Full Active Redundancy (Without Repair)

Continuing with our two-valve example, consider the over-pressure failure mode described in Chapter 2. This is no longer a reliability *series* situation since both valves need to fail open in order for the top event to occur. In this case a *parallel* reliability block diagram is used. Since either or both valves operating correctly is sufficient for system success, then the addition rule in Section 7.2.2 applies. For the two valves it is:

$$R_{\text{system}} = 1 - (1 - R_a)(1 - R_b) \text{ or, in another form}$$
$$R_{\text{system}} = R_a + R_b - R_a R_b$$

In other words, one minus the product of their unreliabilities. Let us assume that the fail open failure rate of a valve is 3×10^{-6} per hour:

$$R_a = R_b = e^{-\lambda t} \quad \text{where } \lambda t = 3 \times 10^{-6} \times 8760 = 0.026$$
$$e^{-\lambda t} = 0.974$$
$$R_{\text{system}} = 1 - (0.026)^2 = \underline{0.999}$$

If there were N items in this redundant configuration such that all may fail except one, then the expression becomes:

$$R_{\text{system}} = 1 - (1 - R_a)(1 - R_b),\ldots,(1-R_n)$$

There is a pitfall at this point that is important to emphasize. The reliability of the system, after substitution of $R = e^{-\lambda t}$, becomes:

$$R_s = 2e^{-\lambda t} - e^{-2\lambda t}$$

It is very important to note that, unlike the series case, this combination of constant failure rate units exhibits a reliability characteristic that is *not* of the form e^{-kt}. In other words, although constant failure rate units are involved, the failure rate of the system is variable. The MTBF can therefore be obtained only from the integral of reliability. In Chapter 2, we saw that

$$\text{MTBF} = \int_0^\infty R(t)\, dt$$

Hence

$$\text{MTBF} = \int_0^\infty (2e^{-\lambda t} - e^{-2\lambda t})$$
$$= 2/\lambda - 1/2\lambda$$
$$= 3/2\lambda$$
$$= \underline{3\theta/2} \text{ where } \theta \text{ is the MTBF of a single unit.}$$

In the above working we substituted θ for $1/\lambda$, which was correct because a unit was being considered for which constant λ applies. The danger now is to assume that the failure rate of the system is $2\lambda/3$. This is not true since the practice of inverting MTBF to obtain failure rate, and vice versa, is valid only *for constant failure rate*.

Figure 7.4 compares reliability against time, and failure rate against time, for series and redundant cases. As can be seen, the failure rate is initially zero because we can argue that there must be an interval short enough such that only one failure can occur. It then increases asymptotically towards the failure rate of one item because we argue that eventually there is a time long enough that one must have failed and therefore only one remains. Reliability in a redundant configuration stays higher than for constant failure rate at the beginning but eventually falls more sharply. The greater the number of redundant units, the longer the period of higher reliability and the sharper the decline. These features of redundancy apply, in principle, to all redundant configurations and only the specific values change.

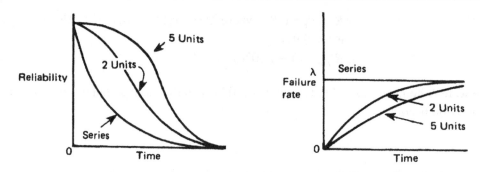

Figure 7.4 Effect of redundancy on reliability and failure rate

7.4.3 Partial Active Redundancy (Without Repair)

Consider three identical units each with reliability R. Let $R + Q = 1$ so that Q is the unreliability (probability of failure in a given time). The binomial expression $(R + Q)^3$ yields the following terms:

$$R^3, 3R^2Q, 3RQ^2, Q^3, \text{which are}$$
$$R^3, 3R^2(1 - R), 3R(1 - R)^2, (1 - R)^3$$

This conveniently describes the probabilities of

$$0, 1, 2, 3 \text{ failures of a single unit.}$$

In Section 7.4.2 the reliability for full redundancy was seen to be:

$$1 - (1 - R)^3$$

This is consistent with the above since it can be seen to be one minus the last term. Since the sum of the terms is unity reliability, it is therefore the sum of the first three terms that, being the probability of zero, one or two failures, is the reliability of a fully redundant system.

In many cases of redundancy, however, the number of units permitted to fail before system failure occurs is less than in full redundancy. In the example of three units, full redundancy requires only one to function, whereas partial redundancy would exist if two units were required with only one allowed to fail. Once again the reliability can be obtained from the binomial expression since it is the probability of zero or one failures that is given by the sum of the first two terms. Hence:

$$R_{system} = R^3 + 3R^2(1 - R)$$
$$= 3R^2 - 2R^3$$

In general, if r items may fail out of n then the reliability is given as the sum of the first $r + 1$ terms of the binomial expansion $(R + Q)^n$. Therefore:

$$R = R^n + nR^{n-1}(1 - R) + \frac{n(n-1)R^{n-2}(1-R)^2}{2!} + \dots$$
$$+ \frac{n(n-1)\dots(n-r+1)R^{n-r}(1-R)^r}{r!}$$

7.4.4 Conditional Active Redundancy

This is best considered by an example. Consider the configuration in Figure 7.5. Three identical digital processing units (A, B and C) have reliability R. They are triplicated to provide redundancy in the event of failure and their identical outputs are fed to a two-out-of-three majority voting gate. If two identical signals are received by the gate then they are reproduced at the output. Assume that the voting gate is sufficiently more reliable than the units so that its probability of failure can be disregarded. Assume also that the individual units can fail either to an open-circuit or a short-circuit output. Random data bit errors are not included in the definition of system failure for the purpose of this example. The question arises as to whether the system has:

partial redundancy   one unit may fail but no more, or
full redundancy   two units may fail.

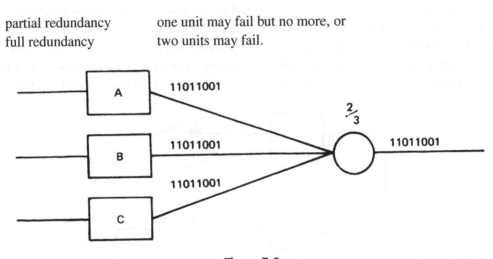

Figure 7.5

The answer is conditional on the mode of failure. If two units fail in a like mode (both outputs logic 1 or logic 0) then the output of the voting gate will be held at the same value and the system will have failed. If, on the other hand, they fail in unlike modes then the remaining unit will produce a correct output from the gate since it always sees an identical binary bit from one of the other units. This conditional situation requires the Bayes theorem introduced in Section 7.2.4. The equation becomes:

$$R_{\text{system}} = R_{\text{given } A} \cdot P_A + R_{\text{given } B} \cdot P_B, \dots, + R_{\text{given } N} \cdot P_N$$

where A to N are mutually exclusive and $\sum_{i=A}^{i=N} P_i = 1$

In this case the solution is:

$$R_{\text{system}} = R_{\text{system given that in the event of failure, 2 units fail alike}} \times P_{\text{failing alike}}$$
$$+ R_{\text{system given that in the event of failure, 2 units fail unalike}} \times P_{\text{failing unalike}}$$

Therefore:

$$R_s = [R^3 + 3R^2(1 - R)]P_A + [1 - (1 - R)^3]P_B$$

since if two units fail alike there is partial redundancy and if two units fail unalike there is full redundancy. Assume that the probability of both failure modes is the same and that $P_A = P_B = 0.5$. The system reliability is therefore:

$$R_s = \frac{R^3 + 3R^2 - 3R^3 + 1 - 1 + 3R - 3R^2 + R^3}{2} = \frac{3R - R^3}{2}$$

7.4.5 Standby Redundancy

So far only active redundancy has been considered where every unit is operating and the system can function despite the loss of one or more units. Standby redundancy involves additional units that are activated only when the operating unit fails. A greater improvement, per added unit, is anticipated than with active redundancy since the standby units operate for less time. Figure 7.6 shows *n* identical units with item 1 active. Should a failure be detected

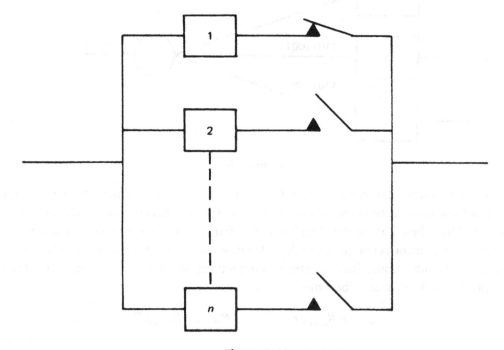

Figure 7.6

then item 2 will be 'switched-in' to take its place. Initially, the following assumptions are made:

1. The means of sensing that a failure has occurred and for switching from the defective to the standby unit is assumed to be failure free.
2. The standby unit(s) are assumed to have identical, constant failure rates to the main unit.
3. The standby units are assumed not to fail while in the idle state.
4. As with the earlier calculation of active redundancy, defective units are assumed to remain so. No repair is effected until the system has failed.

Calculations involving redundancy and repair are covered in the next chapter. The reliability is then given by the first *n* terms of the Poisson expression:

$$R_{system} = R(t) = e^{-\lambda t}\left(1 + \lambda t + \frac{\lambda^2 t^2}{2!} \cdots \frac{\lambda^{(n-1)} t^{(n-1)}}{(n-1)!}\right)$$

which reduces, for two units, to:

$$R_{system} = e^{-\lambda t}(1 + \lambda t)$$

Figure 7.7 shows the more general case of two units with some of the above assumptions removed. In the figure:

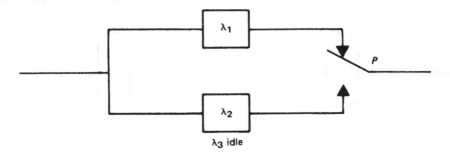

Figure 7.7

λ_1 is the constant failure rate of the main unit,
λ_2 is the constant failure rate of the standby unit when in use,
λ_3 is the constant failure rate of the standby unit in the idle state,
P is the one-shot probability of the switch performing when required.

The reliability is given by:

$$R_{system} = e^{-\lambda_1 t} + \frac{P\lambda_1}{\lambda_2 - \lambda_1 - \lambda_3}\left(e^{-(\lambda_1 + \lambda_3)t} - e^{-\lambda_2 t}\right)$$

It remains only to consider the following failure possibilities. Let λ_4, λ_5 and λ_6 be the failure rates associated with the sums of the following failure modes:

λ_4: dormant failures that inhibit failure sensing or changeover;

λ_5: failures causing the incorrect switching back to the failed unit;

λ_6: false sensing of non-existent failure.

If we think about each of these in turn it will be seen that, from the point of view of the above model:

λ_4 is part of λ_3,

λ_5 is part of λ_2,

λ_6 is part of λ_1.

In the analysis they should therefore be included in the appropriate category.

7.4.6 Load Sharing

The following situation can be deceptive since, at first sight, it appears as active redundancy. Figure 7.8 shows two capacitors connected in series. Given that both must fail short circuit in order for the system to fail, we require a model for the system. It is *not* two units in active redundant configuration because if the first capacitor should fail (short circuit) then the voltage applied to the remaining one will be doubled and its failure rate greatly increased. This situation is known as load sharing and is mathematically identical to a standby arrangement.

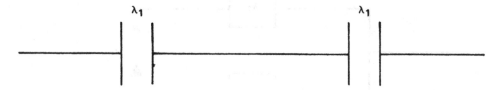

Figure 7.8

Figure 7.9 shows two units in standby configuration. The switchover is assumed to be perfect (which is an appropriate assumption) and the standby unit has an idle failure rate equal to zero with a different (larger) failure rate after switchover. The main unit has a failure rate of twice the single capacitor.

7.5 General Features of Redundancy

7.5.1 Incremental Improvement

As was seen in Figure 7.4, the improvement resulting from redundancy is not spread evenly along the time axis. Since the MTBF is an overall measure obtained by integrating reliability from zero to infinity, it is actually the area under the curve of reliability against time. For

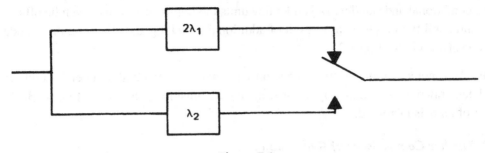

Figure 7.9

short missions (less than one MTBF in duration) the actual improvement in reliability is greater than would be suggested simply by comparing MTBFs. For this reason, the length of mission should be taken into account when evaluating redundancy.

As we saw in Section 7.4, the effect of duplicating a unit by active redundancy is to improve the MTBF by only 50%. This improvement falls off as the number of redundant units increases, as is shown in Figure 7.10. The effect is similar for other redundant configurations

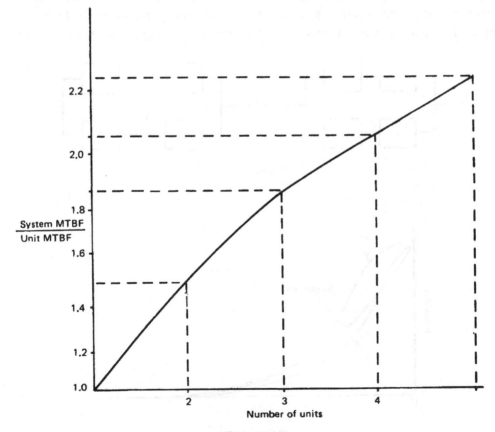

Figure 7.10

such as conditional and standby. Beyond a few units the improvement may even be offset by the unreliability introduced as a result of additional switching and other common mode effects dealt with in Section 8.2.

Figure 7.10 is not a continuous curve since only the points for integral numbers of units exist. It has been drawn, however, merely to illustrate the diminishing enhancement in MTBF as the number of units is increased.

7.5.2 Further Comparisons of Redundancy

Figure 7.11 shows two alternative configurations involving four units in active redundancy: (i) protects against short-circuit failures (ii) protects against short- and open-circuit conditions. As can be seen from Figure 7.12, (ii) has the higher reliability but is harder to implement. If readers care to calculate the MTBF of (i), they will find that it can be less than for a single unit and, as can be seen from the curves, the area under the reliability curve (MTBF) is less. It is of value only for conditions where the short-circuit failure mode is more likely.

Figure 7.13 gives a comparison between units in both standby and active redundancy. For the simple model assuming perfect switching the standby configuration has the higher reliability, although in practice the associated hardware for sensing and switching will erode

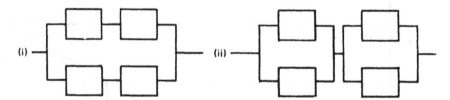

Figure 7.11

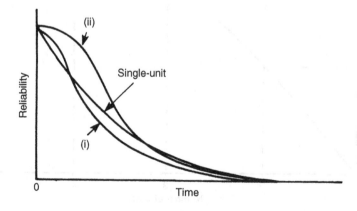

Figure 7.12

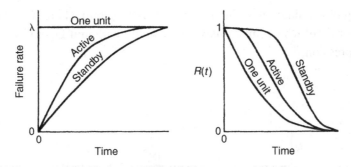

Figure 7.13

the advantage. On the other hand, it is not always easy to achieve active redundancy with true independence between units. In other words, the failure of one unit may cause or at least hasten the failure of another. This common mode effect will be explained in the next chapter (Section 8.2).

7.5.3 Redundancy and Cost

It must always be remembered that redundancy adds:

> capital cost
> weight
> spares
> space
> preventive maintenance cost
> power consumption
> more failures at the unit level (hence more corrective maintenance).

Each of these contributes substantially to cost.

Exercises

1. Provide an expression for the MTBF of the system shown in the following block diagram in Figure 7.14.

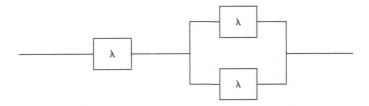

Figure 7.14

2. The following block diagram in Figure 7.15 shows a system whereby unit B may operate with units D or E but where unit A may only operate with unit D, or C with E. Derive the reliability expression.

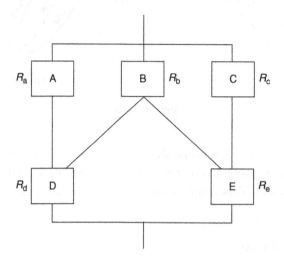

Figure 7.15

CHAPTER 8

Methods of Modeling

In Chapter 1 (Section 1.3) and Chapter 7 (Section 7.1) the limitations of reliability of prediction were emphasized. This chapter describes, in some detail, the methods used.

8.1 Block Diagrams and Repairable Systems

8.1.1 Reliability Block Diagrams

The following is the general approach to block diagram analysis.

8.1.1.1 Establish failure criteria

It is important to define what constitutes a system failure since only then can one determine which failure modes, at the component level, actually cause the system to fail. There may well be more than one type of system failure, in which case a number of predictions giving different reliabilities will be required. This step is absolutely essential if the predictions are to have any significance. It was explained, in Section 2.1, how different system failure modes can involve quite different component failure modes and, indeed, even different series/redundant configurations.

8.1.1.2 Create a reliability block diagram

It is necessary to describe the system as a number of functional blocks which are interconnected according to the effect of each block failure on the overall system reliability.

Figure 8.1 is a series diagram representing a system of two blocks such that the failure of either block prevents operation of the system. Figure 8.2 models the situation where both blocks must fail in order for the system to fail. This is known as a parallel, or redundancy, case. Figure 8.3 shows a combination of series and parallel reliability. It represents a system that will fail if block A fails or if both block B and block C fail. The failure of B or C alone is insufficient to cause system failure.

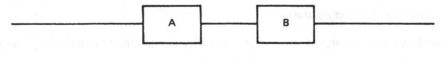

Figure 8.1

Reliability, Maintainability and Risk. DOI: 10.1016/B978-0-08-096902-2.00008-8

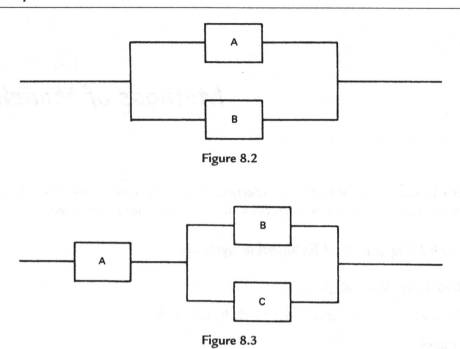

Figure 8.2

Figure 8.3

A number of general rules should be borne in mind when defining the blocks.

1. Each block should represent the maximum number of components in order to simplify the diagram.
2. The function of each block should be easily identified.
3. Blocks should be mutually independent in that failure in one should not affect the probability of failure in another (see Section 8.2, which deals with common cause failure).
4. Blocks should not contain any significant redundancy otherwise the addition of failure rates, within the block, would not be valid.
5. Each replaceable unit should be a whole number of blocks.
6. Each block should contain one technology, that is, electronic or electro-mechanical.
7. There should be only one environment within a block.

8.1.1.3 Failure mode analysis

Failure Mode and Effect Analysis (FMEA) is described later in Chapter 9 (Section 9.3). It provides block failure rates by examining individual component failure modes and failure rates. Given a constant failure rate and no internal redundancy, each block will have a failure rate predicted from the sum of the failure rates on the FMEA worksheet.

8.1.1.4 Calculation of system reliability

Relating the block failure rates to the system reliability is mathematical modeling, the subject of this chapter. In the event that the system reliability prediction fails to meet the objective then improved failure rate (or down time) objectives must be assigned to each block by means of reliability allocation.

8.1.1.5 Reliability allocation

The importance of reliability allocation is stressed in Chapter 11 and an example is calculated. The block failure rates are taken as a measure of the complexity, and improved, suitably weighted objectives are set.

8.1.2 Repairable Systems (Revealed Failures)

In Chapter 7 the basic rules for series and redundant systems were explained. For redundant systems, however, the equations only catered for the case whereby failed redundant units were not repaired. In other words, the reliability was calculated as the reliability of failure, given that failed units stay failed. It is now necessary to address the repairable case.

The traditional approach given in the sixth edition of *Reliability Maintainability and Risk,* together with the majority of textbooks and standards, was challenged in the late 1990s. It is now generally acknowledged that the traditional Markov model does not correctly represent the normal repair activities for redundant repairable systems **when calculating the Probability of Failure on Demand (PFD).** The Journal of The Safety and Reliability Society, *Safety and Reliability* Volume 22, No. 2, Summer 2002, published a paper by W. G. (Bill) Gulland, and another by Kenneth G. L. Simpson, both of which agreed with those findings. The rest of this section therefore provides slightly different formulae for repairable systems than were given in earlier editions.

The problem arises because repair of multiple failures is not necessarily a Markov process (namely that the probability of being in a state can be determined solely from knowledge of the previous state). As a result, Markov modeling gives the correct answer for system failure rate but not for PFD.

For a redundant repairable system without a dedicated repair crew for each unit, the transition from a multiple failure state to an operating state does not depend on the repair of the last failure (as it should for the process to be applicable to a Markov model). In practice, it is related to the repair of the previous failure. For this reason a Markov model is pessimistic because it under-estimates the transition rate from the failed state. It is as if the repair crew abandon the earlier repair **(not so in practice)** to carry out the repair of the latest failure. With a dedicated repair crew per unit then, the repair of the last failure is independent of preceding failures and the process is approximately Markov with the calculations giving the correct answer. However, real-life situations seldom follow this scenario.

It is worth mentioning that, as with all redundant systems, the total system failure rate (or PFD) will usually be dominated by the effect of Common Cause Failure (CCF) dealt with in Section 8.2.

For interest, the traditional (incorrect) Markov model is given in Appendix 13. Coincidentally it does give the correct answer for system failure rate (as mentioned above) if the multiple repair crew assumption is used.

The modeling of systems with revealed failures will be carried out here using simple probability rules.

Consider a simple duplicated (one out of two) redundant system. In other words, one unit out of the two must work for the system to operate. System failure, therefore, requires two failed units. One can argue as follows:

The failure rate, whereby a given unit fails, is, by definition, λ. There are, however, two units and thus the rate for either failing is 2λ. For the system to fail it requires that, coincidentally, the other unit is in the failed state. This is another way of saying it is unavailable. Its unavailability is λ MDT (as we saw in Section 2.6 of Chapter 2).

Therefore, the system failure rate is $2\lambda \times \lambda$ MDT

$$= 2\lambda^2 \text{MDT}$$

The reader can develop the same argument for units with different failure rates and down times to yield

$$\lambda_1 \lambda_2 (\text{MDT}_1 + \text{MDT}_2)$$

The general expression for m units requiring to operate out of a total of n (with the same failure rates, λ, and mean down times, MDT) becomes:

$$n!/[(n-m)!\,(m-1)!] \cdot \lambda^{(n-m+1)} \text{MDT}^{(n-m)}$$

This leads to the following Table 8.1, in which MDT refers to unit MDT.

Table 8.1: System Failure Rates (Revealed)

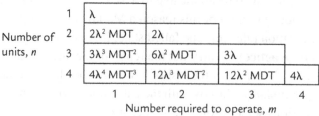

Number of units, n	1	2	3	4
1	λ			
2	$2\lambda^2$ MDT	2λ		
3	$3\lambda^3$ MDT^2	$6\lambda^2$ MDT	3λ	
4	$4\lambda^4$ MDT^3	$12\lambda^3$ MDT^2	$12\lambda^2$ MDT	4λ

Number required to operate, m

Turning to the system unavailability, we need to consider the system mean down time. This is not the same as the unit mean down time (MDT).

Thinking about the simple one out of two case, when the second failure occurs, the repair crew will, on average, be half way through the repair of the first failure. Therefore, the system mean down time will be MDT/2. This argument applies irrespective of whether there is a single or multiple repair crews.

Thus the system unavailability becomes the product of system failure rate and down time = $2\lambda^2$ **MDT** $\times$ **MDT/2**

$$= \lambda^2 \text{MDT}^2$$

The general value of down time for m units requiring to operate out of a total of n becomes MDT/$[n - m + 1]$. Thus, taking Table 8.1 and dividing each cell by the appropriate down time yields Table 8.2 for system unavailability based on:

$$n!/[(n - m + 1)!(m - 1)!] \cdot \lambda^{(n-m+1)} \text{MDT}^{(n-m+1)}$$

in which MDT refers to the MDT of one unit.

Table 8.2: System Unavailabilities (Revealed)

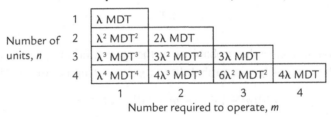

Number of units, n	$m=1$	$m=2$	$m=3$	$m=4$
1	λ MDT			
2	λ^2 MDT^2	2λ MDT		
3	λ^3 MDT^3	$3\lambda^2$ MDT^2	3λ MDT	
4	λ^4 MDT^4	$4\lambda^3$ MDT^3	$6\lambda^2$ MDT^2	4λ MDT

Number required to operate, m

The results shown in Table 8.2 are independent of the number of repair crews for the reason stated, namely that the repairers of the first failure will continue rather than switch to the second failure and start the MDT again.

8.1.3 Repairable Systems (Unrevealed Failures)

In many cases failures are unrevealed, as, for example, the failure of an instrument to measure high pressure in a safety-related monitoring system. The pressure may seldom exceed the alarm limit and thus the inability of the instrument to respond will not be revealed except by some periodic proof test.

Both unit and system down times will be related to the proof-test interval, T. The following arguments are:

If the system is visited every T hours for the inspection of failed units then the average down time AT THE UNIT LEVEL is:

$$T/2 + \text{Repair Time}$$

In general, however, the proof-test interval is large compared with the repair time and we may simplify the above to:

$$\text{Mean Unit Down Time} = T/2$$

In general the mean time to fail of a unit is much greater than the proof-test interval and so, if z events are needed for the **system to fail** then these will be evenly distributed across the interval T and, thus, the system will be in failed state for:

$$\text{Mean System Down Time} = T/(z + 1)$$

For a system where m units are required to operate out of n then $n - m + 1$ must fail for the system to fail and so the **system** down time becomes:

$$T/(n - m + 2)$$

The probability of an individual unit failing between proof tests is simply:

$$\lambda T$$

For the general m out of n case the probability of the system failing prior to the next proof test is approximately the same as for the probability of $n - m + 1$ units failing. This is:

$$n!/[(m - 1)! \, (n - m + 1)!] \cdot (\lambda T)^{(n - m + 1)}$$

The failure rate is obtained by dividing the above by T:

$$n!/[(m - 1)! \, (n - m + 1)!] \cdot \lambda^{(n - m + 1)} \cdot T^{(n - m)}$$

This yields Table 8.3.

Table 8.3: System Failure Rates (Unrevealed)

Number of units, n	1	2	3	4
1	λ			
2	$\lambda^2 T$	2λ		
3	$\lambda^3 T^2$	$3\lambda^2 T$	3λ	
4	$\lambda^4 T^3$	$4\lambda^3 T^2$	$6\lambda^2 T$	4λ

Number required to operate, m

Multiplying these failure rates by the system down time yields Table 8.4.

Table 8.4: System Unavailability (Unrevealed)

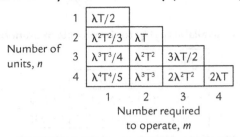

	1	2	3	4
1	$\lambda T/2$			
2	$\lambda^2 T^2/3$	λT		
3	$\lambda^3 T^3/4$	$\lambda^2 T^2$	$3\lambda T/2$	
4	$\lambda^4 T^4/5$	$\lambda^3 T^3$	$2\lambda^2 T^2$	$2\lambda T$

Number of units, n (rows) — Number required to operate, m (columns)

The above formulae apply to proof tests that are carried out simultaneously. In other words, both of the redundant units are inspected at the same time, every T hours. If the proof test were staggered (e.g. January/July) the unavailablity would be improved (decreased). The author addressed this issue in a paper in *Safety & Reliability*, Volume 21, No. 3, Autumn 2001. The result is shown in Table 8.5.

Table 8.5: System Unavailability (Unrevealed – Random Proof Test)

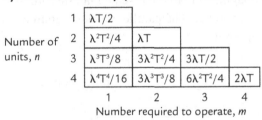

	1	2	3	4
1	$\lambda T/2$			
2	$\lambda^2 T^2/4$	λT		
3	$\lambda^3 T^3/8$	$3\lambda^2 T^2/4$	$3\lambda T/2$	
4	$\lambda^4 T^4/16$	$3\lambda^3 T^3/8$	$6\lambda^2 T^2/4$	$2\lambda T$

Number of units, n (rows) — Number required to operate, m (columns)

Staggered choice of the unit for proof test will lead to yet different results but, this being an unlikely scenario, is not addressed here (an example, for one out of two, is $5\lambda^2 T^2/24$).

The above results are independent of the number of repair crews.

8.1.4 Systems With Cold Standby Units and Repair

Cold standby implies that the redundant units, when not required, are not in use and thus have zero failure rate. Similar logic can be applied to derive expressions for standby scenarios whereby one redundant unit is not activated until needed. A zero failure rate in the dormant state is assumed.

In this case the failure rate becomes $(n-1)\lambda$ because only $m-1$ are available to fail. The failure rate for the second failure is the same because the standby unit has been activated and is thus $(n-1)\lambda$. Thus the probability of the second failure is $(n-1)\lambda$ unit down time. The system failure rate is thus:

$$(n-1)^2\lambda^2 \mathrm{MDT}_{unit}$$

This leads to Table 8.6 for the system unavailabilities.

Table 8.6: System Unavailability (Up to three Units in Standby, i.e. $n=4$, $m=1$)

Number of units, n	1	2	3	4
1	λ MDT			
2	λ^2 MDT^2/2	2λ MDT		
3	λ^3 MDT^3/6	$2\lambda^2$ MDT^2	3λ MDT	
4	λ^4 MDT^4/24	$4\lambda^3$ MDT^3/3	$9\lambda^2$ MDT^2/2	4λ MDT

Number required to operate, m

8.1.5 Modeling Repairable Systems with Both Revealed and Unrevealed Failures

When carrying out the block diagram modeling it will be necessary to take account of units (e.g. a PLC) that have both revealed and unrevealed failures. Since these attract different formulae the approach is to imagine that all the revealed failures are in one hypothetical unit and all the unrevealed failures in another. The parallel block diagram of a one out of two system would become as shown in Figure 8.4.

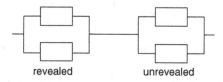

revealed   unrevealed

Figure 8.4: Model showing revealed and unrevealed failures

This is addressed in Section 2 of Chapter 9.

8.1.6 Conventions for Labeling 'Dangerous', 'Safe', Revealed and Unrevealed Failures

Unfortunately many current standards and guidelines make use of the terms 'dangerous' and 'safe' failures. This is potentially misleading, giving the impression that some failures are 'safe'. This is only true in the context of a specific system failure mode leading to a given hazardous event. The so-called 'safe' failures may well be 'dangerous' in another context. Take, for example, the pair of valves in Figure 2.1 in Chapter 2.

If the hazardous event in question is 'downstream overpressure', leading to release of gas, subsequent ignition and a fatality, then the 'fail-open' modes of the valves are indeed 'dangerous'. Similarly the 'fail-shut' modes are 'safe'. However, if the event in question is 'loss of throughput' leading to low pressure, and the potential for a hazardous gas/air mixture, the above defintions are reversed. 'Fail-shut' becomes 'dangerous' and 'fail-open' becomes 'safe'. The author therefore prefers to avoid these subjective terms and refers to the failure mode in question.

Nevertheless the following terms are in common use and the reader should be aware of the above when using them:

λ_{dd} Failure rate of 'dangerous' failures that are revealed (i.e. detected)
λ_{du} Failure rate of 'dangerous' failures that are unrevealed (i.e. not detected)
λ_{sd} Failure rate of 'safe' failures that are revealed (i.e. detected)
λ_{su} Failure rate of 'safe' failures that are unrevealed (i.e. not detected).

8.2 Common Cause (Dependent) Failure

8.2.1 What is CCF?

Common cause failures frequently dominate the unreliability of redundant systems by virtue of defeating the random coincident failure feature, which is the essence of redundancy. Consider the duplicated system in Figure 8.5. The failure rate of the redundant element (in other words the coincident failures) can be calculated using the formula developed in Section 8.1, namely $2\lambda^2$ MDT. Typical figures of 10 per million hours failure rate and twenty-four hours down time lead to a failure rate of $2 \times 10^{-10} \times 24 = 0.0048$ per million hours. However, if only one failure in twenty is of such a nature as to affect both channels and thus defeat the redundancy, it is necessary to add the series element, λ_2, whose failure rate is $5\% \times 10^{-5} = 0.5$ per million hours. The effect is to swamp the redundant part of the prediction. This sensitivity of system failure to CCF places emphasis on the credibility of CCF estimation and thus justifies efforts to improve the models.

Whereas simple models of redundancy (developed in Section 8.1) assume that failures are both random and independent, common cause failure (CCF) modeling takes account of failures that

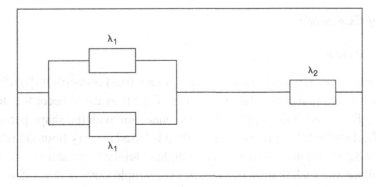

Figure 8.5: Reliability block diagram for CCF

are linked, due to some dependency, and therefore occur simultaneously or, at least, within a sufficiently short interval as to be perceived as simultaneous.

Two examples are:

(a) The presence of water vapor in gas causing both valves in twin streams to seize due to icing. In this case the interval between the two failures might be in the order of days. However, if the proof-test interval for this dormant failure is two weeks then the two failures will, to all intents and purposes, be simultaneous.
(b) Inadequately rated rectifying diodes on identical twin printed circuit boards failing simultaneously due to a voltage transient.

Typically, causes arise from:

(a) *Requirements*: incomplete or conflicting.
(b) *Design*: software, emc, noise.
(c) *Manufacturing*: batch-related component deficiencies.
(d) *Maintenance/operations*: human-induced or test equipment problems.
(e) *Environment*: temperature cycling, electrical interference, etc.

Defenses against CCF involve design and operating features that form the assessment criteria shown in the next section.

The term Common Mode Failure (CMF) is also frequently used and a brief explanation of the difference between CMF and CCF is therefore necessary. Common mode failure refers to co-incident failures of the same mode, in other words failures that have an identical appearance or effect. On the other hand, the term CCF implies that the failures have the same underlying cause. It is possible (although infrequent) for two CMFs not to have a common cause and, conversely, for two CCFs not to manifest themselves in the same mode. In practice the difference is slight and unlikely to affect the data, which rarely contain sufficient detail to justify any difference in the modeling. Since the models described in this section involve assessing defenses against the CAUSES of coincident failure, CCF will be used throughout.

8.2.2 Types of CCF Model

Various approaches to modeling are:

(a) The simple Beta (β) model, which assumes that a fixed proportion (β) of the failures arise from a common cause. The estimation of (β) is assessed according to the system. (Note the Beta used in this context has no connection with the shape parameter used in the Weibull method, Chapter 6.) The method is based on very limited historical data. In Figure 8.5 (λ_1) is the failure rate of a single redundant unit and (λ_2) is the common cause failure rate such that (λ_2) = β (λ_1)for the simple Beta model and also the Partial Beta model, in (b) below.

(b) The Partial Beta model also assumes that a fixed proportion of the failures arise from a common cause. It is more sophisticated than the simple Beta model in that the contributions to Beta are split into groups of design and operating features that are believed to influence the degree of CCF. Thus the Beta factor is made up by adding together the contributions from each of a number of factors within each group. In traditional Partial Beta models the following groups of factors, which represent defenses against CCF, can be found:

- similarity (diversity between redundant units reduces CCF)
- separation (physical distance and barriers reduce CCF)
- complexity (simpler equipment is less prone to CCF)
- analysis (previous FMEA and field data analysis will have reduced CCF)
- procedures (control of modifications and of maintenance activities can reduce CCF)
- training (designers and maintainers can help to reduce CCF by understanding root causes)
- control (environmental controls can reduce susceptibility to CCF, e.g. weather proofing of duplicated instruments)
- tests (environmental tests can remove CCF-prone features of the design, e.g. emc testing).

The Partial Beta model is also represented by the reliability block diagram shown in Figure 8.5. Beta is assumed to be made up of a number of partial βs, each contributed to by the various groups of causes of CCF. β is then estimated by reviewing and scoring each of the contributing factors (e.g. diversity, separation).

(c) The system cut-off model offers a single failure rate for all failures (independent and dependent both combined). It argues that the dependent failure rate dominates the coincident failures. Again, the choice is affected by system features such as diversity and separation. It is the least sophisticated of the models in that it does not base the estimate of system failure rate on the failure rate of the redundant units.

(d) The boundary model uses two limits of failure rate: namely, limit A, which assumes all failures are common cause (λ_u), and limit B, which assumes all failures are random (λ_1). The system failure rate is computed using a model of the following type:

$$\lambda = (\lambda_1{}^n \lambda_u)^{1/(n+1)}$$

where the value of n is chosen according to the degree of diversity between the redundant units. n is an integer, normally from zero to four, which increases with the level of diversity between redundant units. It is chosen in an arbitrary and subjective way. This method is a mathematical device, having no foundation in empirical data, which relies on a subjective assessment of the value of n. It provides no traceable link (as does the Partial Beta method) between the assessment of n and the perceived causes of CCF. Typical values of n for different types of system are:

Configuration	Mode of Operation	Precautions Against CCF	n
Redundant equipment/system	Parallel	No segregation of services or supplies	0
Redundant equipment/system	Parallel	Full segregation of services or supplies	1
Redundant equipment/system	Duty/standby	No segregation of services or supplies	1
Redundant equipment/system	Duty/standby	Full segregation of services or supplies	2
Diverse equipment or system	Parallel	No segregation of services or supplies	2
Diverse equipment or system	Parallel	Full segregation of services or supplies	3
Diverse equipment or system	Duty/standby	No segregation of services or supplies	3
Diverse equipment or system	Duty/standby	Full segregation of services or supplies	4

(e) The multiple greek letter model is similar to the Beta model but assumes that the Beta ratio varies according to the number of coincident failures. Thus two coincident failures and three coincident failures would have different Betas. However, in view of the inaccuracy inherent in the approximate nature of these models it is considered to be too sophisticated and cannot therefore be supported by field data until more detailed information is available.

All the models are, in their nature, approximate but because CCF failure rates (which are in the order of $\beta \lambda$) are much greater than the coincident independent failures (in the order of λ^n), then greater precision in estimating CCF is needed than for the redundant coincident models described in Section 8.1.

8.2.3 The BETAPLUS Model

The BETAPLUS model has been developed from the Partial Beta method, by the author, because:

- it is objective and maximizes traceability in the estimation of Beta. In other words the choice of checklist scores when assessing the design can be recorded and reviewed;
- it is possible for any user of the model to develop the checklists further to take account of any relevant failure causal factors that may be perceived;
- it is possible to calibrate the model against actual failure rates, albeit with very limited data;
- there is a credible relationship between the checklists and the system features being analyzed. The method is thus likely to be acceptable to the non-specialist;
- the additive scoring method allows the partial contributors to β to be weighted separately;
- the β method acknowledges a direct relationship between (λ_2) and (λ_1) as depicted in Figure 8.5;
- it permits an assumed 'non-linearity' between the value of β and the scoring over the range of β.

The Partial Beta model includes the following enhancements:

(a) Categories of factors: Whereas existing methods rely on a single subjective judgement of score in each category, the BETAPLUS method provides specific design and operational related questions to be answered in each category. Specific questions are individually scored in each category (i.e. separation, diversity, complexity, assessment, procedures, competence, environmental control, environmental test) thereby permitting an assessment of the design and its operating and environmental factors. Other Beta methods only involve a single scoring of each category (e.g. a single subjective score for diversity).

(b) *Scoring*: The maximum score for each question has been weighted by calibrating the results of assessments against known field operational data. Programable and non-programable equipment have been accorded slightly different checklists in order to reflect the equipment types (see Appendix 10).

(c) *Taking account of diagnostic coverage*: Since CCF are not simultaneous, an increase in auto-test or proof-test frequency will reduce β since the failures may not occur at precisely the same moment. Thus, more frequent testing will prevent some CCF. Some defenses will protect against the type of failure that increased proof test might identify (for example failures in parallel channels where diversity would be beneficial). Other defenses will protect against the type of failure that increased proof test is unlikely to identify (for example, failures prevented as a result of long-term experience with the type of equipment) and this is reflected in the model.

(d) *Sub-dividing the checklists according to the effect of diagnostics*: Two columns are used for the checklist scores. Column (A) contains the scores for those features of CCF protection that are perceived as being enhanced by an increase of diagnostic frequency (either proof-test or auto-test). Column (B), however, contains the scores for those features thought not to be enhanced by an improvement in diagnostic frequency. In some cases the score has been split between the two columns, where it is thought that some, but not all, aspects of the feature are affected.

(e) *Establishing a model*: The model allows the scoring to be modified by the frequency and coverage of diagnostic test. The (A) column scores are modified by multiplying by a factor (C) derived from diagnostic related considerations. This (C) score is based on the diagnostic frequency and coverage. (C) is in the range one to three. Beta is then estimated from the following RAW SCORE total:

$$S = \text{RAW SCORE} = (A\,C) + B$$

It is assumed that the effect of the diagnostic score (C) on the effectiveness of the (A) features is linear. In other words each failure mode is assumed to be equally likely to be revealed by the diagnostics. Only more detailed data can establish if this is not a valid assumption.

(f) *Non-linearity*: There are currently no CCF data to justify departing from the assumption that as Beta decreases (i.e. improves) then successive improvements become proportionately harder to achieve. Thus the relationship of the Beta factor to the raw score

[(A C)+B] is assumed to be exponential and this non-linearity is reflected in the equation which translates the raw score into a Beta factor.

(g) *Equipment type*: The scoring has been developed separately for programable and non-programable equipment, in order to reflect the slightly different criteria that apply to each type of equipment.

(h) *Calibration*: The model was calibrated against the author's field data.

8.2.3.1 Checklists and scoring of the (A) and (B) factors in the model

Scoring criteria were developed to cover each of the categories (i.e. separation, diversity, complexity, assessment, procedures, competence, environmental control, environmental test). Questions have been assembled to reflect the likely features which defend against CCF. The scores were then adjusted to take account of the relative contributions to CCF in each area, as shown in the author's data. The score values have been weighted to calibrate the model against the data.

When addressing each question, a score, less than the maximum of 100%, may be entered. For example, in the first question, if the judgement is that only 50% of the cables are separated then 50% of the maximum scores (15 and 52) may be entered in each of the (A) and (B) columns (7.5 and 26).

The checklists are presented in two forms (see Appendix 10) because the questions applicable to programable-based equipments will be slightly different to those necessary for non-programable items (e.g. field devices and instrumentation).

The headings (expanded with scores in Appendix 10) are:

(1) **Separation/segregation**
(2) **Diversity**
(3) **Complexity/design/application/maturity/experience**
(4) **Assessment/analysis and feedback of data**
(5) **Procedures/human interface**
(6) **Competence/training/safety culture**
(7) **Environmental control**
(8) **Environmental testing**

8.2.3.2 Assessment of the diagnostic interval factor (C)

In order to establish the (C) score it is necessary to address the effect of the frequency and coverage of proof test or auto-test. The diagnostic coverage, expressed as a percentage, is an estimate of the proportion of failures that would be detected by the proof test or auto-test. This can be estimated by judgement or, more formally, by applying FMEA at the component level to decide whether each failure would be revealed by the diagnostics. Appendix 10 shows the detailed scoring criteria.

An exponential model is proposed to reflect the increasing difficulty in further reducing Beta as the score increases (as discussed in paragraph 8.2.3.f). This is reflected in the following equation:

$$\beta = 0.3 \exp(-3.4S/2624)$$

Because of the nature of this model, additional features (as perceived by any user) can be proposed in each of the categories. The model can then be modified. If subsequent field data indicate a change of relative importance between the categories then adjust the scores in each category so that the category totals reflect the new proportions, also ensuring that the total possible raw score ($S = 2624$) remains unaltered.

The model can best be used iteratively to test the effect of design, operating and maintenance proposals where these would alter the scoring. A Beta value can be assessed for a proposed equipment. Proposed changes can be reflected by altering the scores and recalculating Beta. The increased design or maintenance cost can be reviewed against the costs and/or savings in unavailability by re-running the RAMS predictions using the improved Beta. As with all RAMS predictions the proportional comparison of values rather than the absolute value is of primary value.

8.2.3.3 'M out of N' redundancy/voting

The basic Beta model applies to simple 'one out of two' redundancy. In other words a pair of redundant items where the 'top event' is the failure of both items. However, as the number of voted systems increases (in other words $N > 2$) the proportion of common cause failures varies and the value of β needs to be modified. The reason for this can be understood by thinking about two extreme cases.

'One out of six' voting.

In this case only one out of the six items is required to work and up to five failures can be tolerated. Thus, in the event of a common cause failure, five more failures need to be provoked by the common cause. This is less likely than the 'one out of two' case and β will be smaller. The table suggests a factor of 0.4.

'Five out of six' voting.

In this case five out of the six items are required to work and only one failure can be tolerated. Thus, in the event of a common cause failure, there are five items to which the common cause could apply. This is more likely than the 'one out of two' case and β will be greater. The table suggests a factor of 8.

The model for this factor is explained in 'SINTEF, Reliability Prediction Method for Safety Instrumented Systems STF50 A06031'. Briefly, the argument (for the $N = 3$ case) is:

With three items, A, B and C, the rate for a coincident failure of AB or AC or BC is βλ. The probability of overlap causing ABC (i.e. all three) is assumed to be 0.3βλ. Thus for one out of three (1oo3) redundancy (all three need to fail) β is modified by 0.3.

If AB is given as βλ then:

AB and not C must be βλ less the 0.3βλ = 0.7βλ;

thus as there are three ways to fail, ABnotC, ACnotB and BCnotA,the rate is 2.1βλ;

therefore the rate for two or more failures is 2.1βλ + 0.3βλ = 2.4βλ;

thus for 2oo3 redundancy (any two need to fail) β is modified by 2.4.

The following table shows the results for up to five out of six redundancy:

BETA(MooN) Factor

	M=1	M=2	M=3	M=4	M=5	M=6	M=7	M=8
N=2	1							
N=3	0.3	2.4						
N=4	0.15	0.75	4					
N=5	0.075	0.45	1.2	6				
N=6	*0.0375*	0.26	0.83	1.58	8.1			
N=7	*0.019*	0.15	0.54	1.2	1.9	11		
N=8	*0.009*	0.084	0.35	0.87	1.5	2	14	
N=9	*0.005*	*0.047*	0.22	0.61	1.2	1.8	2	17

However, as can be seen from the bold italic in the above table, the value of the BETA(MooN) factor becomes progressively smaller as N increases with small values of M. The validity of the SINTEF model seems doubtful once the value of BETA(MooN) becomes as small as 0.075 due to the existence of common elements such as cabling and environment, which might well limit the lower value. Thus the BETAPLUS package calculates the BETA(MooN) factor according to the algorithm behind the above table but limits its value to no smaller than 0.075.

It is of interest to note that the BETA(MooN) factor is closest to one at, or near, M=N/2. The BETAPLUS package asks you to input M and N and calculates the modification factor. It caters for values of up to n = 12.

8.3 Fault Tree Analysis

8.3.1 The Fault Tree

A fault tree is a graphical method of describing the combinations of events leading to a defined system failure. In fault tree terminology, the system failure mode is known as the *top event.*

The fault tree involves essentially three logical possibilities and hence two main symbols. These involve *gates* such that the inputs below gates represent failures. Outputs (at the top) of gates represent a propagation of failure depending on the nature of the gate. The three types are:

> *the OR gate*, whereby any input causes the output to occur;
> *the AND gate*, whereby all inputs need to occur for the output to occur;
> *the voted gate,* similar to the *AND* gate, whereby two or more inputs are needed for the output to occur.

Figure 8.6 shows the symbols for the *AND* and *OR* gates and also draws attention to their equivalence to reliability block diagrams. The *AND* gate models the redundant case and is thus equivalent to the parallel block diagram. The *OR* gate models the series case whereby any failure causes the top event. An example of a 'two out of three' voted gate is shown in Figure 8.10.

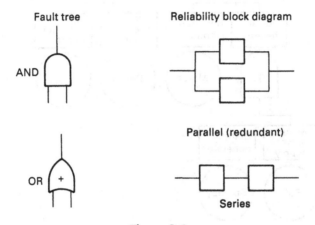

Figure 8.6

For simple trees the same equations given in Section 8.1 on reliability block diagrams can be used and the difference is merely in the graphical method of modeling. In probability terms the *AND* gate involves multiplying probabilities of failure and the *OR* gate the addition rules given in Chapter 7. Whereas reliability block diagrams model paths of success, the fault tree models the paths of failure to the top event.

A fault tree is constructed as shown in Figure 8.7, in which two additional symbols can be seen. The rectangular box serves as a place for the description of the gate below it. Circles, always at the furthest point down any route, represent the basic events that serve as the enabling inputs to the tree.

8.3.2 Calculations

Having modeled the failure logic for a system as a fault tree the next step is to evaluate the frequency of the top event. As with block diagram analysis, this can be performed, for simple trees, using the formulae from Section 8.1. More complex trees will be considered later.

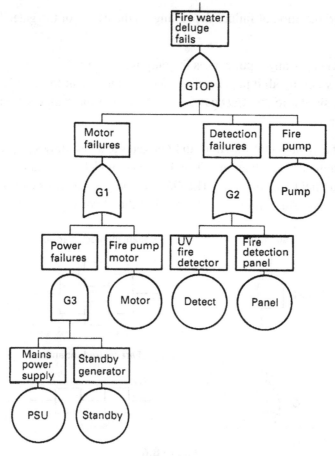

Figure 8.7

The example shown in Figure 8.7 could be evaluated as follows. Assume the following basic event data:

	Failure Rate (PMH)	MDT (Hours)
PSU	100	24
Standby	500	168
Motor	50	168
Detector	5	168
Panel	10	24
Pump	60	24

The failure rate 'output' of the AND gate G3 can be obtained from the formula $\lambda_1 \times \lambda_2 \times (\text{MDT}_1 + \text{MDT}_2)$. Where an AND gate is actually a voted gate, as, for example, two out of three, then again the formulae from Section 8.1 can be used. The outputs of the OR gates G1 and GTOP can be obtained by adding the failure rates of the inputs. Figure 8.8 has the failure rate and MDT values shown.

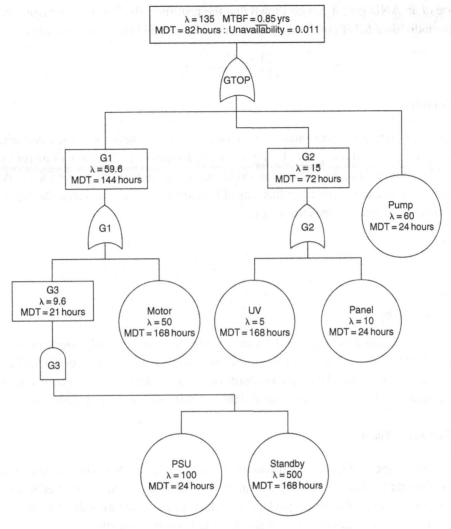

Figure 8.8

It often arises that the output of an OR gate serves as an input to another gate. In this case the MDT associated with the input would be needed for the equation. If the MDTs of the two inputs to the lower gate are not identical then it is necessary to compute an equivalent MDT. In Figure 8.8 this has been done for G1 even though the equivalent MDT is not needed elsewhere. It is the weighted average of the two MDTs weighted by failure rate. In this case,

$$\frac{(21 \times 9.6) + (168 \times 50)}{(9.6 + 50)} = 144 \text{ h}$$

In the case of an AND gate it can be shown that the resulting MDT is obtained from the multiple of the individual MDTs divided by their sum. Thus for G3 the result becomes

$$\frac{(24 \times 168)}{(24 + 168)} = 21h$$

8.3.3 Cutsets

A problem arises, however, in evaluating more complex trees where the same basic initiating event occurs in more than one place. Using the above formulae, as has been done for Figure 8.8, would lead to inaccuracies because an event may occur in more than one *Cutset*. A Cutset is the name given to each of the combinations of base events which can cause the top event. In the example of Figure 8.7 the cutsets are:

> pump
> motor
> panel
> detector
> PSU and standby.

The first four are referred to as first-order Cutsets since they involve only single events which alone trigger the top event. The remaining cutset is known as a second-order Cutset because it comprises a pair of events. There are no third- or higher-order Cutsets in this example. The relative frequency of cutsets is of interest and this is addressed in the next section.

8.3.4 Computer Tools

Manually evaluating complex trees, particularly with basic events that occur more than once, is not easy and would be time-consuming. Fortunately, with today's computer speed and memory capacity, a large number of software packages (such as TTREE) are available for fault tree analysis. They are quite user-friendly and the degree of competition ensures that efforts continue to enhance the various packages in terms of facilities and user-friendliness.

The majority of packages are sufficiently simple to use that even the example in Figure 8.7 would be undertaken considerably more quickly by computer. The time taken to draw the tree manually would exceed that needed to input the necessary logic and failure rate data to the package. There are two methods of inputting the tree logic:

1. *Gate logic*, which is best described by writing the gate logic for Figure 8.7 as follows:

> GTOP   + G1 G2 PUMP
> G1   + G3 MOTOR
> G3   * PSU STANDBY
> G2   + DETECT PANEL

+ represents an OR gate and * an AND gate. Each gate, declared on the right-hand side, subsequently appears on the left-hand side until all gates have been described in terms of all the basic events on the right. Modern packages are capable of identifying an illogical tree. Thus, gates that remain undescribed or are unreachable will cause the program to report an error.

2. *A graphical tree*, which is constructed on the computer screen by use of cursors or a mouse to pick up standard gate symbols and to assemble them into an appropriate tree.

Failure rate and mean down time data are then requested for each of the basic events. The option exists to describe an event by a fixed probability as an alternative to stating a rate and down time. This enables fault trees to contain 'one-shot' events such as lightning and human error.

Most computer packages reduce the tree to Cutsets (known as minimal Cutsets), which are then quantified. Some packages compute by the simulation technique described in Section 9.5.

The outputs consist of:

Graphics to a plotter or printer (e.g. Figures 8.7, 8.9, 8.10)
MTBF, availability, rate (for the top event and for individual Cutsets)
Ranked cutsets
Importance measures.

Cutset ranking involves listing the Cutsets in ranked order of one of the variables of interest – say, failure rate. In Figure 8.8 the Cutset whose failure rate contributes most to the top event is the PUMP (50%). The least contribution is from the combined failure of UV DETECTOR and PANEL. The ranking of Cutsets is thus:

pump	(44%)
motor	(37%)
PSU and standby	(7%)
panel	(7%)
UV detector	(4%)

There are various applications of the *importance* concept but, in general, they involve ascribing a number either to the basic events or to Cutsets which describes the extent to which they contribute to the top event. In the example, the PUMP MTBF is 1.9 years whereas the overall top event MTBF is 0.85 year. Its contribution to the overall failure rate is thus 44%. Thus an importance measure of 44% is one way of describing the PUMP either as a basic event or, as is the case here, a single-order Cutset.

If, on the other hand, the cutsets were to be ranked in order of unavailability the picture might be different, since the down times are not all the same. In Exercise 3, at the end of Chapter 9, the reader can compare the ranking by unavailability with the above ranking by failure rate.

An extract from a typical cutset ranking (taken from the Technis TTREE computer package – see end of book) is shown below.

Barlow–Proschan Measure of Cutset Importance					
Rank 1	Importance 0.353		MTBF hours 0.283E+08		MTBF years 0.323E+04
	Basic		**Failure**	**Mean Fault**	**Constant**
	Event	**Type**	**Rate**	**Duration**	**Probability**
	FLANGE	I/E	0.180E-03	20.0 (MDT)	
	HE1	E			0.100E-01
	CPU	I/E	0.500E-05	0.800E+04 (PTI)	
Rank 2	Importance 0.179		MTBF hours 0.558E+08		MTBF years 0.636E+04
	Basic		**Failure**	**Mean Fault**	**Constant**
	Event	**Type**	**Rate**	**Duration**	**Probability**
	FLANGE	I/E	0.180E-03	20.0 (MDT)	
	HE1	E			0.100E-01
	HE2	E			0.100E-01
Rank 3	Importance 0.110		MTBF hours 0.905E≤08		MTBF years 0.103E≤05
	Basic		**Failure**	**Mean Fault**	**Constant**
	Event	**Type**	**Rate**	**Duration**	**Probability**
	FLANGE	I/E	0.180E-03	20.0 (MDT)	
	CCF1	I/E	0.780E-06	0.800E+04 (PTI)	
	CPU	I/E	0.500E-05	0.800E+04 (PTI)	
Rank 4	Importance 0.574E-01		MTBF hours 0.174E+09		MTBF years 0.199E+05
	Basic		**Failure**	**Mean Fault**	**Constant**
	Event	**Type**	**Rate**	**Duration**	**Probability**
	FLANGE	I/E	0.180E-03	20.0 (MDT)	
	HE1	E			0.100E-01
	CCF3	I/E	0.800E-06	0.800E+04 (PTI)	

The names Barlow–Proschan refer to the cutsets being ranked in order of MTBF (rather than unavailability). Ranking by unavailability is referred to by the names Fussell–Vesely. It can be seen that the first three cutsets account for 64.2% (i.e. 35.3% + 17.9% + 11%) of the total failure rate. Note, also, that the event 'flange' is common to all three. Thus, suggestions for improving the reliability will take account of this. It can also be seen that the next (fourth) cutset has an importance of only 5.7%. This tells us that from the fourth cutset downwards, no one cutset is particularly significant. One other point to note is the column for down time, which (in the TTREE package) distinguishes between revealed failures, having a mean down time (MDT), and unrevealed failures, subject to a Proof-Test Interval (PTI).

8.3.5 Allowing for CCF

Figure 8.9 shows the reliability block diagram of Figure 8.5 in fault tree form. The common cause failure can be seen to defeat the redundancy by introducing an OR gate above the redundant G1 gate.

Figure 8.10 shows another example, this time of two out of three redundancy, where a voted gate is used.

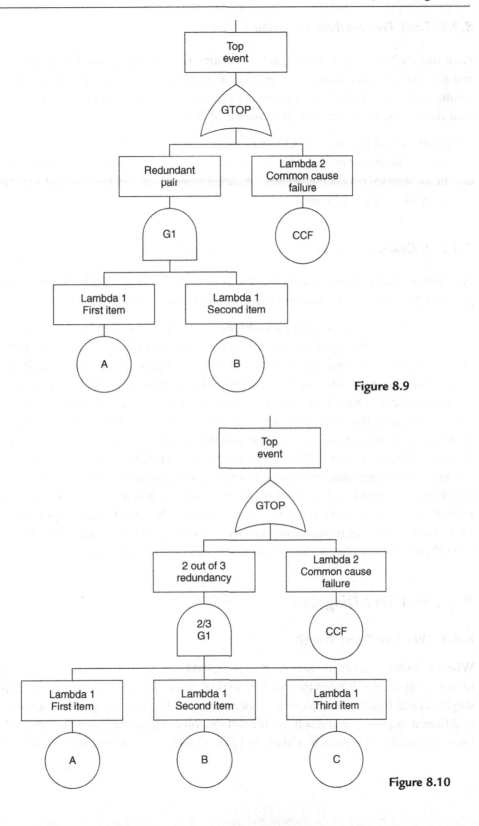

Figure 8.9

Figure 8.10

8.3.6 Fault Tree Analysis in Design

Fault tree analysis, with fast computer evaluation (i.e. seconds), enables successive design and maintenance alternatives to be evaluated. The effect of changes to redundancy or to maintenance intervals can be tested against the previous run. Again, evaluating the relative changes is of more value than obtaining the absolute MTBF.

Frequently a fault tree analysis identifies that the reduction of down time for a few component items has a significant effect on the top event MTBF. This can often be achieved by a reduction in the interval between preventive maintenance, which has the effect of reducing the repair time for dormant failures.

8.3.7 A Cautionary Note

Problems can arise in interpreting the results of fault tree analysis which contain only fixed probability events or a mixture of fixed probability and rate and time events.

If a tree combines fixed probabilities with rates and times then beware of the tree structure. If there are routes to the top of the tree (i.e. cutsets) which involve only fixed probabilities and, in addition, there are other routes involving rates and times then it is possible that the tree logic is flawed. This is illustrated by the example in Figure 8.11. G1 describes the scenario whereby leakage, which has a rate of occurrence, meets a source of ignition. Its contribution to the top event is thus a rate at which explosion may occur. Conversely G2 describes the human error of incorrectly opening a valve and then meeting some other source of ignition. In this case, the contribution to the top event is purely a probability. It is in fact the probability of an explosion for each maintenance activity. It can be seen that the tree is not realistic and that a probability cannot be added to a rate. In this case, a solution would be to add an additional event to G2 as shown in Figure 8.12. G2 now models the rate at which explosion occurs by virtue of including the maintenance activity as a rate (e.g. twice a year for eight hours). G1 and G2 are now modeling failure in the same units (i.e. rate and time).

8.4 Event Tree Diagrams

8.4.1 Why Use Event Trees?

Whereas fault tree analysis (Section 8.3) is probably the most widely used technique for quantitative analysis, it is limited to AND/OR logical combinations of events that contribute to a single defined failure (the top event). Systems where the same component failures occurring in different sequences can result in different outcomes cannot so easily be modeled by fault trees. The fault tree approach is likely to be pessimistic since a fault tree acknowledges the

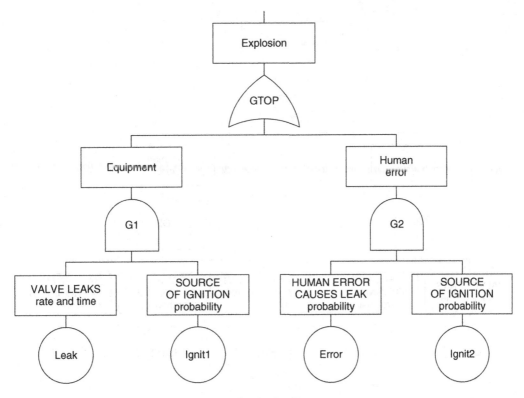

Figure 8.11

occurrence of both combinations of the inputs to an AND gate whereas an event tree or cause consequence model can, if appropriate, permit only one sequence of the inputs.

8.4.2 The Event Tree Model

Event trees or Cause Consequence Diagrams (CCDs) resemble decision trees, which show the likely train of events between an initiating event and any number of outcomes. The main element in a CCD is the decision box, which contains a question/condition with YES/NO outcomes. The options are connected by paths, either to other decision boxes or to outcomes. Comment boxes can be added at any stage in order to enhance the clarity of the model.

Using a simple example, the equivalence of fault tree and event tree analysis can be demonstrated. Figures 8.13 and 8.14 compare the fault tree AND and OR logic cases with their equivalent CCD diagrams. In both cases there is only one Cutset in the fault tree.

These correspond to the 'system fails' and 'no alarm' paths through the CCD diagrams in Figure 8.14(a) and (b) respectively.

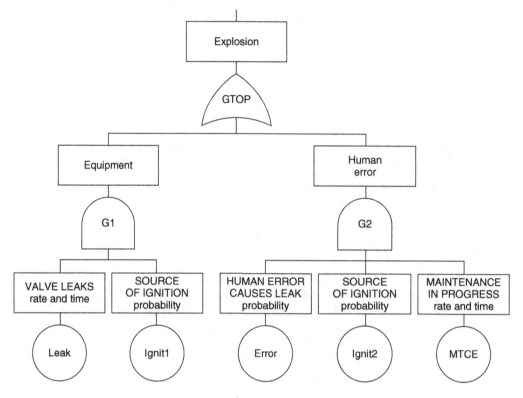

Figure 8.12

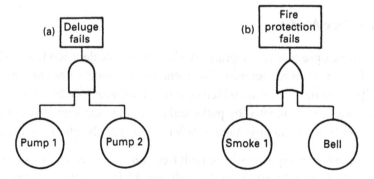

Figure 8.13

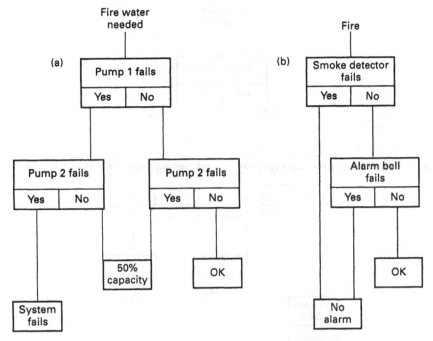

Figure 8.14: (a) Pump 1 *and* 2. (b) Smoke detector *or* alarm bell

Simple CCDs, with no feedback (explained later), can often be modeled using equivalent fault trees but in cases of sequential operation the CCD may be easier to perceive.

8.4.3 Quantification

A simple event tree, with no feedback loops, can be evaluated by simple multiplication of YES/NO probabilities where combined activities are modeled through the various paths.

Figure 8.15 shows the fire water deluge example using a pump failure rate of 50 per million hours with a mean down time of fifty hours. The unavailability of each pump is thus obtained from:

$$50 \times 10^{-6} \times 50 = 0.0025$$

The probability of a pump not being available on demand is thus 0.0025 and the probabilities of both 100% system failure and 50% capacity on demand are calculated.

The system fail route involves the square of 0.0025. The 50% capacity route involves two ingredients of 0.0025×0.9975. The satisfactory outcome is, therefore, the square of 0.9975.

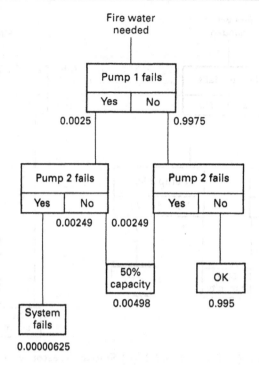

Figure 8.15

8.4.4 Differences

The main difference between the two models (fault tree and event tree) is that the event tree models the order in which the elements fail. For systems involving sequential operation it may well be easier to model the failure possibilities by event tree rather than to attempt a fault tree.

In the above example the event tree actually evaluated two possible outcomes instead of the single outcome (no deluge water) in the corresponding fault tree. As was seen in the example, the probabilities of each outcome were required and were derived from the failure rate and down time of each event.

The following table summarizes the main differences between event tree and fault tree models.

Cause Consequence	Fault Tree
Easier to follow for non-specialist	Less obvious logic
Permits several outcomes	Permits one top event
Permits sequential events	Static logic (implies sequence is irrelevant)
Permits intuitive exploration of outcomes	Top-down model requires inference
Permits feedback (e.g. waiting states)	No feedback
Fixed probabilities	Fixed probabilities and rates and times

8.4.5 Feedback Loops

There is a complication that renders event trees difficult to evaluate manually. In the examples quoted so far, the exit decisions from each box have not been permitted to revisit existing boxes. In that case the simple multiplication of probabilities is not adequate.

Feedback loops are required for continuous processes or where a waiting process applies such that an outcome is reached only when some set of circumstances arises. Figure 8.16 shows a case where a feedback loop is needed where it is necessary to model the situation that a flammable liquid may or may not ignite before a relief valve closes. Either numerical integration or simulation (Section 9.5) is needed to quantify this model and a computer solution is preferred.

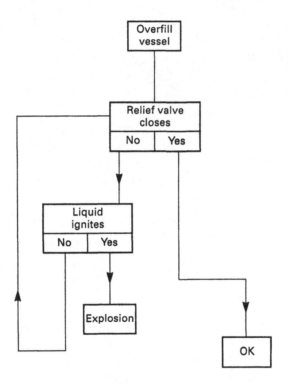

Figure 8.16

8.6.3 *Feedback Loops*

There is a complication that renders event trees difficult to use. Chains brought in to be considered so far assume 'quasi-serial' paths, they have not been a matter to recycle, or still process in such case the simple multiplication of probabilities to be caught.

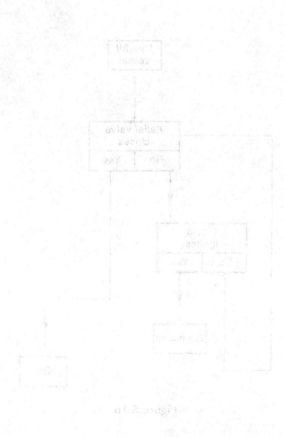

(Figure 8.1a)

CHAPTER 9

Quantifying the Reliability Models

9.1 The Reliability Prediction Method

This section summarizes how the methods described in Chapters 7 and 8 are brought together to quantify RAMS, and Figure 9.1 gives an overall picture of the prediction process. It has already been emphasized that each specific system failure mode has to be addressed separately and thus targets are required for each mode ('top event' in fault tree terminology). Reliability prediction requires the choice of suitable failure rate data, which has been dealt with in detail in Chapter 4. Down times also need to be assessed and it will be shown, in Section 9.2, how repair times and diagnostic intervals both contribute to down time. The probability of human error (Section 9.4) may also need to be assessed where fault trees or event trees contain human events (e.g. operator fails to act following an alarm).

One or more of the techniques described in Chapter 8 will have been chosen to model a scenario. The choice between block diagram and fault tree modeling is very much a matter for the analyst and will depend upon:

- which technique the analyst prefers or is more familiar with;
- what tools are available (i.e. fault tree program, spreadsheet);
- the complexity of the system to be modeled;
- the most appropriate graphical representation for the failure logic.

Chapter 8 showed how to evaluate the model in terms of random coincident failures. Common cause failures then need to be assessed as was shown in Section 8.2. These can be added into the models either as:

- series elements in reliability block diagrams (Section 8.1)
- 'OR' gates in fault trees (Section 8.3.5).

Traditionally this process provides a single predicted RAMS figure. However, the work described in Section 4.4 allows the possibility of expressing the prediction as a confidence range and showed how to establish the confidence range for mixed data sources. Section 9.6 shows how comparisons with the original targets might be made.

Reliability, Maintainability and Risk. DOI: 10.1016/B978-0-08-096902-2.00009-X

Figure 9.1 also reminds us that the opportunity exists to revise the targets should they be found to be unrealistic. It also emphasizes that the credibility of the whole process is dependent on field data being collected to update the data sources being used. The following sections address specific items that need to be quantified.

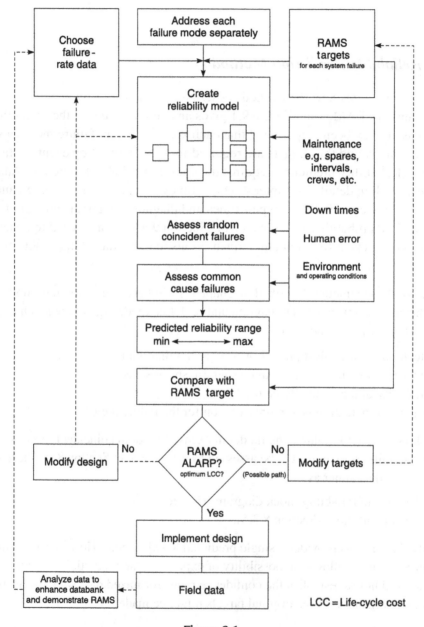

Figure 9.1

9.2 Allowing for Diagnostic Intervals

9.2.1 Establishing Diagnostic Coverage

We saw, in Section 8.1.3, how the down time of unrevealed failures could be assessed. Essentially it is obtained from a fraction of the proof-test interval (i.e. half, at the unit level) together with the MTTR (mean time to repair).

Some databases include information about MTTRs and those that do have been mentioned in Section 4.2.

In many cases there is both auto-test, whereby a programable element in the system carries out diagnostic checks to discover unrevealed failures, as well as a manual proof test. In practice the auto-test will take place at some relatively short interval (e.g. eight minutes) and the proof test at a longer interval (e.g. a year).

The question arises as to how the reliability model can take account of the fact that failures revealed by the auto-test enjoy a shorter down time than those that remain unrevealed until the proof test. The ratio of the former to the total is a measure of the diagnostic coverage and is expressed as a percentage of failures revealed by the test.

There are three ways in which diagnostic coverage can be assessed:

1. *By test*: in other words failures are simulated and the number of diagnosed failures counted.
2. *From field data*: as for test, but using maintenance records.
3. *By FMEA*: in other words the circuit is examined (by FMEA described in Section 9.3), ascertaining, for each potential component failure mode, whether it would be revealed by the diagnostic program.

Clearly 60% diagnostic coverage could be demonstrated fairly easily by either method. Test (or field) data would require a sample of only a few failures to establish a ratio of 60%. For 90% coverage however, the data sample would now need to exceed 20 failures and the cost and time begin to become onerous. For 99% coverage the sample size would now need to exceed 200 failures and this is likely to be impracticable.

9.2.2 Modeling

Consider a dual redundant configuration with components subject to 90% auto-test coverage. Let the auto-test interval be 4 hours and the manual proof-test interval be 4380 hours. Assume that the manual test reveals 100% of the remaining failures. The reliability block diagram needs to split the model into two parts in order to calculate separately in respect of the auto-diagnosed and manually diagnosed failures.

Figure 9.2 shows the parallel and common cause elements twice and applies the equations from Section 8.1 to each element. The total failure rate of the item, for the failure mode in question, is λ and the failure rates of the elements are shown below.

The equivalent fault tree is shown in Figure 9.3.

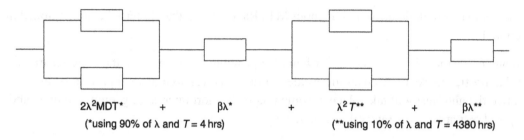

$$2\lambda^2 MDT^* \quad + \quad \beta\lambda^* \qquad + \qquad \lambda^2 T^{**} \quad + \quad \beta\lambda^{**}$$

(*using 90% of λ and $T = 4$ hrs)   (**using 10% of λ and $T = 4380$ hrs)

Figure 9.2

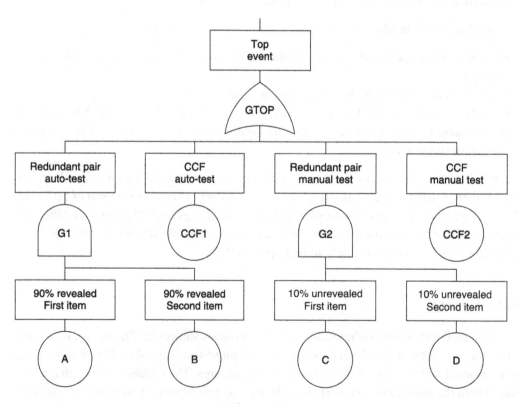

Figure 9.3

9.2.3 Partial Stroke Testing

This technique has become widely used in applications whereby the total closure of a 'shut-down valve' would cause undue disruption to a process. It involves starting to close the valve (thereby proving the 'health' of the actuator) but sensing the movement (either by means of additional actuator position switches or by observing small changes in pressure) and then discontinuing the closure. It is argued that the diagnostic coverage of such an action, whilst not 100%, will still be significant.

Current opinion is that partial stroke testing reveals approximately 75% of unrevealed failures in valves (50% in the case of tight shut-off devices) whereas full testing reveals approximately 95% of unrevealed failures. In that case the following calculation demonstrates how the proability of failure of successful closure can be calculated, taking account of all the variables.

Let λ be the failure rate of the valve.
Let PSI be the interval between partial stroke tests (assume 75% diagnostic coverage).
Let PTI be the interval between full proof tests (assume 95% diagnostic coverage).
Let DI be the interval between actual demands on the valve, say 10 years (assume 100% diagnostic coverage).

$$\text{PFD} = (75\% \; \lambda \; \text{PSI}/2) + (1 - 71\%) \; 95\% \; \lambda \; \text{PTI}/2 + 1.2\% \; \lambda \; \text{DI}/2$$

9.2.4 Safe Failure Fraction

This is a metric based on diagnostic coverage. It has been introduced as a result of standards in the safety-related systems area. It combines the proportion of revealed 'dangerous' failures with those that are not 'dangerous'. It is dealt with in Chapter 22.

9.3 FMEA (Failure Mode and Effect Analysis)

The fault trees, block diagrams and event tree models, described earlier, require failure rates for the individual blocks and enabling events. FMEA involves studying a circuit or mechanical assembly to decide how its component parts contribute to the overall failure mode in question.

This process is known as FMEA and consists of assessing the effect of each component part failing in each possible mode. The process consists of defining the overall failure modes (there will usually be more than one) and then listing each component failure mode that contributes to it. Failure rates are then ascribed to each component-level failure mode and the totals for each of the overall modes are obtained.

The process of writing each component and its failures into rows and columns is tedious but PC programs are now available to simplify the process. Figure 9.4 is a sample output from the FARADIP.THREE package. Each component is entered by entering the 'Reference', 'Name', 'Failure Rate', 'Modes' and 'Mode Percentages'. The table, which can be imported into most word-processing packages, is then printed with failure rate totals for each mode.

The concept of FMEA can be seen in Figure 9.4 by looking at the column headings 'Failure Mode 1' and 'Failure Mode 2'. Specific component modes have been assessed as those giving rise to the two overall modes (Failure to detect and spurious output) for the circuit being analyzed.

Note that the total of the two overall mode failure rates is less than the parts count total. This is because the parts count total is the total failure rate of all the components including all of their failure modes, whereas the specific modes being analyzed do not cover all failures. In other words, there are component failure modes that do not cause either of the modes being considered.

FARADIP3 5.2   Data output

DETECTOR CIRCUIT

Environment factor   1.00   Quality factor   1.00

Component Ref	Component Name	Component Failure Rate	Total Failure Rate	Failure Mode 1	Mode 1 Factor	Failure Rate Mode 1	Failure Mode 2	Mode 2 Factor	Failure Rate Mode 2
IC1	CPU	0.0300	0.0300	20%	0.2000	0.0060	20%	0.2000	0.0060
IC2	CMOS MSI	0.0100	0.0100	LOW	0.5000	0.0050	HIGH	0.5000	0.0050
D21	LP DIODE	0.0050	0.0050	O/C	0.1000	0.0005	S/C	0.7500	0.0037
TR30	NPN	0.0400	0.0400	S/C	0.3000	0.0120	O/C	0.3000	0.0120
Y1	CRYSTAL	0.0700	0.0700	ALL	1.000	0.0700	NONE	0.0000	0.0000
C9	TANTALUM	0.0007	0.0007	S/C	0.5500	0.0004	O/C	0.3000	0.0002
*25	RFILM	0.0002	0.0050	O/C	0.8000	0.0040	NONE	0.0000	0.0000
UV1	UVDETECT	8.000	8.000	FAIL	0.4000	3.200	SPUR	0.6000	4.800
*150	CONNS	0.0001	0.0150	50%	0.5000	0.0075	50%	0.5000	0.0075
SW1	uSWITCH	0.1000	0.1000	O/C	0.3000	0.0300	S/C	0.1000	0.0100
PCB	BOARD	0.0100	0.0100	50%	0.5000	0.0050	50%	0.5000	0.0050
R5COIL	COIL	0.0500	0.0500	O/C	1.000	0.0500	NONE	0.0000	0.0000
R5CON	CONTACT	0.4500	0.4500	O/C	0.7500	0.3375	S/C	0.1000	0.0450
X1	TRANSF'R	0.0300	0.0300	100%	1.000	0.0300	NONE	0.0000	0.0000
F1	FUSE	0.0010	0.0010	O/C	0.1000	0.0001	NONE	0.0000	0.0000

Parts count
Failure rate = 8.817 per million hours
MTBF   = 12.95 years

FAILURE TO DETECT
Mode 1 failure rate = 3.758 per million hours
Mode 1 MTBF   = 30.38 years

SPURIOUS OUTPUT
Mode 2 failure rate = 4.894 per million hours
Mode 2 MTBF   = 23.32 years

Figure 9.4

The FMEA process does not enable one to take account of any redundancy within the assembly which is being analyzed. In practice, this is not usually a problem, since small elements of redundancy can often be ignored, their contribution to the series elements being negligible.

In the previous section we looked at the need to assess the diagnostic coverage. In Figure 9.4 (an example of a detector circuit) it might well be that failures involving inability to detect fire are diagnosed by some auto-test provided by the circuitry and its software. This can be assessed, during the FMEA, by asking if each component failure is diagnosed. For this reason the term FMEDA (failure mode and effect and diagnostic analysis) is sometimes used. Figure 9.5 shows the FMEA, for that failure mode, with an additional column for diagnosis – 100% indicates that the failure is diagnosed and 0% indicates that it is not. Notice the 90% for the UV detector suggesting a 'smart' instrument purchased with a claim of 90% diagnostic capability; 50% has been entered for doubtful items such as connections, the PC board and the CPU device.

The assessment of safe failure fraction (Chapter 22) can also be obtained from this and it is given, along with the diagnostic coverage, as can be seen in Figure 9.5.

FARADIP3 5.2   Data output

DETECTOR CIRCUIT

FAILURE TO DETECT

Environment factor   1.00   Quality factor   1.00

Component Ref	Component Name	Component Failure Rate	Total Failure Rate	Failure Mode 1	Mode 1 Factor	Failure Rate Mode 1	Diagnostic Coverage Mode 1
IC1	CPU	0.0300	0.0300	20%	0.2000	0.0060	50.00
IC2	CMOS MSI	0.0100	0.0100	LOW	0.5000	0.0050	100.0
D21	LP DIODE	0.0050	0.0050	O/C	0.1000	0.0005	0.0000
TR30	NPN	0.0400	0.0400	S/C	0.3000	0.0120	100.0
Y1	CRYSTAL	0.0700	0.0700	ALL	1.000	0.0700	100.0
C9	TANTALUM	0.0007	0.0007	S/C	0.5500	0.0004	0.0000
*25	RFILM	0.0002	0.0050	O/C	0.8000	0.0040	100.0
UV1	UVDETECT	8.000	8.000	FAIL	0.4000	3.200	90.00
*150	CONNS	0.0001	0.0150	50%	0.5000	0.0075	50.00
SW1	uSWITCH	0.1000	0.1000	O/C	0.3000	0.0300	100.0
PCB	BOARD	0.0100	0.0100	50%	0.5000	0.0050	50.00
R5COIL	COIL	0.0500	0.0500	O/C	1.000	0.0500	100.0
R5CON	CONTACT	0.4500	0.4500	O/C	0.7500	0.3375	100.0
X1	TRANSF'R	0.0300	0.0300	100%	1.000	0.0300	100.0
F1	FUSE	0.0010	0.0010	O/C	0.1000	0.0001	100.0

Parts count
Failure rate = 8.817 per million hours
MTBF   = 12.95 years

FAILURE TO DETECT
Mode 1 failure rate   = 3.758 per million hours
Mode 1 MTBF   = 30.38 years
Mode 1 diagnostic cover   = 91.22%
Mode 1 safe failure fraction = 96.26%

Figure 9.5

9.4 Human Factors

9.4.1 Background

It can be argued that the majority of well-known major incidents, such as Three Mile Island, Bhopal, Chernobyl, Zeebrugge and Clapham, are related to the interaction of complex systems with human beings. In short, the implication is that human error was involved, to a larger or greater extent, in these and similar incidents. For many years there has been an interest in modeling these factors so that quantified reliability and risk assessments can take account of the contribution of human error to the system failure.

As with other forms of reliability and risk assessment, the first requirement is for failure rate/ probability data to use in the fault tree or alternative model. Thus, human error probabilities for various activities are needed. In the early 1960s there were attempts to develop a database of human error rates and these led to models whereby probabilities of human error could be estimated by assessing the relevant error producing factors such as stress, training, unfamiliarity, complexity and so on. These human error probabilities include not only simple failure to carry out a given task but diagnostic tasks where errors in reasoning, as well as action, are involved. A great deal of data has been collected over the years across the industry sectors (rail in particular). Difficulties include:

- Low probabilities require large amounts of experience in order for a meaningful statistic to emerge.
- Data collection concentrates on recording the event rather than analyzing the causes.
- Many large organizations have not been prepared to commit the necessary resources to collect data.

Over the years interest has grown in exploring the underlying reasons, as well as probabilities, of human error. In this way, assessments can involve not only quantification of the hazardous event but also an assessment of the changes needed to bring about a reduction in error.

9.4.2 Models

There are several models, each developed by separate groups of analysts working in this field. Whenever several models are available for quantifying an event, the need arises to compare them and to decide which is the most suitable for the task in hand. Factors for comparison could be:

- *Accuracy*: there are difficulties in the lack of suitable data for comparison and validation.
- *Consistency*: between different analysts studying the same scenario.

- *Usefulness*: in identifying factors to change in order to reduce the human error rate.
- *Resources*: needed to implement the study.

One such comparison was conducted by a subgroup of the Human Factors in Reliability Group, and their report *Human Reliability Assessor's Guide* (SRDA R11), which addresses eight of the better-known models, is available from SRD, AEA Technology. The report is dated June 1995 but nevertheless continues to be a useful reference.

The following descriptions of three of the available models will provide some understanding of the approach. A full application of each technique, however, would require more detailed study.

9.4.3 HEART (Human Error Assessment and Reduction Technique)

This is a deterministic and fairly straightforward method developed by J. C. Williams during the early 1980s. It involves choosing a human error probability from a table of error rates and then modifying it by multiplication factors identified from a table of error-producing conditions. It is considered to be of particular use during design since it identifies error-producing conditions and therefore encourages improvements. It is a quick and flexible technique requiring few resources. The error rate table contains nine basic error task types. It is:

Task	Probability of Error
Totally unfamiliar, perform at speed, no idea of outcome	0.55
Restore system to new or original state on a single attempt without supervision or procedures checks	0.26
Complex task requiring high level of comprehension and skill	0.16
Fairly simple task performed rapidly or given scant attention	0.09
Routine, highly practised, rapid task involving relatively low level of skill	0.02
Restore system to new state following procedure checks	0.003
Totally familiar task, performed several times per hour, well motivated, highly trained staff, time to correct errors	0.0004
Respond correctly when there is augmented supervisory system providing interpretation	0.00002
Miscellaneous task – no description available	0.03

The procedure then describes 38 'error-producing conditions' to each of which a maximum multiplier is ascribed. Any number of these can be chosen and, in turn, multiplied by a number between zero and one in order to take account of the analyst's assessment of what proportion of the maximum to use. The modified multipliers are then used to modify the above probability. Examples are:

Error-Producing Condition	Maximum Multiplier
Unfamiliar with infrequent and important situation	×17.
Shortage of time for error detection	×11.
No obvious means of reversing an unintended action	×8.
Need to learn an opposing philosophy	×6.
Mismatch between real and perceived task	×4.
Newly qualified operator	×3.
Little or no independent checks	×3.
Incentive to use more dangerous procedures	×2.
Unreliable instrumentation	×1.6
Emotional stress	×1.3
Low morale	×1.2
Inconsistent displays and procedures	×1.2
Disruption of sleep cycles	×1.1

The following example illustrates the way the tables are used to calculate human error probability.

Assume that an inexperienced operator is required to restore a plant bypass, using strict procedures but that are different to his normal practice. Assume that he is not well aware of the hazards, it is late in the shift and that there is an atmosphere of unease due to worries about impending plant closure.

The probability of error, chosen from the first table, might appropriately be **0.003.**

Five error-producing conditions might be chosen from the second table as can be seen in the following table.

For each condition the analyst assigns a 'proportion of the effect' from judgement (in the range zero to one).

The table is then drawn up using the calculation:

$$[(EPC - 1) \times (Proportion)] + 1$$

The final human error probability is the multiple of the calculated values in the table times the original 0.003.

Factor	EPC	Proportion Effect	$[(EPC - 1) \times (Proportion)] + 1$
Inexperience	3.	0.4	$[(3 - 1) \times (0.4)] + 1 = 1.8$
Opposite technique	6.	1.	6.
Low awareness of risk	4.	0.8	3.4
Conflicting objectives	2.5	0.8	2.2
Low morale	1.2	0.6	1.12

Hence <u>ERROR RATE</u> $0.003 \times 1.8 \times 6 \times 3.4 \times 2.2 \times 1.12 = \underline{0.27}$

Similar calculations can be performed at percentile bounds. The full table provides 5th and 95th percentile bands for the error-rate table.

Note that since the probability of failure cannot exceed one and, therefore, for calculations taking the prediction above one it will be assumed that the error WILL almost certainly occur.

9.4.4 THERP (Technique for Human Error Rate Prediction)

This was developed by A. D. Swain and H. E. Guttmann and is widely used. The full procedure covers the definition of system failures of interest, through error rate estimation, to recommending changes to improve the system. The analyst needs to break each task into steps and then identify each error that can occur. Errors are divisible into types as follows:

> omission of a step or an entire task
> selects a wrong command or control
> incorrectly positions a control
> wrong sequence of actions
> incorrect timing (early/late)
> incorrect quantity.

The sequence of steps is represented in a tree so that error probabilities can be multiplied along the paths for a particular outcome.

Once again (as with HEART), there is a table of error probabilities from which basic error rates for tasks are obtained. These are then modified by 'shaping parameters' that take account of stress, experience and other factors known to affect the error rates.

The analysis takes account of dependence of a step upon other steps. In other words, the failure of a particular action (step) may alter the error probability of a succeeding step.

9.4.5 TESEO (Empirical Technique to Estimate Operator Errors)

This was developed by G. C. Bellow and V. Colombari from an analysis of available literature sources in 1980. It is applied to the plant control operator situation and involves an easily applied model whereby five factors are identified for each task and the error probability is obtained by multiplying together the factors as follows:

Activity	
Simple	0.001
Requires attention	0.01
Non-routine	0.1
Time stress (in seconds available)	
2 (routine), 3 (non-routine)	10
10 (routine), 30 (non-routine)	1.
20 (routine)	0.5
45 (non-routine)	0.3
60 (non-routine)	0.1
Operator	
Expert	0.5
Average	1.
Poorly trained	3.
Anxiety	
Emergency	3.
Potential emergency	2.
Normal	1.
Ergonomic (i.e. plant interface)	
Excellent	0.7
Good	1.
Average	3–7
Very poor	10.

Figure 9.6 is an example of applying HEART and TESEO to a specific task. The questions have been abbreviated onto a spreadsheet for performing the predictions.

9.4.6 Other Methods

There are other methods including:

SLIM (Success Likelihood Index Method)
APJ (Absolute Probability Judgement)
paired comparisons
IDA (the Influence Diagram Approach)
HCR (Human Cognitive Reliability correlation).

These are well described in the Human Factors in Reliability Group (HFRG) document mentioned above.

9.4.7 Human Error Rates

Frequently there are insufficient resources to use the above models. In those cases a simple error rate per task is needed. Appendix 6 provides tables of such error probabilities from a number of sources but it must be emphasized that error probabilities can vary by an order of magnitude according to the application specific factors. Thus, Appendix 6 should only be taken as a general guide and an overview of the picture.

DESCRIBE TASK – OPERATOR DOES NOT USE GOGGLES				
HEART: ENTER NOMINAL ERROR. ENTER SCORE OR 1. ENTER PROPORTION [0 – 1] EFFECT OR 1.				
	SCORE	%		
(17) UNFAMILIAR, IMPORTANT SITUATION, INFREQUENT	17	1	1	
(11) SHORT TIME FOR ERROR DETECT/CORRECT	11	0.3	4	
(10) LOW SIGNAL/NOISE	10	1	1	
(9) EASY TO SUPPRESS INFORMATION	9	1	1	
(8) NO MEANS TO CONVEY SPATIAL INFO	8	1	1	
(8) MISMATCH OPERATOR/DESIGNER MODEL OF WORLD	8	1	1	
(8) NO OBVIOUS MEANS OF REVERSING UNINTENDED ACT	8	0.2	2.4	
(6) SIMULTANEOUS NON-REDUNDANT INFORMATION	6	1	1	
(6) NEED TO UNLEARN AND LEARN NEW PHILOSOPHY	6	1	1	
(5.5) NEED TO TRANSFER KNOWLEDGE TASK/TASK	5.5	1	1	
(5) AMBIGUITY IN REQUIRED PERFORMANCE STD'S	5	1	1	
(4) MISMATCH PERCEIVED/REAL RISK	4	1	1	
(4) POOR/AMBIGUOUS SYSTEM FEEDBACK	4	0.1	1.3	
(4) NO CONFORMATION OF ACTION BY SYSTEM	4	0.3	1.9	
(3) OPERATOR INEXPERIENCED	3	1	1	
(3) POOR INFO FROM PROC'S & PERSONS	3	1	1	
(3) NO INDEPENDENT CHECKING	3	1	1	
(2.5) IMMEDIATE V. LONG-TERM OBJECTIVES	2.5	1	1	
(2.5) NO DIVERSE INFO FOR VERIFICATION	2.5	1	1	
(2) EDUCATIONAL MISMATCH WITH TASK	2	1	1	
(2) INCENTIVE TO USE OTHER DANGEROUS PROC	2	1	1	
(1.8) NO MENTAL/PHYSICAL EXERCISE	1.8	1	1	
(1.6) UNRELIABLE INSTRUMENTS (known)	1.6	1	1	
(1.6) JUDGEMENTS NEEDED > CAPABILITY	1.6	1	1	
(1.6) UNCLEAR RESPONSIBILITY	1.6	1	1	
(1.4) NO WAY TO KEEP TRACK OF PROGRESS	1.4	1	1	
(1.4) DANGER THAT PHYSICAL CAPAB'S EXCEEDED	1.4	1	1	
(1.4) TASK HAS NO INTRINSIC MEANING	1.4	1	1	
(1.3) EMOTIONAL STRESS	1.3	1	1	
(1.2) ILL HEALTH	1.2	1	1	
(1.2) MORALE	1.2	1	1	
(1.2) INCONSISTENT DISPLAYS	1.2	1	1	
(1.15) POOR ENVIRONMENT	1.15	1	1	
(1.1) REPETITIOUS	1.1	0.9	1.09	
(1.1) DISRUPT SLEEP CYCLES	1.1	1	1	
(1.06) PACING BY INTERRUPTION BY OTHERS	1.06	1	1	
(1.03) MORE STAFF THAN NECESSARY	1.03	1	1	
(1.02) AGE	1.02	1	1	
	145.66		25.84608	
NOMINAL =	0.0004			
ERROR =	**0.010338**			
(0.5) UNFAMILIAR, SPEED, CONSEQUENCES??				
(0.26) RESTORE SYS TO ORIGINAL STATE IN ONE GO. NO PROCS/SUPY				
(0.16) COMPLEX, HIGH COMPREHENSION/SKILL				
(0.09) SIMPLE, RAPIDLY, SCANT ATTENTION				
(0.02) ROUTINE, PRACTISED, RAPID LOW SKILL				
(0.003) RESTORE SYS TO ORIG + PROCS + SUPY CHECKS				
(0.0004) V FAMILIAR, n/HR, MOTIVE, TIME TO CORRECT, etc				
(0.00002) RESP TO SYS COMMAND WHEN AUTO SUPY INFO IS THERE				
(0.03) MISC (DON'T KNOW)				
TESEO: ENTER ESTIMATES				SCORE
SIMPLE(0.001), REQUIRES ATT(0.01), NON-ROUTINE(0.1)				0.01
TIME STRESS (secs), [2R3NR = 10][10R30NR = 1][20R = 0.5][45NR = 0.3][60NR = 0.1]*				1
EXPERT(0.5), AVERAGE(1), POORLY TRAINED(3)				1
EMERGENCY(3), POTENTIAL EMRG(2), NORMAL(1)				1
ERGONOMIC EXCELLENT(0.7), GOOD(1), AV(3–7)VPOOR(10)				1
*2R MEANS 2s ROUTINE, etc.				
ERROR =	**0.01**			

Figure 9.6

One approach, when using error rates in a fault tree or other quantified method, is to select a pessimistic value (the circumstances might suggest 0.01 or even 0.1) for the task error rate. If, in the overall incident probability computed by the fault tree, the contribution from that human event is negligible then the problem can be considered unimportant. If, however, the event dominates the overall system failure rate then it would be wise to re-run the fault tree (or simulation) using an error rate an order less pessimistic (e.g. 0.001). If the event still dominates the analysis then there is a clear need for remedial action by means of an operational or design change. If the event no longer dominates at the lower level of error probability then there is a grey area that will require judgement to be applied according to the circumstances. In any case, a more detailed analysis is suggested.

A factor that should be kept in mind when choosing error rates is that human errors are not independent. Error rates are likely to increase as a result of previous errors. For instance, an audible alarm is more likely to be ignored if a previous warning gauge has been recently misread.

In the 1980s it was recognized that a human error database would be desirable. In the USA the NUCLARR database (see also Section 4.2.2.5) was developed and this consists of about 50% human error data although this is heavily dependent on expert judgement rather than solid empirical data. In the UK, there is the CORE-DATA (Computerized Operator Reliability and Error Database) developed at the University of Birmingham.

9.4.8 Trends in Rigor of Assessment

Traditionally, the tendency has been to add additional levels of protection rather than address the underlying causes of error. More recently there is a focus of interest in analyzing the underlying causes of human error and seeking appropriate procedures and defenses to minimize or eliminate them.

Regulatory bodies, such as the UK Health and Safety Executive, have taken (rightly) a great interest in this area and questions are frequently asked about the role of human error in the hazard assessments that are a necessary part of the submissions required from operators of major installations (see Chapter 21).

For multiple-hazard studies, the author has found it useful to concentrate all of the human error rate assessments into one composite report. This has the advantage of encouraging consistency in the assesment of similar activities in different areas. It should contain, for each task:

- a description of the task including, documentation, training, competency, frequency of carrying out the task;
- HEART, TESEO or other assessments (preferably at least two for comparison);

- anecdotal data about the number of such tasks performed and the number of errors;
- an overall argument (combining the above) for the error probablity chosen.

The UK HSE refer to a '7-step approach' to addressing human factors. The steps are:

Step 1: consider the main site hazards;
Step 2: identify manual activities that affect these hazards;
Step 3: outline the key steps in these activities;
Step 4: identify potential human failures in these steps;
Step 5: identify factors that make these failures more likely;
Step 6: manage the failures using hierarchy of control;
Step 7: manage error recovery.

9.5 Simulation

9.5.1 The Technique

Block diagram, fault tree and cause consequence analyses were treated, in Chapters 7–9, as deterministic methods. In other words, given that the model is correct then, for given data, there is only one numerical answer from the model. If two components are in series (i.e. the fault tree 'OR' gate) then, if each has a failure rate of 5 per million hours, the overall failure rate is 10 per million hours – no more, no less. Another approach is to perform a computer-based simulation, sometimes known as Monte Carlo analysis, in which random numbers are used to sample from probability distributions.

In the above example, two random distributions each with a rate of 5 per million would be set up. Successive time slots would be modeled by sampling from the two distributions in order to ascertain if *either* distribution yielded a failure in that interval.

One approach, known as event-driven simulation, inverts the distribution to represent time as a function of the probability of a failure occurring. The random-number generator is used to provide a probability of failure, which is used to calculate the time to the next failure. The events generated in this manner are then logged in a 'diary' and the system failure distribution is derived from the component failure 'diary'. As an example assume we wish to simulate a simple exponential distribution then the probability of failing in time *t* is given by:

$$R(t) = e^{-\lambda t}$$

Taking logs and then inverting the expression we can say that:

$$t = (\log_e R)/\lambda$$

Since R is a number between zero and one the random-number generator can be used to provide this value, which is divided by λ to provide the next value of t. The same approach is adopted for more complex expressions such as the Weibull.

A simulation would be run many thousands of times and the overall rate of system failure counted. This might be 10 per million or 9.99998 or 10.0012 and, in any case, will yield slightly different results for each trial. The longer each simulation runs, and the more runs attempted, the closer will the ultimate answer approach 10 per million hours. This may seem a laborious method for assessing what can be obtained more easily from deterministic methods. Fault tree, cause consequence and simple block diagram methods are, however, limited to simple AND/OR logic and *constant failure rates* and straightforward *mean down times*.

Frequently problems arise due to complicated failure and repair scenarios where the effect of failure and the redundancy depend upon demand profiles and the number of repair teams. Also, it may be required to take account of failure rates and down times that are not constant. The assessment may therefore involve:

- LogNormal down times
- Weibull down times
- Weibull models for not-constant failure rates
- standby items with probabilities of successful start
- items with profiles where the MTBF varies with some process throughput
- spares requirements
- maintenance skill types and quantities
- logistical delays
- ability to make up lost availability within defined rules and limits.

It is not easy to take account of the above items using the techniques explained in this chapter and, for them, simulation provides a quick and cost-effective method.

One drawback to the technique is that the lower the probability of the events, the greater the number of trials that are necessary in order to obtain a satisfactory result. The other limitation, involving program cost and computer run times, has been steadily eroded over the years due to ever-increasing PC power. There are a number of cost-effective packages that can rival the deterministic techniques.

A recent development in reliability simulation (see Section 9.5.2) is the use of genetic algorithms. This technique enables modeling options (e.g. additional redundant streams) to be specified in the simulation. The algorithm then develops and tests the combinations of possibilities depending on the relative success of the outcomes.

There are a variety of algorithms for carrying out this approach and they are used in the various PC packages that are available. Some specific packages are described in the following sections. There are many similarities between them.

9.5.2 Some Packages

9.5.2.1 Optagon

This package was developed by GL Noble Denton (formerly Advantica and before that British Gas Research and Development) to link availability, reliability, operational and commercial information together in order to allow overall performance to be optimized. It was first developed in 1990 and since then has been used on an extensive number of applications representing interests in the UK and worldwide. The package is primarily intended for modeling the performance of oil and gas production assets where there are complexities that require a high level of functionality. These complexities include:

- multiple revenue streams and product streams;
- tank storage and shipping (e.g. LNG import and export facilities);
- gas substitution arrangements;
- operational strategy (e.g. over-producing to make up for production losses);
- resource constraints (e.g. repair team limitations, spares holding);
- variable demand profile (e.g. seasonal) with random fluctuations;
- partial states of operation;
- system configuration changes and equipment throughput variations;
- well profiles;
- conditional logic and equipment dependencies.

The package uses the Monte Carlo approach to overcome the modeling complexities. This approach consists of explicitly modeling the system being studied, subjecting it to a typical set of events over its lifetime, and empirically observing how well it performs. The typical events that are directed at the model are generated stochastically and this means that any individual simulation of the system's lifetime cannot be taken as a guide to its average performance – the model may have been subject to events that were more or less favorable than the average. Instead, it is necessary to carry out a large number of individual simulations. The performance of the model over many simulations gives an indication of how it is likely to perform on average, and how widely the range of possible performance is spread. The application of this approach in OPTAGON enables not only the prediction of mean values for availability statistics such as production availability, shortfall, operational availability, etc., but also enables the possible range of likely performance to be estimated using P5, P10, P50 (median), P90, P95. For example, the P10 value for shortfall is the value for which there is a 10% probability that the shortfall will be less than this value and a 90% probability that the shortfall will be greater. This can give more useful information than purely quoting mean values as it provides an indication of the likely range of performance.

A particular feature is the use of the genetic algorithms already mentioned. These apply the Darwinian principle of natural selection by searching for the optimal solution emerging from the successive simulation runs. It is achieved by expressing the characteristics of a system (such as the complexities listed earlier) in the form of a binary string. The string (known as a gene-string) can then be created by random number generation. A weighting process is used to favor those genes that lead to the more optimistic outcomes by increasing the probability of their choice in successive simulations. Further information can be obtained from the website http://www.gl-nobledenton.com/en/consulting/optagon.php.

9.5.2.2 MAROS, TARO, TRAIL and OFFOP

The following packages are available from DNV: www.dnv.com/services/software/products/safeti/safetipf.

MAROS is a RAM tool with extensive features for modeling networks, maintenance, operations, and demand scenarios. The product has been specifically developed to model issues unique to the oil and gas industry. An 'event-driven' simulation algorithm is used to create life-cycle scenarios of the system under investigation accounting for its reliability, maintainability and operating policies. The simulator is inherently incorporated with an algorithm to analyze the impact of each individual scenario on how the system performs. By studying performance results, and how they respond to altering specific parameters in the design or its logistics, it is possible to optimize the system with respect to a given constraint. Features include:

- intuitive graphical user interface (network and RBD based);
- failure mode, failure and repair data and maintenance task data import from Excel;
- reservoir and demand profiles over variable time periods (with Excel import);
- flaring related environmental constraints (time-or volume-based limits);
- compensation mechanisms (boosting, linepacking, storage, third-party substitution);
- maintenance and logistics, shift constraints, mobilisation delays, spares analysis;
- highly interactive results viewer with drill-down (results output in Access form);
- results with executive summary and detailed results breakdown (efficiency, system, Category and equipment criticality, production, contract loss, outages, etc.).

TARO stands for Total Asset Reliability and Optimization and was developed for modeling refining and petrochemical plants in which the production efficiency is a complex interaction between reliability, blending and yield rules, flow routing (including recycle), and intermediate storage options. Used extensively to identify and improve low unit availability and utilization. It caters for:

- multiple feed stock supply;
- intermediate product streams and routing issues;
- storage volumes – feed, intermediate and product;
- unit overcapacities (maximum capacity versus normal throughput);
- unit turndown capabilities (minimum flow/slowdowns);

- maintenance resources and availability of spares;
- unit re-start times (maintenance and process related).

TRAIL is a life-cycle simulator designed specifically for the railway industry. TRAIL uses discrete event technology to represent the life-cycle operation of railway infrastructure systems. Its main aim is to model and improve timetable and delay performance on large rail networks. A TRAIL simulation requires the items of infrastructure to be divided into sections. Each section is generally defined as a continuous piece of track with no interchanges, or a set of points. Each section is described in terms of its components and their failure and repair characteristics. Each train journey must be defined within the simulation; and involves detailing:

- routes in terms of their connecting sections and time to traverse each section;
- timetables for all journeys, and, in addition, penalties for delays are described;
- any alternative paths available in the event of failure.

OFFOP is a comprehensive decision analysis tool designed for the planning, costing, risk analysis and operations research of offshore operations. The progress and eventual completion of many offshore activities are routinely subject to delays resulting from inhibitive environmental conditions, e.g. extreme sea states, fog, etc. The software analyzes the progress of work activity from sea-going vessels, such as is required for many offshore installation and maintenance programs, using historical environmental data (wind speed, wave height, wave period, etc.).

9.5.2.3 RAM4, RAMP and SAM

These packages are from W. S. Atkins.

RAM4, a Reliability Block Diagram (RBD) tool, has been developed for modeling continuously running systems, and is mission oriented. Typical applications include communications, defense, control systems, signaling and space systems. RAM4 is sponsored by the Ministry of Defence, and is a recommended modeling tool for MoD projects.

RAM4 gives design and reliability engineers a powerful, easy to use analysis tool for the assessment of the reliability of complex repairable and non-repairable systems – whether electronic, mechanical or a mixture of both – as described by RBDs. The program has been continually enhanced and currently includes features such as:

- finite spares and maintenance resources
- preventive maintenance schedules
- logistic delays
- startup delays for standby items
- equipment wear out (Weibull failure distribution)
- common mode failures

- active and standby redundancy
- queuing for repairs
- simulation lifetime.

(www.atkinsglobal.com/defence)

RAMP is a powerful availability modeling tool for the analysis of process systems. RAMP is similar to RAM4 in that it is a RBD-based tool. However, as well as modeling system availability, RAMP models may be created to investigate the effects of scheduled and unscheduled downtime on process throughput, establishing relationships between equipment reliability, configuration, maintenance strategy, spares and resources. RAMP built-in features are similar to those of RAM4 but additionally include:

- time varying delivery and demand/nomination profiles
- bulk and buffer stores.

(www.atkinsglobal.com/areas_of_business/energy/oil_and_gas)

SAM (the Systems Availability Model) is a program designed to assess the RAM characteristics of multiple systems used over operating scenarios that place varying demands upon those systems, such as those encountered in complex military or industrial installations. It is the ability of SAM to overlay system dependencies onto complex mission profiles that makes it a powerful and flexible RAM modeling tool. A mission profile may be built up of a variety of activities, each demanding use of different combinations of equipments, rather than a fixed time at risk approach adopted by many simpler modeling tools.

(www.atkinsglobal.com/defence)

9.5.2.4 ITEM ToolKit

This is also a Monte Carlo package based on reliability block diagrams. It copes with revealed and unrevealed failures, preventive and corrective maintenance regimes, ageing and maintenance queuing. The usual standby and start-up scenarios are modeled and non-random distributions for failure rate and down time can be modeled. System performance is simulated over a number of life-cycles to predict unavailability, number of system failures and required spares levels.

The RBD software module offers easy construction of network diagrams using 'point and click' or 'drag and drop' techniques. Once complete, Boolean algebra expressions are used to determine minimal cutsets or the minimum combination of failures required to cause a system failure. As well as 15 built-in failure models, Markov models can be used for standby systems with respect to maintenance arrangements. A RBD calculates system failure frequency and unavailability. The RBD module also features a capability to estimate performance degradation due to the unavailability of one or more blocks in the system. This Helian model feature allows the user to allocate processing capacities to each block in the model, and computes the reduced throughput of the system in case of partial system failure.

At the time of writing (2010) ToolKit is soon to be replaced by iQT, which will include both ToolKit and iQRAS together.

The website address is www.itemsoft.com.

9.6 Comparing Predictions with Targets

In the light of the work described in Section 4.4 we saw that it is possible to attempt some correlation between predicted and field reliability and that the confidence in the prediction depends upon the data used.

These studies indicate that the results are equally likely to be optimistic as pessimistic. Thus one interpretation is that we can be 50% confident that the field result will be equal to or better than the predicted RAMS value. However, a higher degree of confidence may be desired, particularly if the RAMS prediction is for a safety-related failure mode. If industry-specific data have been used for the prediction and 90% confidence is required then, consulting the tables in Section 4.4, a failure rate of four times the predicted value would be used.

■ Exercises

1. The reliability of the two-valve example of Figure 2.1 was calculated, for two failure modes, in Sections 7.3 and 7.4. Imagine that, to improve the security of supply, a twin parallel stream is added as follows:
   Construct reliability block diagrams for:
      (a) loss of supply;
      (b) failure to control downstream over-pressure;
   and recalculate the reliabilities for one year.
2. For this twin-stream case, imagine that the system is inspected every two weeks for valves that have failed shut. (a) How does this affect the system failure rate in respect of loss of supply? (b) Now calculate the effect of addressing CCF (say 5%).
3. In Section 8.3, the cutsets were ranked by failure rate. Repeat this ranking by unavailability.

CHAPTER 10

Risk Assessment (QRA)

10.1 Frequency and Consequence

Having identified a hazard, the term 'risk analysis' is often used to embrace two types of assessment:

- The frequency (or probability) of the event.
- The consequences of the event.

Thus, for a process plant the assessments could be:

- The probability of an accidental release of a given quantity of toxic (or flammable) material might be 1 in 10 000 years.
- The consequence, following a study of the toxic (or thermal radiation) effects and having regard to the population density, might be 'a 1 in 10 chance of the event propagating to 40 fatalities'.

The term QRA (Quantified Risk Assessment) refers to assessing the frequency of an event and its measurable consequences (e.g. fatalities, damage).

The analysis of consequence is a specialist area within each industry and may be based on chemical, electrical, gas or nuclear technology. Prediction of frequency, however, is essentially the same activity as reliability prediction, the methods for which have been described in Chapters 7–9. Events are dependent on:

- random hardware (component) failures
- systematic failures including software
- human error.

Quantitative risk assessment has received increasing attention over the years, particularly following Lord Cullen's inquiry into the Piper Alpha disaster.

Risk analysis also involves factors such as lightning, collision, weather factors, flood, etc., and these are dealt with in Section 10.4.

Reliability, Maintainability and Risk. DOI: 10.1016/B978-0-08-096902-2.00010-6

10.2 Perception of Risk, ALARP and Cost per Life Saved

10.2.1 Maximum Tolerable Risk (Individual Risk)

When quantifying the risk of fatality the meanings of the words 'tolerable', 'acceptable' and 'unacceptable' are important. There is, of course, no such thing as zero risk and it is necessary to think about what is 'tolerable' or even 'acceptable'. Appendix 7 shows some typical risk figures, which put the matter into perspective.

In this context the word 'acceptable' is generally taken to mean that we accept the probability of fatality as reasonable, having regard to the circumstances, and would not seek to expend much effort in reducing it further.

Tolerable, on the other hand, implies that whilst we are prepared to live with the particular risk level, we would continue to review its causes and the defenses we might take with a view to reducing it further. Cost would probably come into the picture in that any potential reduction in risk would be compared with the cost needed to achieve it (see ALARP in 10.2.3 below).

Unacceptable means that we would not tolerate that level of risk and would not participate in the activity in question nor permit others to operate a process that exhibited it.

'Individual risk' is the frequency of fatality for a hypothetical person in respect of a specific hazard. This is different from 'societal risk', which takes account of multiple fatalities. Society has a greater aversion to multiple fatalities than single ones in that killing 10 people in a single incident is perceived as worse than 10 separate single fatalities.

At the lower end of the risk scale, a 'broadly acceptable' risk is nearly always defined. This is the risk below which one would not, normally, seek further risk reduction. It is approximately two orders of magnitude less than the total of random risks to which one is exposed in everyday life.

There is a body of opinion that multiple fatalities should also affect the choice of 'maximum tolerable individual risk'. The targets in Table 10.1 reflect an attempt (current UK gas industry guidance) to take account of societal risk concerns in a relatively simple way and is thus a hybrid compromise. More complex calculations for societal risk (involving F–N curves) are sometimes addressed by specialists, as are adjustments for particularly vulnerable sections of the community (the disabled, children etc). See Section 10.2.4 below.

Table 10.1: Target Single- and Multiple-Fatality Individual Risks

Maximum Tolerable Individual Risk (per annum)			
	1–2 Fatalities	3–5 Fatalities	6 or More Fatalities
Employee	10^{-4}	$3\ 10^{-5}$	10^{-5}
Public	10^{-5}	$3\ 10^{-6}$	10^{-6}
Broadly Acceptable Risk (per annum)			
Employee and public	10^{-6}	$3\ 10^{-7}$	10^{-7}

The location, i.e. site or part of a site for which a risk is being addressed, may be exposed to multiple potential sources of risk. The question arises as to how many potential separate hazards an individual (or group) in any one place and time is exposed to. Therefore, in the event of exposure to several hazards at one time, one should seek to allow for this by specifying a more stringent target for each hazard. For example, a study addressing a multi-risk installation might need to take account of an order of magnitude of sources of risk. On the other hand, an assessment of a simple district pressure regulator valve for the local distribution of natural gas implies a limited number of sources of risk (perhaps only one).

A typical assessment confined to employees on a site might use the recommended 10^{-4} pa maximum tolerable risk (for 1–2 fatalities) but might identify 10 sources of risk to an individual in a particular place. Thus, an average of 10^{-5} pa would be used as the maximum tolerable risk for each of the 10 safety functions involved. By the same token, the broadly acceptable risk would be factored from 10^{-6} pa to 10^{-7} pa.

In any event, the final choice of maximum tolerable risk (in any scenario) forms part of the 'safety argument' put forward by a system user. There are no absolute rules but the foregoing provides an overview of current practice. *The Safety Critical Systems Handbook*, 3rd Edition, (D.J. Smith and K.G.L. Simpson) addresses this area in greater depth.

10.2.2 Maximum Tolerable Failure Rate

This involves factoring the maximum tolerable risk according to totally external levels of protection and to circumstances that limit the propagation to fatality of the event. Table 10.2 gives examples of the elements that might be considered. These are not necessarily limited to the items described below and the analyst(s) must be open-ended in identifying and assessing the factors involved.

The maximum tolerable failure rate is then targeted by taking the maximum tolerable risk and factoring it according to the items assessed. Thus, for the examples given in Table 10.2 (assuming a 10^{-5} pa involuntary risk):

$$\text{Maximum Tolerable Failure Rate} = 10^{-5}\,\text{pa} / (0.6 \times 0.2 \times 0.7 \times 0.25 \times 0.9 \times 0.25)$$
$$= 2.1\ 10^{-3}\,\text{pa}$$

■ Example:

A gas release (e.g. a natural gas holder over-fill) is judged to be a scenario leading to a single on-site fatality and three off-site fatalities. Both on and off site, person(s) are believed to be exposed to that one risk from the installation.

■

Table 10.2: Factors Leading to the Maximum Tolerable Failure Rate

Factor Involving the Propagation of the Incident or Describing an Independent Level of Protection	Probability (example)	This Column is Used to Record Arguments, Justifications, References etc to Support the Probability Used
The profile of time at risk	60%	Quantifying whether the scenario can develop. This may be <100% as for example if: • flow, temp, pressure etc profiles are only sufficient at specific times, for the risk to apply • the process is only in use for specific periods.
Unavailability of separate mitigation fails (i.e. another level of protection)	20%	Mitigation outside the scope of this study and not included in the subsequent modeling that assesses if the system meets the risk target. Examples are: • a down stream measurement (temp, pressure etc.) leading to manual intervention • a physical item of protection (for example, vessel; bund) not included in the study.
Probability of the scenario developing	70%	Examples are: • the vessel/line will succumb to the over-temp, over-pressure etc. • the release has an impact on the passing vehicle.
Person(s) exposed (i.e. being at risk)	25%	Proportion of time during which some person or persons are close enough to be at risk should the event propagate. Since a person may be exposed to a range of risks during the working week, this factor should not be erroneously reduced to the proportion of time exposed to the risk in question. If that were repeated across the spectrum of risks then each would be assigned an artificially optimistic target. The working week is approximately 25% of the time and thus that is the factor which would be anticipated for an on-site risk. In the same way, an off-site risk may only apply to a given individual for a short time.
Probability of subsequent ignition	90%	Quantifying if the released material ignites/explodes.
Fatality ensues	25%	The likelihood that the event, having developed, actually leads to fatality.

10.2.2.1 On site

Proportion of time system can offer the risk	75%	40 weeks pa
Probability of ignition	5%	Judgement
Person at risk	25%	Working week i.e. 42 hrs/168 hrs
Probability of fatality	75%	Judgment

From Table 10.1, the maximum tolerable risk is 10^{-4} pa. Thus, the maximum tolerable failure rate (leading to the event) is calculated as:

$$10^{-4} \text{pa}/(0.75 \times 0.05 \times 0.25 \times 0.75) = 1.4 \ 10^{-2} \text{pa}$$

10.2.2.2 Off site

Proportion of time system can offer the risk	75%	40 weeks pa
Probability of ignition	5%	Judgement
Person(s) at risk	33%	Commercial premises adjoin
Probability of three fatalities	10%	Offices well protected by embankments

From Table 10.1 the maximum tolerable risk is $3\ 10^{-6}$ pa. Thus the maximum tolerable failure rate (leading to the event) is calculated as:

$$3\ 10^{-6}\,\text{pa}/(0.75 \times 0.05 \times 0.33 \times 0.1) = 2.4\ 10^{-3}\,\text{pa}$$

Thus, **$2.4\ 10^{-3}$ pa**, being the more stringent of the two, is taken as the maximum tolerable failure rate target.

Chapter 22 (Section 22.2.1) carries on to establish safety-integrity targets from these maximum tolerable failure rates.

10.2.3 ALARP and Cost per Life Saved

The principle of ALARP (as low as reasonably practicable) describes the way in which risk is treated legally and by the HSE (Health and Safety Executive) in the UK. The concept is that all reasonable measures will be taken in respect of risks that lie in the 'tolerable' zone to reduce them further until the cost of further risk reduction is grossly disproportionate to the benefit. In any case it is always necessary, whatever the cost–benefit arguments, to demonstrate the application of good practice. Figure 10.1 shows the so-called ALARP triangle, which illustrates these regions.

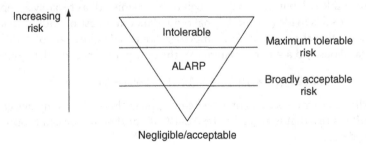

Figure 10.1: ALARP triangle

The question arises as to the 'cost per life saved' criterion to be used. Organizations are reluctant to state grossly disproportionate levels of CPL. Currently, figures in the range of £500 000 to £4 000 000 are common. Where a risk has the potential for multiple fatalities then higher sums may be used.

However, a value must be chosen by the plant operator for each assessment. The value selected must take account of any uncertainty inherent in the assessment and may have to take account of any company-specific issues such as the number of similar installations. The greater the potential number of lives lost and the greater the aversion to the scenario then the larger is the choice of the cost per life saved criteria. Values that have been quoted include:

1. Approximately £1 000 000, where there is a recognized scenario, a voluntary aspect to the exposure, a sense of having personal control, small numbers of casualties per incident. An example would be PASSENGER ROAD TRANSPORT.
2. Approximately £2 000 000–£4 000 000, where the risk is not under personal control and therefore an involuntary risk. An example would be TRANSPORT OF DANGEROUS GOODS.
3. Approximately £5 000 000–£15 000 000, where there are large numbers of fatalities, there is uncertainty as to the frequency and no personal control by the victim. An example would be MULTIPLE OFFSHORE FATALITIES.

This is a controversial area and figures can be subject to rapid revision in the light of catastrophic incidents and subsequent media publicity.

The following is an example:

The maximum tolerable risk of fatality associated with a particular system failure mode might be 10^{-4} per annum. The failure rate, for that mode, which risk assessment shows is associated with that frequency, is say 10^{-3} failures per annum (i.e. there is a 1:10 chance of the failure leading to fatality). If the broadly acceptable risk is 10^{-6} per annum then it follows that it will be achieved with a failure rate 10 times less, 10^{-5} per annum.

Let the predicted failure rate (using industry-specific data) for the system failure mode in question be 8×10^{-4} per annum (in other words a fatality risk of 8×10^{-5} per annum). This is better than the maximum tolerable risk but not small enough to be 'dismissed' as broadly acceptable. Therefore, a design proposal is made (e.g. additional redundancy at a cost of £5000) to improve the failure rate. Assume that the outcome is an improved predicted failure rate of 4×10^{-4} per annum. Assuming two fatalities and a 40-year system life, the cost per life saved calculation is:

$$£5000/([8 \times 10^{-5} - 4 \times 10^{-5}] \, 2 \times 40) = £1.5 \, million$$

If this exceeds the cost per life saved criterion being applied then the existing design would be considered to offer a risk that is argued to be ALARP. If not then the design proposal would need to be considered.

However, in many assessments, no specific risk reduction measure has yet been proposed and thus no cost per life saved can be calculated. However, the above equation can be used, rather than to calculate the CPL, to calculate the cost that should be contemplated, given some CPL criteria.

> **The following is an example:**
>
> The frequency of some hazardous failure maps to a risk of 6.5×10^{-6} pa. It is less than the 'maximum tolerable risk' but not small enough to be considered 'broadly acceptable' and is therefore in the ALARP region.
>
> If a cost per life saved criterion of £4 000 000 is used then the expenditure on any proposal that might reduce the risk to 10^{-6} pa can be calculated (assuming two fatalities and a 30-year plant life) as:
>
> $$£4\,000\,000 = £\text{proposed}/([6.5 \times 10^{-6} - 1 \times 10^{-6}] \times 2 \times 30)$$
>
> Thus £proposed = 1320
> Any proposal involving less than £1320 that would reduce the risk to 10^{-6} pa should be considered. This might well be possible if proof-test intervals are reduced.

The foregoing example provides a useful way of indicating whether risk reduction is or is not feasible within the cost indicated.

The concept of 'grossly disproportionate' also arises. In order to demonstrate that ALARP has been achieved, the cost of implementing a measure to reduce risk shall be grossly disproportionate to the benefit. Note: Guidance such as SPC/Permissioning/9 and SPC/Permissioning/12 is available from the HSE. The suggestion is that the cost per life saved criterion is multiplied by a gross disproportion factor of between 1 and 2 towards the bottom of the ALARP region (i.e. just above the 'broadly acceptable' level) and 10 towards the top of the ALARP region (i.e. just below the 'intolerable' level).

10.2.4 Societal Risk

It is important to note that individual risk and societal risk calculations are fundamentally different. Thus the starting points for maximum tolerable risk, in the case of a single fatality, do not immediately coincide. It should be stressed that, for societal risk, it is a fatal event frequency (irrespective of any one individual) and, for individual risk, a frequency of death to an individual – not the same thing.

For societal risk the distinction between voluntary and involuntary is largely irrelevant since voluntary risk usually involves specific individuals in a specific place and is thus dealt with by individual risk. Societal risk usually involves involuntary, randomly involved individuals and thus the involuntary concept applies.

Scenarios, such as sites, usually imply a risk to the same (more or less) groups of individuals (be it on site or off site) at any time. 'Distributed' risks, such as, for example, pipelines across wide areas, rail journeys, tunnels with rapidly changing identities of individuals, are the scenarios for which the involuntary risk approach becomes limited and the societal risk approach more appropriate.

First one assesses the number of potential fatalities. This may not be a single number at all times of the day. The following example shows how a weighted average can be arrived at:

for 4 hours per day, 60 persons are at risk;
for 20 hours per week, 17 persons are at risk;
for 24 hours per day, 1 person is at risk.
Weighted average of exposure is:

$$4/24 \times 60 + 17/168 \times 10 + 24/24 \times 1 = 12 \text{ fatalities:}.$$

It is now necessary to address the maximum tolerable risk. unlike the individual Risk criteria (Table 10.1), which address the probability as applying to an individual, the criterion becomes the frequency of a fatal event (irrespective of the individuals concerned). Figure 10.2 suggests criteria based on the number of potential fatalities.

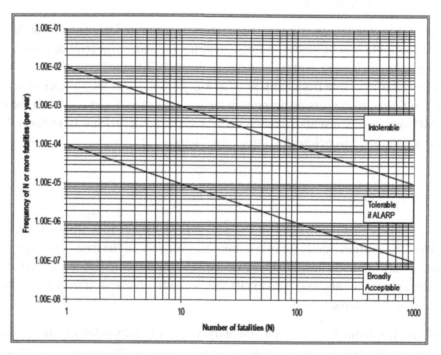

Figure 10.2: Societal risk criteria.

Although expressed in log by log format, it is a linear relationship that can be summarized as:

maximum tolerable risk (societal) = 10^{-2} pa/N;
broadly acceptable risk (societal) = 10^{-4} pa/N;
where N is the number of potential fatalities.

The propagation to fatality of an event is calculated as for involuntary risk BUT ignoring the element that addresses what proportion of the time any one is at risk, it having been dealt with already in the societal risk concept.

ALARP is dealt with much as for individual risk. The cost per life saved (CPL) is:

CPL = £proposed/(predicted societal risk – broadly acceptable risk) × N × plant-life

Disproportionality was mentioned in Section 10.2.3 above and the following are typical criteria:

Number of Potential Fatalities	Cost per Life Saved Criterion
1–2	£2 000 000
3–5	£4 000 000
6–10	£6 000 000
11–50	£10 000 000
>50	£20 000 000

This raises the question as to which approach (individual or societal) should prevail in any given scenario. Examples quoted above (e.g. site killing specific people versus pipeline to which numerous ever-changing identities of persons are exposed) are fairly straightforward.

Some scenarios might need the application of BOTH individual and societal calculations and for ALARP to be satisfied in both cases. The following is an example of such a case.

A pipeline passes through a tunnel which is utilized 24 hours per day such that at any time 100 randomly selected persons are at risk from pipeline rupture. It is assessed that there would be potentially 100 fatalities given that an incident has a 75% chance of propagating to fatality. However, there are also three specific maintenance personnel at any time, each being present for 35 hours per week (20%). It is assessed that all three might be potentially simultaneous fatalities given that an incident has a 50% chance of propagating to their fatality. There are no other simultaneous risks perceived. A reliability/integrity study has established a predicted frequency of pipeline breach of $5 \ 10^{-5}$ pa. The pipeline will remain in situ for 25 years.

Individual Risk Study
From Table 10.1 a voluntary (3 fatality) max tolerable risk of $3 \ 10^{-5}$ pa is chosen.
The broadly acceptable risk is $3 \ 10^{-7}$ pa.
The maximum tolerable failure rate for the pipeline is thus $3 \ 10^{-5}$ pa/(50% × 20%) = $3 \ 10^{-4}$ pa
The predicted failure rate for the pipeline is $5 \ 10^{-5}$ pa (from above)
Thus the predicted individual risk is $3 \ 10^{-5}$ pa × $5 \ 10^{-5}/3 \ 10^{-4}$ = $5 \ 10^{-6}$ pa
The cost per life saved criterion (above) is £4 000 000 (three fatalities)
ALARP is tested as follows:
£4 000 000 = £proposed/($5 \ 10^{-6}$ pa – $3 \ 10^{-7}$ pa) × 3 fatalities × 25 yrs
Thus any expenditure within a budget of **£1400**/that might reduce the risk to the broadly acceptable level should be considered. Since no realistic risk reduction can be obtained within this sum it might be argued that ALARP is satisfied.

(Cont.)

Societal Risk Study

From Figure 10.2 the max tolerable risk of 10^{-2} fatalities pa is factored by 100 fatalities to yield a frequency target of 10^{-4} pa.

The broadly acceptable risk is by the same token therefore 10^{-6} pa.

The maximum tolerable failure rate for the pipeline is thus 10^{-4} pa/(75%) = $1.3 \ 10^{-4}$ pa

The predicted failure rate for the pipeline is $5 \ 10^{-5}$ pa (from above)

Thus the predicted risk frequency is 10^{-4} pa × $5 \ 10^{-5}/1.3 \ 10^{-4}$ = $3.8 \ 10^{-5}$ pa

The cost per life saved criterion (above) is £20 000 000 (100 fatalities)

ALARP is tested as follows:

£20 000 000 = £proposed/ ($3.8 \ 10^{-5}$ pa – 10^{-6} pa) × 100 fatalities × 25 yrs

Thus any expenditure within a budget of **£1.85 million** that might reduce the risk to the broadly acceptable level should be considered.

Conclusion

From the individual risk standpoint ALARP is argued to be satisfied.

From the societal risk standpoint ALARP is not satisfied and risk reduction should be studied within the budget indicated.

10.2.5 Production/Damage Loss

The same technique may be applied to production loss as well as to safety. The unavailability of a process can be predicted and used to assess the cost of outage that can be anticipated. Thus, a process for which outage costs £50 000 per day, and having a predicted unavailability of 5×10^{-3}, will lose:

$$£50 000 \times 5 \times 10^{-3} \times 365 = £91 250 \text{ per annum.}$$

Suppose that a proposed modification, involving redundant items, is assessed to cost £100 000 and that it improves the unavailability from 5×10^{-3} to 8×10^{-4}.

The saving is thus:

$$£50 000 \times (5 \times 10^{-3} - 8 \times 10^{-4}) \times 365 = £76 650 \text{ per annum.}$$

There is a saving of £14 600 pa and it could therefore be argued that the proposed modification is acceptable in that the proposed expenditure of £100 000 offers nearly 15% return on the investment.

10.3 Hazard Identification

Before an event (failure) can be quantified it must first be identified and there are a number of formal procedures for this process. Hazard Identification (HAZID) is used to identify the possible hazards, HAZOP (Hazard and Operability Studies) is used to establish how the hazards might arise in a process, whereas HAZAN (Hazard Analysis) refers to the process of analyzing the outcome of a hazard otherwise known as consequence analysis.

This is carried out at various levels of detail from the earliest stages of design throughout the project design cycle.

Preliminary hazard analysis, at the early system design phase, identifies safety critical areas, identifies and begins to quantify hazards and begins to set safety targets. It may include:

> previous experience (historical information)
> review of hazardous materials, energy sources, etc.
> interfaces with operators, public, etc.
> applicable legislation, standards and regulations
> hazards from the environment
> impact on the environment
> software implications
> safety-related equipment.

More detailed hazard analysis follows in the detailed design stages. Now that specific hardware details are known and drawings exist, studies can address the effects of failure at component and software level. FMEA and fault tree techniques (Chapter 8) as well as HAZOP and consequence analyses are applicable here.

10.3.1 HAZOP

HAZOP is a technique developed in the 1970s by loss prevention engineers working for Imperial Chemical Industries at Tees side, UK. The purpose of a HAZOP is to identify hazards in a process. At one time this was done by individuals or groups of experts at a project meeting. This slightly blinkered approach tended to focus on the more obvious hazards and those that related to the specific expertise of the participants. In contrast to this, HAZOP involves a deliberately chosen balanced team using a systematic approach. The method is to systematically brainstorm the plant, part by part, and to review how deviations from the normal design quantities and performance parameters would affect the situation. Appropriate remedial action is then agreed.

One definition of HAZOP has been given as:

> A Study carried out by a *Multidisciplinary Team,* who apply *Guidewords* to identify *Deviations* from the Design Intent of a system and its Procedures. The team attempt to identify the *Causes and Consequences* of these Deviations and the Protective Systems installed to minimize them and thus to make *Recommendations* which lead to Risk Reduction.

This requires a full description of the design (up-to-date engineering drawings, line diagrams, etc.) and a full working knowledge of the operating arrangements. A HAZOP is thus usually conducted by a team that includes designers and operators (including plant, process and instrumentation) as well as the safety (HAZOP) engineer.

A typical small process plant might be 'HAZOPed' by a team consisting of a:

  Chemical Engineer
  Mechanical Engineer
  Instrument Engineer
  Loss Prevention (or Safety or Reliability) Engineer
  Chemist
  Production Engineer/Manager
  Project Manager.

A key feature is the HAZOP team leader, who must have experience of HAZOP and be full-time in the sense that he attends the whole study, whereas some members may be part-time. An essential requirement for the leader is experience of HAZOP in other industries so as to bring as wide as possible a view to the probing process. Detailed recording of problems and actions is essential during the meeting. Follow-up and review of actions must also be formal. There must therefore be a full-time team secretary who records all findings and actions.

The procedure will involve:

  define the scope and objectives of the HAZOP
  define the documentation required
  select the team
  prepare for the HAZOP (pre-reading)
  carry out and record the HAZOP
  implement the follow-up action
  record results.

In order to formalize the analysis, a 'guideword' methodology has evolved in order to point the analysts at the types of deviation. The guidewords are applied to each of the process parameters such as flow, temperature, pressure, etc. under normal operational as well as start-up and shut-down modes. Account should be taken of safety systems that are allowed, under specified circumstances, to be temporarily defeated. The following table describes the approach:

Guideword	Meaning	Explanation
NO or NOT	The parameter is zero	Something does not happen but no other effect
MORE THAN or LESS THAN	There are increases or decreases in the process parameter	Flows and temperatures are not normal
AS WELL AS	Qualitative increase	Some additional effect
PART OF	Qualitative increase	Partial effect (not all)
THE REVERSE	Opposite	Reverse flow or material
OTHER THAN	Substitution	Totally different effect

Each deviation of a parameter must have a credible *cause*, typically a component or human error related failure or a deviation elsewhere in the plant. Examples of typical *causes* might be:

Deviation	Cause
More flow	Line rupture
	Control valve fail 'open'
Less flow	Control valve fail 'closed'
	Leaking vessel or heat exchanger
No flow	Blockage
	Rupture
Reverse flow	Siphoning
	Check-valve failure
More pressure	Restricted flow
	Boiling
Less/no pressure	Excessive flow out
	Insufficient flow in
More level	Operator error
	Vessel leak
Less/no level	Drain left open
	High barometric pressure
More temperature	Loss of cooling
	Latent heat release
Less temperature	Joule-Thomson cooling
	Adiabatic expansion
Part composition	Loss of ratio control
	Dosing pump failure
More composition	Carry-over
	By-products

Causes lead to *consequences*, which need to be assessed. When a parameter has varied beyond the design intent then it might lead to vessel rupture, fire, explosion, toxic release, etc.

The *likelihood* may also be assessed. The reliability prediction techniques described earlier in this book can be used to predict the frequency of specific events. However, these techniques may be reserved for the more severe hazards. In order to prioritize, a more qualitative approach at the HAZOP stage might be to assign, using team judgement only, say five grades of likelihood as for example:

1. not more than once in the plant life
2. up to once in 10 years
3. up to once in 5 years
4. up to once a year
5. more frequent than annually.

A similar approach can be adopted for classifying *severity* pending more formal quantification of the more severe consequences. The ranking might be:

1. no impact on plant or personnel
2. damage to equipment only or minor releases
3. injuries to unit personnel (contained on site)
4. major damage, limited off-site consequences
5. major damage and extensive off-site consequences.

One approach is to use a *risk matrix* to combine the *likelihood* and *severity* assessments in order to prioritize items for a more quantified approach and for further action. One such approach is:

	Severity 1	Severity 2	Severity 3	Severity 4	Severity 5
Likelihood 1	1	2	3	4	5
Likelihood 2	2	4	6	7	8
Likelihood 3	3	6	7	8	9
Likelihood 4	4	7	8	9	10
Likelihood 5	5	8	9	10	10

where '10' is the highest ranking of consequence and '1' is the lowest.

HAZOP was originally applied to finalized plant design drawings. However, changes arising at this stage can be costly and the technique has been modified for progressive stages of application throughout the design cycle. As well as being a design tool HAZOP can be equally successfully applied to an existing plant and can lead to worthwhile modifications to the maintenance procedures.

Typical phases of the life cycle at which HAZOP might be applied are:

Conceptual design
Detailed design
Approved for construction
'As-built'
Proposed modifications
Response to regulatory requirements.

HAZOP can be applied to a wide number of types of equipment including:

process plants
transport systems
data and programable systems
buildings and structures
electricity generation and distribution
mechanical equipment
military equipment.

In summary, a HAZOP study not only reveals potential hazards but leads to a far deeper understanding of a plant and its operations.

Appendix 11 provides a somewhat simple example of a HAZOP.

10.3.2 HAZID

Whereas HAZOP is an open-ended approach, HAZID is a checklist technique. At an early stage, such as the feasibility study for a hazardous plant, HAZID enables the major hazards to be identified. At the conceptual stage a more detailed HAZID would involve designing out some of the major problems.

Often, HAZID uses a questionnaire approach and each organization tends to develop and evolve its own list, based on experience. Appendix 12 gives an example of such a list and is reproduced by kind permission of the Institution of Gas Engineers (guidance document SR24).

10.3.3 HAZAN (Consequence Analysis)

This technique is applied to selected hazards following the HAZOP and HAZID activities. It is usually the high-consequence activities such as major spillage of flammable or toxic materials or explosion that are chosen. High-consequence scenarios usually tend to be the low-probability hazards.

Consequence analysis requires a detailed knowledge of the materials/hazards involved in order to predict the outcome of the various failures. knowledge of the physics and chemistry of the outcomes is necessary in order to construct mathematical models necessary to calculate the effects on objects and human beings. Some examples are:

> flammable and toxic releases (heat radiation, food/water pollution and poisoning)
> structural collapse
> vehicle, ships and other impact (on structures and humans)
> nuclear contamination
> explosion (pressure vessels and chemicals)
> large-scale water release (vessels, pipes and dams).

Reference to specific literature, in each case, is necessary.

10.4 Factors to Quantify

The main factors that may need to be quantified in order to assess the frequency of an event are as follows.

10.4.1 Reliability

Chapters 7–9 cover this element in detail.

10.4.2 Lightning and Thunderstorms

It is important to differentiate between thunderstorm-related damage, which affects electrical equipment by virtue of induction or earth currents, and actual lightning strikes. The former is approximately one order (ten times) more frequent.

BS 6651: 1990 indicates an average of 10 thunderstorm days per annum in the UK. This varies, according to the geography and geology of the area, between 3 and 21 days per annum. Thunderstorm damage (usually electrical) will thus be related to this statistic. Some informal data suggest damage figures such as:

*   five incidents per square kilometer per annum where electrical equipment is used in outdoor or unprotected accommodation;
*   0.02 incidents per microwave tower.

Lightning strike, however, is a smaller probability and the rate per annum is derived by multiplying the effective area in square kilometers by the strikes per annum per square kilometer in Figure 10.3 (reproduced by kind permission of the British Standards Institution). The average is in the area of 0.3–0.5 per annum.

The effective area is obtained by subtending an angle of 45° around the building or object in question. Figure 10.4 illustrates the effect upon one elevation of a square building of side 10 m and height 2 m. The effective length is thus 14 m (10 + 2 + 2). BS 6651: 1990, from which Figure 10.4 is reproduced, contains a fuller method of assessment.

It must not be inferred, automatically, that a strike implies damage. This will depend upon the substance being struck, the degree of lightning protection and the nature of the equipment contained therein.

10.4.3 Aircraft Impact

Aircraft crash is a high-consequence but a low-probability event. The data are well recorded and a methodology has evolved for calculating the probability of impact from a crashing aircraft according to the location. This type of study stemmed from concerns when carrying out risk assessments of a nuclear plant but can be used for any other safety study where impact damage is relevant.

This subject is extensively covered in *The Calculation of Aircraft Risk in the UK*, J. P. Byrne, AEA Technology, Report 150, 1997 (HSE Books ISBN 0 7176 1447 6).

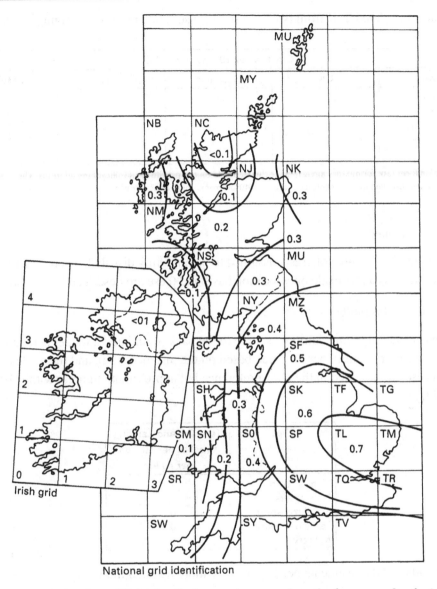

Figure 10.3: Number of lightning flashes to the ground per km² per year for the UK

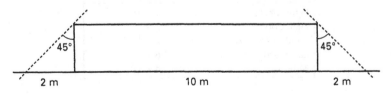

Figure 10.4: Lightning – effect of building elevation

Crashes are considered as if from two (additive) causes and approximate figures are:

Crashes per 10^{-5} pa per sq km				
Type	UK	Scotland	Wales	UK Mainland
Light aircraft	3.7	1	0.8	2.5
Helicopters				1.2
Small transport				0.1
Large transport				0.2
Military				5.*

*There is no uniform rate for military aircraft since this depends on the proximity of military installations. The range is 0.5–10 and spans so-called low concentration areas (>40 km from a high concentration) to high concentration areas.

10.4.3.1 Background

This is the 'ambient' source of crash, assumed to be randomly distributed across the UK. More accurate estimates can be made on a location basis and these are described in the above report.

10.4.3.2 Airfield Proximity

These are considered as an additional source to the background and a model is required that provides a crash probability per unit area based on the distance from, and the angle to, the runway. These are provided in the above reference but approximate figures (within 10 km of the airfield) are:

Type	Take Off and Landing Crashes per 10^{-5} pa per sq km
Light aircraft	1.2
Helicopters	–
Small transport	0.1
Large transport	0.1
Military	0.1

Large airfields (>50 movements per day) require more detailed modeling.

Expressed as crashes per movement we have:

Type	Crash per Movement
Light aircraft	1×10^{-6}
Helicopters	2×10^{-6}
Small transport	2×10^{-6}
Large transport	0.6×10^{-6}
Military	4×10^{-6}

10.4.4 Earthquake

Earthquake intensities are defined according to Mercalli and the modified scale can be summarized as follows:

Intensity	Effect
I	Not felt.
II	Felt by persons at rest on upper floors.
III	Felt indoors. Hanging objects swing. Vibration similar to light trucks passing. May not be recognized as an earthquake.
IV	Hanging objects swing. Vibration like passing of heavy trucks or jolt sensation like heavy ball striking wall. Parked motor cars rock. Windows, dishes and doors rattle. Glasses and crockery clink. Wooden walls may creak.
V	Felt outdoors. Sleepers awakened. Liquids disturbed and some spilled. Small unstable objects displaced or upset. Pendulum clocks affected. Doors, pictures, etc. move.
VI	Felt by all. People frightened, run outdoors and walk unsteadily. Windows, dishes, glassware broken. Items and books off shelves. Pictures off walls and furniture moved or overturned. Weak plaster and masonry D cracked. Small bells ring, trees or bushes visibly shaken or heard to rustle.
VII	Difficult to stand. Noticed by drivers of motor cars. Hanging objects quiver. Furniture broken, damage to masonry D including cracks. Weak chimneys broken at roof line. Plaster, loose bricks, stones, tiles, etc. fall and some cracks to masonry C. Waves on ponds. Large bells ring.
VIII	Steering of motor cars affected. Damage or partial collapse of masonry C. Some damage to masonry B but not A. Fall of stucco and some masonry walls. Twisting and falling chimneys, factory stacks, elevated tanks and monuments. Frame houses moved on foundations if not secured. Branches broken from trees and cracks in wet ground.
IX	General panic. Masonry D destroyed and C heavily damaged (some collapse) and B seriously damaged. Reservoirs and underground pipes damaged. Ground noticeably cracked.
X	Most masonry and some bridges destroyed. Dams and dikes damaged. Landslides. Railway lines slightly bent.
XI	Rails bent. Underground pipelines destroyed.
XII	Total damage. Large rocks displaced. Objects in air.

The masonry types referred to are:

D	Weak materials, poor workmanship.
C	Ordinary materials and workmanship but not reinforced.
B	Good workmanship and mortar. Reinforced.
A	Good workmanship and mortar and laterally reinforced using steel, concrete, etc.

The range of interest is V to VIII, since below V the effect is unlikely to be of concern and above VIII the probability of that intensity in the UK is negligible.

The following table of frequencies is assumed to apply across the UK:

Intensity	Annual Probability
V	12×10^{-3}
VI	3.5×10^{-3}
VII	0.7×10^{-3}
VIII	0.075×10^{-3}

A useful reference is *Elementary Seismology*, by C. F. Richter (Freeman).

For interest, the average annual incidence of earthquakes (worldwide) is:

Intensity (or Greater)	
IV	13 000
V	1 319
VI	134
VII	17
VIII	1

10.4.5 Meteorological Factors

The Meteorological Office publishes a range of documents giving empirical data by place and year, covering:

- extreme wind speeds and directions
- barometric pressure
- snow depth
- temperature
- precipitation.

These can be obtained from HMSO (Her Majesty's Stationery Office) and may be consulted in modeling the probability of extreme conditions that have been identified as being capable of causing the event in question.

Potential flooding areas can be obtained from www.environment-agency.gov.uk.

10.4.6 Other Consequences

As a result of extensive measurements of real events, models have been developed to assess various consequences. The earlier sections have outlined specific examples such as lightning, earthquake and aircraft impact. Other events, which are similarly covered in the appropriate literature and by a wide range of computer programs, include:

chemical release
gas explosion

fire and blast
ship collision
pipeline corrosion
pipeline rupture
jet dispersion
thermal radiation
pipeline impact
vapor cloud/pool dispersion.

PART 4

Achieving Reliability
and Maintainability

CHAPTER 11

Design and Assurance Techniques

This chapter outlines the activities and techniques, in design and operation, that are used to optimize reliability.

11.1 Specifying and Allocating the Requirement

The main objective of a reliability and maintainability program is to ensure adequate performance consistent with minimal maintenance costs. This can be achieved only if, in the first place, objectives are set and then described by suitable parameters. The intended use and environment of a system must be accurately stated in order to set realistic objectives and, in the case of contract design, the customer requirements must be clearly stated. It may well be that the customer has not considered these points and guidance may be necessary in persuading him or her to set appropriate targets with regard to the technology, environment and overall cost envisaged. Appropriate parameters have then to be chosen.

System reliability and maintainability will be specified, perhaps in terms of MTBF and MTTR, and values have then to be assigned to each separate unit. Thought must be given to the *allocation* of these values throughout the system such that the overall objective is achieved without over-specifying the requirement for one unit while under-specifying for another. Figure 11.1 shows a simple system comprising two units connected in such a way that neither may fail if the system is to perform. We saw in Chapter 7 that the system MTBF is given by:

$$\theta_s = \frac{\theta_1 \theta_2}{\theta_1 + \theta_2}$$

If the design objective for θ_s is 1000 hrs then this may be met by setting θ_1 and θ_2 both at 2000 hrs. An initial analysis of the two units, however, could reveal that unit 1 is twice as complex as, and hence likely to have half the MTBF of, unit 2. If the reliability is allocated

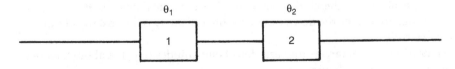

Figure 11.1

Reliability, Maintainability and Risk. DOI: 10.1016/B978-0-08-096902-2.00011-8

equally, as suggested, then the design task will be comparatively easy for unit 2 and unreasonably difficult for unit 1. Ideally, the allocation of MTBF should be weighted so that:

$$2\theta_1 = \theta_2$$

Hence

$$\theta S = \frac{2\theta_1^2}{3\theta_1} = \frac{2\theta_1}{3} = 1000\,hrs$$

Therefore

$$\theta_1 = 1500\,hrs$$

and

$$\theta_2 = 3000\,hrs$$

In this way the overall objective is achieved with the optimum design requirement being placed on each unit. The same philosophy should be applied to the allocation of repair times such that more attention is given to repair times in the high-failure-rate areas.

System reliability and maintainability are not necessarily defined by a single MTBF and MTTR. It was emphasized in Chapter 2 that it is essential to treat each failure mode separately and, perhaps, to describe it by means of different parameters. For example, the requirement for an item of control equipment might be stated as follows:

- *Spurious failure*: whereby a plant shutdown is carried out despite no valid shutdown condition:

  MTBF – 10 years

- *Failure to respond*: whereby a valid shutdown condition does not lead to a plant shutdown (NB: a dormant failure):

  Probability of failure on demand, which is, in fact, the unavailability = 0.0001

  (NB: The unavailability is therefore 0.0001 and thus the availability is 0.9999. The MTBF is therefore determined by the down time since unavailability is approximated from failure rate × down time.)

  See apportionment of targets for hazardous failure modes (e.g. the above 'failure to respond'), dealt with in Chapter 22.

11.2 Stress Analysis

Component failure rates are very sensitive to the stresses applied. Stresses, which can be classified as environmental or self-generated, include:

Temperature Shock Vibration Humidity Ingress of foreign bodies	Environmental
Power dissipation Applied voltage and current Self-generated vibration Wear	Self-generated

The sum of these stresses can be pictured as constantly varying, with peaks and troughs, and to be superimposed on a distribution of strength levels for a group of devices. A failure is seen as the result of stress exceeding strength. The average strength of the group of devices will increase during the early failures period owing to the elimination, from the population, of the weaker items.

Random failures are assumed to occur because of the overlap of chance peaks in the stress distribution with the weaknesses in the population. It is for this reason that screening and burn-in are highly effective in decreasing component failure rates. During wearout, strength declines owing to physical and chemical processes. An overall change in the average stress will cause more of the peaks to exceed the strength values and more failures will result. Figure 11.2 illustrates this concept, showing a range of strength illustrated as a bold curve

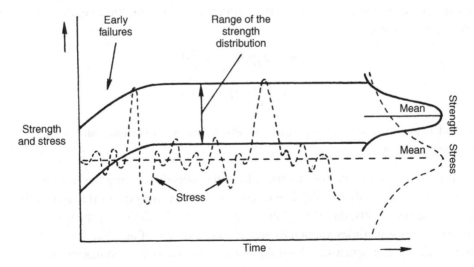

Figure 11.2 Strength and stress

overlapping with a distribution of stress shown by the dotted curve. At the left-hand end of the diagram the strength is shown increasing as the burn-in failures are eliminated. Although not shown, wearout would be illustrated by the strength curves falling again at the right-hand end.

For specific stress parameters, calculations are carried out on the distributions of values. The devices in question can be tested to destruction in order to establish the range of strengths. The distribution of stresses is then obtained and the two compared. In Figure 11.2 the two curves are shown to overlap significantly in order to illustrate the concept, whereas in practice that overlap is likely to be at the extreme tails of two distributions. The data obtained may well describe the central shapes of each distribution but there is no guarantee that the tails will follow the model that has been assumed. The result would then be a wildly inaccurate estimate of the failure probability. The stress/strength concept is therefore a useful model to understand failure mechanisms, but only in particular cases can it be used to make quantitative predictions.

The principle of operating a component part below the rated stress level of a parameter in order to obtain a longer or more reliable life is well known. It is particularly effective in electronics where under-rating of voltage and temperature produces spectacular improvements in reliability. Stresses can be divided into two broad categories – environmental and operating.

Operating stresses are present when a device is active. Examples are voltage, current, self-generated temperature and self-induced vibration. These have a marked effect on the frequency of random failures as well as hastening wearout. Figure 11.3 shows the relationship of failure rate to the voltage and temperature stresses for a typical wet aluminum capacitor.

Note that a 5 to 1 improvement in failure rate is obtained by either a reduction in voltage stress from 0.9 to 0.3 or a 30 °C reduction in temperature. The relationship of failure rate to stress in electronic components is often described by a form of the Arrhenius equation, which relates chemical reaction rate to temperature. Applied to random failure rate, the following two forms are often used:

$$\lambda_2 = \lambda_1 \exp K \left(\frac{1}{T_1} - \frac{1}{T_2} \right)$$

$$\lambda_2 = \lambda_1 \left(\frac{V_2}{V_1} \right)^n G(T_2 - T_1)$$

V_2, V_1, T_2 and T_1 are voltage and temperature levels. λ_2 and λ_1 are failure rates at those levels. K, G and n are constants.

It is dangerous to use these types of empirical formulae outside the range over which they have been validated. Unpredicted physical or chemical effects may occur that render them inappropriate and the results, therefore, can be misleading. Mechanically, the principle of excess material is sometimes applied to increase the strength of an item. It must be remembered that this can sometimes have the reverse effect and the elimination of large sections in a structure can increase the strength and hence reliability.

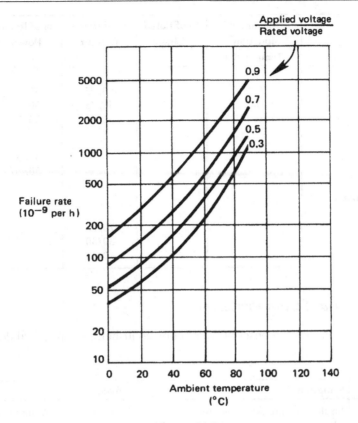

Figure 11.3

A number of general derating rules have been developed for electronic items. They are summarized in the following table as percentages of the rated stress level of the component. In most cases two figures are quoted, these being the rating levels for high reliability and good practice respectively. The temperatures are for hermetic packages and 20 °C should be deducted for plastic encapsulation.

	Maximum Junction Temp. (°C)	% of Rated Voltage	% of Rated Current	% of Rated Power	Fanout
Microelectronics					
Linear	100/110	70/80	75/80		
Hybrid	100				
Digital TTL	120/130	75/85			75/80
Digital MOS	100/105	75/85			75/80
Transistor					
Si signal	110/115	60/80	75/85	50/75	
Si power	125/130	60/80	60/80	30/50	
FET junction	125	75/85		50/70	
FET MOS	85/90	50/75		30/50	

(Contd.)

	Maximum Junction Temp. (°C)	% of Rated Voltage	% of Rated Current	% of Rated Power	Fanout
Diode					
Si signal	110/115	50/75	50/75	50/75	
Si power/SCR	110/115	50/70	50/75	30/50	
Zener	110/115		50/75	50/75	
Resistor					
Comp. and Film				50/60	
Wire wound				50/70	
Capacitor		40/50			
Switch and Relay contact					
Resistive/ capacitive			70/75		
Inductive			30/40		
Rotating			10/20		

11.3 Environmental Stress Protection

Environmental stress hastens the onset of wearout by contributing to physical deterioration. Included are:

Stress	Symptom	Action
High temperature	Insulation materials deteriorate Chemical reactions accelerate	Dissipate heat. Minimize thermal contact. Use fins. Increase conductor sizes on PCBs. Provide conduction paths
Low temperature	Mechanical contraction damage Insulation materials deteriorate	Apply heat and thermal insulation
Thermal shock	Mechanical damage within LSI components	Shielding
Mechanical shock	Component and connector damage	Mechanical design. Use of mountings
Vibration	Hastens wearout and causes connector failure	Mechanical design
Humidity	Coupled with temperature cycling causes 'pumping' – filling up with water	Sealing. Use of silica gel
Salt atmosphere	Corrosion and insulation degradation	Mechanical protection
Electromagnetic radiation	Interference to electrical signals	Shielding and part selection
Dust	Long-term degradation of insulation. Increased contact resistance	Sealing. Self-cleaning contacts
Biological effects	Decayed insulation material	Mechanical and chemical protection
Acoustic noise	Electrical interference due to microphonic effects	Mechanical buffers
Reactive gases	Corrosion of contacts	Physical seals

11.4 Failure Mechanisms

11.4.1 Types of Failure Mechanism

The majority of failures are attributable to one of the following physical or chemical phenomena.

Alloy formation: formation of alloys between gold, aluminum and silicon causes what is known as 'purple plague' and 'black plague' in silicon devices.

Biological effects: moulds and insects can cause failures. Tropical environments are particularly attractive for moulds and insects, and electronic devices and wiring can be affected.

Chemical and electrolytic changes: electrolytic corrosion can occur wherever a potential difference together with an ionizable film are present. The electrolytic effect causes interaction between the salt ions and the metallic surfaces, which act as electrodes. Salt-laden atmospheres cause corrosion of contacts and connectors. Chemical and physical changes to electrolytes and lubricants both lead to degradation failures.

Contamination: dirt, particularly carbon or ferrous particles, causes electrical failure. The former deposited on insulation between conductors leads to breakdown and the latter to insulation breakdown and direct short circuits. Non-conducting material such as ash and fibrous waste can cause open-circuit failure in contacts.

Depolymerization: this is a degrading of insulation resistance caused by a type of liquefaction in synthetic materials.

Electrical contact failures: failures of switch and relay contacts occur owing to weak springs, contact arcing, spark erosion and plating wear. In addition, failures due to contamination, as mentioned above, are possible. Printed-board connectors will fail owing to loss of contact pressure, mechanical wear from repeated insertions and contamination.

Evaporation: filament devices age owing to evaporation of the filament molecules.

Fatigue: this is a physical/crystalline change in metals that leads to spring failure, fracture of structural members, etc.

Film deposition: all plugs, sockets, connectors and switches with non-precious metal surfaces are likely to form an oxide film, which is a poor conductor. This film therefore leads to high-resistance failures unless a self-cleaning wiping action is used.

Friction: friction is one of the most common causes of failure in motors, switches, gears, belts, styli, etc.

Ionization of gases: at normal atmospheric pressure a.c. voltages of approximately 300 V across gas bubbles in dielectrics give rise to ionization, which causes both electrical noise and ultimate breakdown. This reduces to 200 V at low pressure.

Ion migration: if two silver surfaces are separated by a moisture-covered insulating material then, providing an ionizable salt is present as is usually the case, ion migration causes a silver 'tree' across the insulator.

Magnetic degradation: modern magnetic materials are quite stable. However, degraded magnetic properties do occur as a result of mechanical vibration or strong a.c. electric fields.

Mechanical stresses: bump and vibration stresses affect switches, insulators, fuse mountings, component lugs, printed-board tracks, etc.

Metallic effects: metallic particles are a common cause of failure as mentioned above. Tin and cadmium can grow 'whiskers', leading to noise and low-resistance failures.

Moisture gain or loss: moisture can enter equipment through pin holes by moisture vapor diffusion. This is accelerated by conditions of temperature cycling under high humidity. Loss of moisture by diffusion through seals in electrolytic capacitors causes reduced capacitance.

Molecular migration: many liquids can diffuse through insulating plastics.

Stress relaxation: cold flow ('creep') occurs in metallic parts and various dielectrics under mechanical stress. This leads to mechanical failure. This is not the same as fatigue, which is caused by repeated movement (deformation) of a material.

Temperature cycling: this can be the cause of stress fluctuations, leading to fatigue or to moisture build-up.

11.4.2 Failures in Semiconductor Components

The majority of semiconductor device failures are attributable to the wafer-fabrication process. The tendency to create chips with ever-decreasing cross-sectional areas increases the probability that impurities, localized heating, flaws, etc., will lead to failure by deterioration, probably of the Arrhenius type (Section 11.2). Table 11.1 shows a typical proportion of failure modes.

As microelectronics packaging density increases, small chip geometries entail much higher current densities. This suggests a greater need for derating in the application of such

Table 11.1

	Specific			
	Linear (%)	TTL (%)	CMOS (%)	In general (%)
Metalization	18	50	25	
Diffusion	1	1	9	55
Oxide	1	4	16	
Bond – die	10	10	–	25
Bond – wire	9	15	15	
Packaging/ hermeticity	5	14	10	
Surface contamination	55	5	25	20
Cracked die	1	1	–	

devices. Another complication is provided by changing materials to improve performance and to overcome the chip density problem. An example is the replacement of aluminum interconnection with lower-resistance copper to cut propagation delays. The overall effect is likely to accelerate the long-term wearout characteristic. Whereas in the 1970s chip lifetimes were thought to be of the order of hundreds of years, more recent estimates are an order less.

Percentage failure modes of microelectronic and discrete devices are given in Appendix 5.

11.4.3 Discrete Components

The most likely causes of failure in resistors and capacitors are shown in Tables 11.2 and 11.3. Short-circuit failure is rare in resistors. For composition resistors, fixed and variable, the division tends to be 50% degradation failures and 50% open circuit. For film and wire-wound resistors the majority of failures are of the open-circuit type.

Table 11.2

Resistor Type	Short	Open	Drift
Film	Insulation breakdown due to humidity. Protuberances of adjacent spirals	Mechanical breakdown of spiral due to r.f. Thin spiral	–
Wire wound	Over-voltage	Mechanical breakdown due to r.f. Failure of winding termination	
Composition			r.f. produces capacitance or dielectric loss
Variable (wire and composition)		Wiper arm wear. Excess current over a small segment owing to selecting low value	Noise
			Mechanical movement

11.5 Complexity and Parts

11.5.1 Reduction of Complexity

Higher scales of integration in electronic technology enable circuit functions previously requiring many hundreds (or thousands) of devices to be performed by a single component. Hardware failure is restricted to either the device or its connections (sometimes 40 pins) to the remaining circuitry. A reduction in total device population and quantity leads, in general, to higher reliability.

Standard circuit configurations help to minimize component populations and allow the use of proven reliable circuits. Regular planned design reviews provide an opportunity to assess the economy of circuitry for the intended function. Digital circuits provide an opportunity for

Table 11.3

Capacitor Type	Short	Open	Drift
Mica	Water absorption. Silver ion migration	Mechanical vibration	
Electrolytic solid tantalum	Solder balls caused by external heat from soldering	Internal connection	
		Failures due to shock or vibration	
Electrolytic non-solid tantalum	Electrolyte leakage due to temperature cycling	External welds	
Electrolytic aluminum oxide		Lead dissolved in electrolyte	Low capacitance due to aluminum oxide combining with electrolyte
Paper	Moisture. Rupture	Poor internal connections	
Plastic	Internal solder flow Instantaneous breakdown in plastic causing s/c	Poor internal connections	
Ceramic	Silver ion migration	Mechanical stress. Heat rupture internal	
Air (variable)	Loose plates. Foreign bodies	Ruptured internal connections	

reduction in complexity by means of logical manipulation of the expressions involved. This enables fewer logic functions to be used in order to provide a given result.

11.5.2 Part Selection

Since hardware reliability is largely determined by the component parts, their reliability and fitness for purpose cannot be over-emphasized. The choice often arises between standard parts with proven performance which just meet the requirement and special parts that are totally applicable but unproven. Consideration of design support services when selecting a component source may be of prime importance when the application is a new design. General considerations should be:

- function needed and the environment in which it is to be used;
- critical aspects of the part such as, for example, limited life, procurement time, contribution to overall failure rate, cost, etc;
- availability: number of different sources;
- stress: given the application of the component the stresses applied to it and the expected failure rate. The effect of burn-in and screening on actual performance.

11.5.3 Redundancy

This involves the use of additional active units or of standby units. Reliability may be enhanced by this technique, which can be applied in a variety of configurations:

- *Active redundancy*:
     Full: with duplicated units, all operating, one surviving unit ensures non-failure.
     Partial: a specified number of the units may fail as, for example, two out of four engines on an aircraft. Majority voting systems often fall into this category.
     Conditional: a form of redundancy that occurs according to the failure mode.
- *Standby redundancy*: involves extra units that are not brought into use until the failure of the main unit is sensed.
- *Load sharing*: active redundancy where the failure of one unit places a greater stress on the remaining units.
- *Redundancy and repair*: where redundant units are subject to immediate or periodic repair, the system reliability is influenced both by the unit reliability and the repair times.

The decision to use redundancy must be based on an analysis of the trade-offs involved. It may prove to be the only available method when other techniques have been exhausted. Its application is not without penalties since it increases weight, space and cost and the increase in number of parts results in an increase in maintenance and spares holding costs. Remember, as we saw in Chapter 2, redundancy can increase the reliability for one failure mode but at the expense of another. In general, the reliability gain obtained from additional elements decreases beyond a few duplicated elements owing to either the common mode effects (Section 8.2) or to the reliability of devices needed to implement the particular configuration employed. Chapters 7–9 deal, in detail, with the quantitative effects of redundancy.

11.6 Burn-In and Screening

For an established design the early failures portion of the bathtub curve represents the populations of items having inherent weaknesses due to minute variations and defects in the manufacturing process. Furthermore, it is increasingly held that electronic failures – even in the constant failure rate part of the curve – are due to microscopic defects in the physical build of the item. The effects of physical and chemical processes with time cause failures to occur in both the early failures and constant failure rate portions of the bathtub. Burn-in and screening are thus effective means of enhancing component reliability:

- *Burn-in* is the process of operating items at elevated stress levels (particularly temperature, humidity and voltage) in order to accelerate the processes leading to failure. The populations of defective items are thus reduced.
- *Screening* is an enhancement to quality control whereby additional detailed visual and electrical/mechanical tests seek to reveal defective features that would otherwise increase the population of 'weak' items.

The relationship between various defined levels of burn-in and screening and the eventual failure rate levels is recognized and has, in the case of electronic components, become formalized. For microelectronic devices US MIL STD 883 provides a uniform set of test, screening and burn-in procedures. These include tests for moisture resistance, high temperature, shock, dimensions, electrical load and so on. The effect is to eliminate the defective items mentioned above. The tests are graded into three classes in order to take account of the need for different reliability requirements at appropriate cost levels. These levels are:

Class C: the least stringent which requires 100% internal visual inspection. There are electrical tests at 25 °C but no burn-in.

Class B: in addition to the requirements of Class C there is 160 hrs of burn-in at 125 °C and electrical tests at temperature extremes (high and low).

Class S: in addition to the tests in Class B there is longer burn-in (240 hrs) and more severe tests including 72 hrs reverse bias at 150 °C.

The overall standardization and QA programs described in US-MIL-M-38510 call for the MIL 883 tests procedures. The UK counterpart to the system of controls is BS 9000, which functions as a four-tier hierarchy of specifications from the general requirements at the top, through generic requirements, to detail component manufacture and test details at the bottom. Approximate equivalents for the screening levels are:

MIL 883	BS 9400	Relative Cost (Approx.)
S	A	10
B	B	5
C	C	3
–	D	1
		0.5 (plastic)

11.7 Maintenance Strategies

This is dealt with, under reliability centered maintenance, in Chapter 16. It involves:

- routine maintenance (adjustment, overhaul)
- preventive discard (replacement)
- condition monitoring (identifying degradation)
- proof testing for dormant redundant failures.

CHAPTER 12

Design Review, Test and Reliability Growth

12.1 Review Techniques

Design review is the process of comparing the design, at points during design and development, with the requirements of earlier stages. Examples are a review of:

- the functional specification against the requirements specification;
- circuit or mechanical assembly performance against the functional specification;
- predicted reliability/availability against targets in the requirements specification;
- some software source code against the software specification.

Two common misconceptions about design review are:

- that they are schedule progress meetings;
- that they are to appraise the designer.

They are, in fact, to verify the design, as it exists at a particular time against the requirements. It is a measure, as is test, but carried out by document review and predictive calculations. The results of tests may well be an input to the review but the review itself is an intellectual process rather than a test.

It is a feedback loop that verifies each stage of design and provides confidence to proceed to the next. Review is a formal activity and should not be carried out casually. The following points are therefore important when conducting reviews:

- They must be carried out against a defined baseline of documents. In other words, the design must be frozen at specific points, known as baselines, that are defined by a list of documents and drawings each at a specific issue status.
- Results must be recorded and remedial actions formally followed up.
- All documents must be available in advance and checklists prepared of the points to be reviewed.
- Functions and responsibilities of the participants must be defined.
- The review must be chaired by a person independent of the design.
- The purpose must be specific and stated in advance in terms of what is to be measured. Consequently, the expected results should be laid down.

Reliability, Maintainability and Risk. DOI: 10.1016/B978-0-0809-6902-2.00012-X

The natural points in the design cycle that lend themselves to review are:

1. *Requirements specification*: this is the one point in the design cycle above which there is no higher specification against which to compare. It is thus the hardest review in terms of deciding if the outcome is satisfactory. Nevertheless, features such as completeness, unambiguity and consistency can be considered. A requirement specification should not prejudge the design and therefore it can be checked that it states *what* is required rather than *how* it is to be achieved.
2. *Functional specification*: this can be reviewed against the requirements specification and each function checked off for accuracy or omission.
3. *Detailed design*: this may involve a number of reviews depending on how many detailed design documents/modules are created. At this level, horizontal, as well as vertical, considerations arise. In addition to measuring the documents' compliance with the preceding stages, it is necessary to examine their links to other specifications/modules/drawings/diagrams, etc. Reliability predictions and risk assessments, as well as early test results, are used as inputs to measure the assessed conformance to higher requirements.
4. *Software*: code reviews are a particular type of review and are covered in Section 17.4.5.
5. *Test results*: although test follows later in the design cycle, it too can be the subject of review. It is necessary to review the test specifications against the design documents (e.g. functional specification). Test results can also be reviewed against the test specification.

A feature of review is the checklist. This provides some structure for the review and can be used for recording the results. Also, checklists are a means of adding questions based on experience and can be evolved, as lessons are learned from reviews. Section 17.6 provides specific checklists for software reviews. It is important, however, not to allow checklists to constrain the review process since they are only an aide-memoire.

12.2 Categories of Testing

There are four categories of testing:

1. *Design testing*: laboratory and prototype tests aimed at proving that a design will meet the specification. Initially prototype functional tests aim at proving the design. This will extend to pre-production models which undergo environmental and reliability tests and may overlap with:
2. *Qualification testing*: total proving cycle using production models over the full range of the environmental and functional specification. This involves extensive marginal tests, climatic and shock tests, reliability and maintainability tests and the accumulation of some field data. It must not be confused with development or production testing. The purpose of qualification testing is to ensure that a product meets all the requirements laid

down in the engineering specification. This should not be confused with product testing, which takes place after manufacture. Items to be verified are:

Function: specified performance at defined limits and margins.

Environment: ambient temperature and humidity for use, storage, etc. Performance at the extremes of the specified environment should be included.

Life: at specified performance levels and under storage conditions.

Reliability: observed MTBF under all conditions.

Maintainability: MTTR/MDT for defined test equipment, spares, manual and staff.

Maintenance: is the routine and corrective maintenance requirement compatible with use?

Packaging and transport: test under real conditions including shock tests.

Physical characteristics: size, weight, power consumption, etc.

Ergonomics: consider interface with operators and maintenance personnel.

Testability: consider test equipment and time required for production models.

Safety: use an approved test house such as BSI or the British Electrotechnical Approvals Board.

3. *Production testing and commissioning*: verification of conformance by testing modules and complete equipment. Some reliability proving and burn-in may be involved. Generally, failures will be attributable to component procurement, production methods, etc. Design-related queries will arise but should diminish in quantity as production continues.

4. *Demonstration testing*: an acceptance test whereby equipment is tested to agreed criteria and passes or fails according to the number of failures.

These involve the following types of test.

12.2.1 Environmental Testing

This proves that equipment functions to specification (for a sustained period) and is not degraded or damaged by defined extremes of its environment. The test can cover a wide range of parameters and it is important to agree a specification that is realistic. It is tempting, when in doubt, to widen the limits of temperature, humidity and shock in order to be extra sure of covering the likely range which the equipment will experience. The resulting cost of over-design, even for a few degrees of temperature, may be totally unjustified.

The possibilities are numerous and include:

Electrical:
    electric fields
    magnetic fields
    radiation.

Climatic:

temperature extremes / temperature cycling (internal and external may be specified)
humidity extremes
temperature cycling at high humidity
thermal shock – rapid change of temperature
wind – both physical force and cooling effect
wind and precipitation
direct sunlight
atmospheric pressure extremes.

Mechanical:

vibration at given frequency – a resonant search is often carried out
vibration at simultaneous random frequencies – used because resonances at different
frequencies can occur simultaneously
mechanical shock – bump
acceleration.

Chemical and hazardous atmospheres:

corrosive atmosphere – covers acids, alkalis, salt, greases, etc.
foreign bodies – ferrous, carbon, silicate, general dust, etc.
biological – defined growth or insect infestation
reactive gases
flammable atmospheres.

12.2.2 Marginal Testing

This involves proving the various system functions at the extreme limits of the electrical and
mechanical parameters and includes:

Electrical:

mains supply voltage
mains supply frequency
insulation limits
earth testing
high voltage interference – radiated. Typical test apparatus consists of a spark plug,
induction coil and break contact
mains-borne interference
line error rate – refers to the incidence of binary bits being incorrectly transmitted in
a digital system. Usually expressed as in 1 in 10^{-n} bits
line noise tests – analog circuits
electrostatic discharge – e.g. 10 kV from 150 pF through 150 Ω to conductive surfaces
functional load tests – loading a system with artificial traffic to simulate full
utilization (e.g. call traffic simulation in a telephone exchange)

input/output signal limits – limits of frequency and power

output load limits – sustained voltage at maximum load current and testing that current does not increase even if load is increased as far as a short circuit.

Mechanical:

dimensional limits – maximum and minimum limits as per drawing

pressure limits – covers hydraulic and pneumatic systems

load – compressive and tensile forces and torque.

12.2.3 High-Reliability Testing

The major problem in verifying high reliability, emphasized in Chapter 5, is the difficulty of accumulating sufficient data, even with no failures, to demonstrate statistically the value required. If an MTBF of, say, 10^6 hrs is to be verified, and 500 items are available for test, then 2000 elapsed hours of testing (3 months of continuous test) are required to accumulate sufficient time for even the minimum test that involves no failures. In this way, the MTBF is demonstrated with 63% confidence. Nearly two and a half times the amount of testing is required to raise this to 90%.

The usual response to this problem is to accelerate the failure mechanisms by increasing the stress levels. This involves the assumption that relationships between failure rate and stress levels hold good over the range in question. Interpolation between points in a known range presents little problem, whereas extrapolation beyond a known relationship is of dubious value. Experimental data can be used to derive the constants found in the equations shown in Section 11.2. In order to establish if the Arrhenius relationship applies, a plot of $\log_e$ failure rate against the reciprocal of temperature is made. A straight line indicates that it holds for the temperature range in question. In some cases parameters such as ambient temperature and power are not independent, as in transistors where the junction temperature is a function of both. Accelerated testing gives a high confidence that the failure rate at normal stress levels is, at least, less than that observed at the elevated stresses.

Where MTBF is expressed in cycles or operations, as with relays, pistons, rods and cams, the test may be accelerated without a change in the physics of the failure mechanism. For example, 100 contactors can be operated to accumulate 3×10^8 operations in one month although, in normal use, it might well take several years to accumulate the same amount of data.

12.2.4 Testing for Packaging and Transport

There is little virtue in investing large sums in design and manufacture if inherently reliable products are to be damaged by inadequate packaging and handling. The packaging needs to match the characteristics and weaknesses of the contents with the hazards it is likely to meet. The major causes of defects during packaging, storage and transport are:

1. Inadequate or unsuitable packaging materials for the transport involved.
    Transport, climatic and vibration conditions not foreseen.
    Storage conditions and handling not foreseen:
    -   requires consideration of waterproofing, hoops, bands, lagging, hermetic seals, desiccant, ventilation holes, etc.
2. Inadequate marking – see BS 2770 pictorial handling instructions.
3. Failure to treat for prevention of corrosion:
    -   various cleaning methods for the removal of oil, rust and miscellaneous contamination followed by preventive treatments and coatings.
4. Degradation of packaging materials owing to method of storage prior to use.
5. Inadequate adjustments or padding prior to packaging.
    Lack of handling care during transport:
    -   requires adequate work instructions, packing lists, training, etc.

Choosing the most appropriate packaging involves considerations of cost, availability and size, for which reason a compromise is usually sought. Crates, rigid and collapsible boxes, cartons, wallets, tri-wall wrapping, chipboard cases, sealed wrapping, fabricated and moulded spacers, corner blocks and cushions, bubble wrapping, etc. are a few of the many alternatives available to meet any particular packaging specification.

Environmental testing involving vibration and shock tests together with climatic tests is necessary to qualify a packaging arrangement. This work is undertaken by a number of test houses and may save large sums if it ultimately prevents damaged goods being received since the cost of defects rises tenfold and more, once equipment has left the factory. As well as specified environmental tests, the product should be transported over a range of typical journeys and then retested to assess the effectiveness of the proposed pack.

12.2.5 Multiparameter Testing

More often than not, the number of separate (but not independent) variables involved in a test makes it impossible for the effect of each to be individually assessed. To hold, in turn, all but one parameter constant and record its effect and then to analyze and relate all the parametric results would be very expensive in terms of test and analysis time. In any case, this has the drawback of restricting the field of data. Imagine that, in a three-variable situation, the limits are represented by the corners of a cube as in Figure 12.1, then each test would be confined to a straight line through the cube.

One effective approach involves making measurements of the system performance at various points, including the limits, of the cube. For example, in a facsimile transmission system the three variables might be the line error rate, line bandwidth and degree of data compression. For each combination the system parameters would be character error rate on received copy

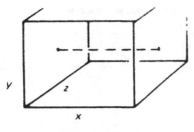

Figure 12.1

and transmission time. Analysis of the cube would reveal the best combination of results and system parameters for a cost-effective solution.

12.2.6 Step-Stress Testing

For electrical parameters, there is some predictable degree of correlation between stress and failure as was addressed by the Arrhenius relationships in Chapter 11. However, for mechanical failures (and indeed many electrical failures) there is no obvious model based on empirical data. In these cases accelerated step testing offers a limited form of assessment based on extrapolation.

This involves successively increasing one or more parameters. The stress parameters chosen (e.g. temperature, mechanical load) are increased by increments at defined time intervals. Thus, for example, a mechanical component could be tested at its nominal temperature and loading for a period of time. Both temperature and load would then be increased by a defined amount for a further equal period. Successive increments of stress would then be applied after each period.

The median rank cumulative failure percentages would then be plotted against the failure times (loglog against log) and a line obtained that (assuming the majority of failures occurred at the higher stresses) can be extrapolated back to the normal stress condition. The target probability of failure for some defined time period, at normal stress, will be a single point on the graph paper.

If the target point falls well to the left of the line then there is SOME evidence (NOT necessarily conclusive) that the design is adequate. Advantages and disadvantages of such a judgement are:

>   ADVANTAGES:
>   gives some indication of failure-free life
>   gives some confidence in the design.
>   DISADVANTAGES:
>   the assumption of linearity of the plot may not be valid
>   does not address all combinations of stresses
>   inaccuracies in the plot.

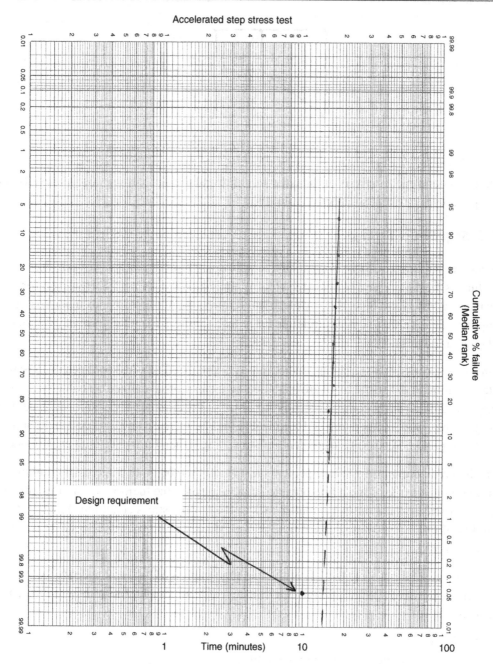

Figure 12.2: Accelerated step test plot

Consider the following example in which design requirement is for a reliability of 0.995 for a period of 10 minutes under a load of $2 \times 10^8\,Nm^{-2}$ at 450 °C. These two values are applied for 10 minutes and then increased by $0.1 \times 10^8\,Nm^{-2}$ and by 10 °C after each subsequent 30 seconds.

Failure No.	Median Rank %	Time Mins	Load $10^8\,Nm^{-2}$	Temperature °C
1	6.7	15.1	3.1	520
2	16	15.2	3.1	520
3	26	16.4	3.4	550
4	36	16.5	3.4	550
5	45	16.7	3.4	550
6	55	16.9	3.4	550
7	64	17	3.5	560
8	74	17.6	3.6	570
9	84	17.9	3.6	570
10	93	18	3.6	570

These results are plotted in Figure 12.2, from which it can be seen that the 'design point' is to the left of the extrapolated line. The inference is that the design is such that the requirement will be met. However, the above limitations must be kept in mind.

12.3 Reliability Growth Modeling

This concerns the improvement in reliability, during use, that comes from field data feedback resulting in modifications. Improvements depend on ensuring that field data actually lead to design modifications. Reliability growth, then, is the process of eliminating design-related failures. It must not be confused with the decreasing failure rate described by the bathtub curve.

Figure 12.3 illustrates this point by showing two bathtub curves for the same item of equipment. Both show an early decreasing failure rate whereas the later model, owing to reliability growth, shows higher reliability in the random failures part of the curve.

12.3.1 The CUSUM Technique

A simple but powerful method of plotting growth is the use of CUSUM (Cumulative Sum Chart) plots. In this technique an anticipated target MTBF is chosen and the deviations are plotted against time. The effect is to show the MTBF by the slope of the plot, which is more sensitive to changes in reliability.

The following example shows the number of failures after each 100 hrs of running of a generator. The CUSUM is plotted in Figure 12.4.

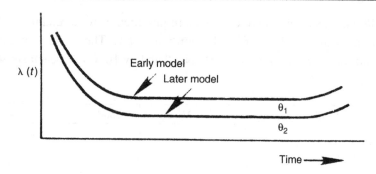

Figure 12.3:

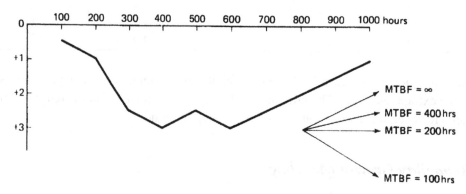

Figure 12.4: CUSUM plot

Cumulative Hours	Failures	Anticipated Failures if MTBF Were 200 hrs	Deviation	CUSUM
100	1	0.5	+0.5	+0.5
200	1	0.5	+0.5	+1.
300	2	0.5	+1.5	+2.5
400	1	0.5	+0.5	+3.
500	0	0.5	−0.5	+2.5
600	1	0.5	+0.5	+3.
700	0	0.5	−0.5	+2.5
800	0	0.5	−0.5	+2.
900	0	0.5	−0.5	+1.5
1000	0	0.5	−0.5	+1.

The CUSUM is plotted for an objective MTBF of 200 hrs. It shows that for the first 400 hrs the MTBF was in the order of half the requirement. From 400 to 600 hrs there was an improvement to about 200 hrs MTBF and thereafter there is evidence of reliability growth. The plot is sensitive to the changes in trend, as can be seen from the above.

The reader will note that the axis of the deviation has been inverted so that negative variations produce an upward trend. This is often done in reliability CUSUM work in order to reflect improving MTBFs by an upward curve, and vice versa.

12.3.2 Duane Plots

Whereas a CUSUM provides a clear picture of past events, it is sometimes required to establish a relationship between MTBF and time for the purposes of predicting reliability growth. The best-known model is that described by J. T. Duane in 1962. It assumes an empirical relationship whereby the improvement in MTBF is proportional to T^α where T is the total equipment time and α is a growth factor.

This can be expressed in the form:

$$\theta = kT^\alpha$$

Which means that with two sample data points:

$$\theta_2/\theta_1 = (T_2/T_1)^\alpha$$

Hence, if any two pairs of values of T and MTBF are known the equations can be solved to obtain k and α. The amount of T required to reach a given desired MTBF can then be predicted, with the assumption that the growth rate does not change. Typically α is between 0.1 and 0.65.

Figure 12.5 shows Duane plots of cumulative MTBF against cumulative time on log axes. The middle line ($\alpha = 0.35$) is a Duane plot based on field trial data for a product with a target MTBF of 50 years. As can be seen, the Duane plot predicts that 25 000 cumulative hours would be required, at the curent reliablity growth, to achieve the target. The ($\alpha = 0.5$) line shows that, were the growth rate to be substantially increased, one might anticipate reaching the target after 2500 cumulative hours. The ($\alpha = 0.2$) line is shown purely to illustrate the approximate limits.

A drawback to the Duane plot is that it does not readily show changes in the growth rate since the data are effectively smoothed. This effect becomes more pronounced as the plot progresses since, by using cumulative time, any sudden deviations are damped.

It is a useful technique during a field trial for predicting, at the current growth rate, how many field hours need to be accumulated in order to reach some target MTBF. In Figure 12.5, if the $\alpha = 0.2$ line was obtained from field data after, say, 800 cumulative field years then, if the objective MTBF were 500 years, the indication is that 10 000 cumulative years are needed at that growth rate. The alternative would be to accelerate the reliability growth by more active follow-up of the failure analysis.

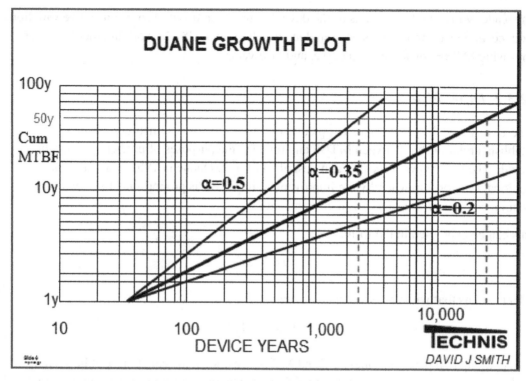

Figure 12.5: Duane plots

■ Exercises

1. One hundred items are placed on simulated life test. Failures occur at:
   17, 37, 45, 81, 88, 110, 122, 147, 208, 232, 235, 263, 272, 317, 325, 354, 355, 403 hrs.
   A 3000 hr MTBF is hoped for. Construct a CUSUM, in 3000 cumulative hour increments, to display these results.

2. Fifty items are put on field trial for three months and have generated 20 failures. A further 50 are added to the trial and, after a further three months, the total number of failures has risen to 35.
   Calculate the parameters and determine when the MTBF will reach 12 000 hrs.
   If the growth factor is increased to 0.6, when will an MTBF of 12 000 hrs be reached?
   Hint: as shown in Figure 12.6, the recalculation of k should be carried out using the second data pair in order to construct the line as shown.

■

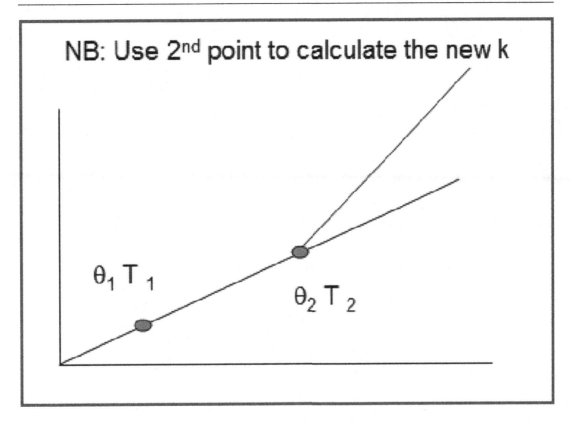

Figure 12.6: Duane plot

CHAPTER 13

Field Data Collection and Feedback

13.1 Reasons for Data Collection

Failure data can be collected from prototype and production models or from the field. In either case a formal failure-reporting document is necessary in order to ensure that the feedback is both consistent and adequate. Field information is far more valuable since it concerns failures and repair actions that have taken place under real operating conditions. Since recording field incidents relies on people, it is subject to errors, omissions and misinterpretation. It is therefore important to collect all field data using a formal document. Information of this type has a number of uses, the main two being feedback, resulting in modifications to prevent further defects, and the acquisition of statistical reliability and repair data. In detail, then, they:

- Indicate design and manufacture deficiencies and can be used to support reliability growth programs (Section 12.3)
- Provide quality and reliability trends
- Identify wearout and decreasing failure rates
- Provide subcontractor ratings
- Contribute statistical data for future reliability and repair time predictions
- Assist second-line maintenance (workshop)
- Enable spares provisioning to be refined
- Allow routine maintenance intervals to be revised
- Enable the field element of quality costs to be identified.

A failure-reporting system should be established for every project and product. Customer cooperation with a reporting system is essential if feedback from the field is required and this could well be sought, at the contract stage, in return for some other concession.

13.2 Information and Difficulties

A failure report form must collect information covering the following:

- Repair time – active and passive
- Type of fault – primary or secondary, random or induced, etc.

Reliability, Maintainability and Risk. DOI: 10.1016/B978-0-08-096902-2.00013-1

- Nature of fault – open or short circuit, drift condition, wearout, design deficiency
- Fault location – exact position and detail of LRA or component
- Environmental conditions – where these are variable, record conditions at time of fault if possible
- Action taken – exact nature of replacement or repair
- Personnel involved
- Equipment used
- Spares used
- Unit running time (from installation until the failure).

The main problems associated with failure recording are:

1. *Inventories*: whilst failure reports identify the numbers and types of failure they rarely provide a source of information as to the total numbers of the item in question and their installation dates and running times.

2. *Motivation*: if the field service engineer can see no purpose in recording information it is likely that items will be either omitted or incorrectly recorded. The purpose of fault reporting and the ways in which it can be used to simplify the task need to be explained. If the engineer is frustrated by unrealistic time standards, poor working conditions and inadequate instructions, then the failure report is the first task which will be skimped or omitted. A regular circulation of field data summaries to the field engineer is the best (possibly the only) way of encouraging feedback. It will help him to see the overall field picture and advice on diagnosing the more awkward faults will be appreciated.

3. *Verification*: once the failure report has left the person who completes it the possibility of subsequent checking is remote. If repair times or diagnoses are suspect then it is likely that they will go undetected or be unverified. Where failure data are obtained from customer's staff, the possibility of challenging information becomes even more remote.

4. *Cost*: failure reporting is costly in terms of both the time to complete failure-report forms and the hours of interpretation of the information. For this reason, both supplier and customer are often reluctant to agree to a comprehensive reporting system. If the information is correctly interpreted and design or manufacturing action taken to remove failure sources, then the cost of the activity is likely to be offset by the savings and the idea must be 'sold' on this basis.

5. *Recording non-failures*: the situation arises where a failure is recorded although none exists. This can occur in two ways. First, there is the habit of locating faults by replacing suspect but not necessarily failed components. When the fault disappears the first (wrongly removed) component is not replaced and is hence recorded as a failure. Failure rate data are therefore artificially inflated and spares depleted. Second, there is the interpretation of secondary failures as primary failures. A failed component may cause stress conditions upon another which may, as a result, fail. Diagnosis may reveal both

failures but not always which one occurred first. Again, failure rates become wrongly inflated. More complex maintenance instructions and the use of higher-grade personnel will help reduce these problems at a cost.

6. *Times to failure*: these are necessary in order to establish wearout. See next section.

13.3 Times to Failure

In most cases fault data schemes yield the numbers of failures/defects of equipment. Establishing the inventories, and the installation dates of items, is also necessary if the cumulative times are also to be determined. This is not always easy as plant records are often incomplete (or out of date) and the exact installation dates of items have sometimes to be guessed.

Nevertheless, establishing the number of failures and the cumulative time enables failure rates to be inferred as was described in Chapter 5.

Although this failure rate information provides a valuable input to reliability prediction and to optimum spares provisioning (Chapter 16), it does not enable the wearout and burn-in characteristics of an item to be described. In Chapter 6, the Weibull methodology for describing variable failure rates was described and in Chapter 16 it is shown how to use this information to optimize replacement intervals.

For this to happen it is essential that each item is separately identified (usually by a tag number) and that each failure is attributed to a specific item. Weibull models are usually, although not always, applicable at the level of a specific failure mode rather than to the failures as a whole. A description of failure mode is therefore important and the physical mechanism, rather than the outcome, should be described. For example the phrase 'out of adjustment' really describes the effect of a failure whereas 'replaced leaking diaphragm' more specifically describes the mode.

Furthermore, if an item is removed, replaced or refurbished as new then this needs to be identified (by tag number) in order for the correct start times to be identified for each subsequent failure time. In other words if an item that has been in situ for five years had a new diaphragm fitted one year ago then, for diaphragm failures, the time to failure dates from the latter. On the other hand failures of another mode might well be treated as times dating from the former.

Another complication is in the use of operating time rather than calendar time. In some ways the latter is more convenient if the data are to be used for generic use. In some cases, however, especially where the mode is related to wear and the operating time is short compared with calendar time, then operating hours will be more meaningful. In any case consistency is the rule.

If this information is available then it will be possible to list:

- Individual times to failure (calendar or operating);
- Times for items that did not fail;
- Times for items that were removed without failing.

In summary the following are needed:

- Installed (or replaced/refurbished) dates and tag numbers;
- Failure dates and tag numbers;
- Failure modes (by physical failure mechanism);
- Running times/profiles unless calendar time is be used.

13.4 Spreadsheets and Databases

Many data-collection schemes arrange for the data to be manually transferred from the written form into a computer. In order to facilitate data sorting and analysis it is very useful if the information can be in a coded form. This requires some form of codes database for the field maintenance personnel in order that the various entries can be made by means of simple alphanumerics. This has the advantage that field reports are more likely to be complete since there is a code available for each box on the form. Furthermore, the codes then provide definitive classifications for subsequent sorting. Headings include:

Equipment code

Preferably a hierarchical coding scheme that defines the plant, subsystem and item as, for example, RC1-66-03-5555, where:

Code	Meaning
R	Southampton plant
C1	Compression system
66	Power generation
03	Switchgear
5555	Actual item

How found

The reason for the defect being discovered as, say, a two-digit code:

Code	Meaning
01	Plant shutdown
02	Preventive maintenance
03	Operating problem
etc.	

Type of fault

The failure mode, for example:

Code	Meaning
01	Short circuit
02	Open circuit
03	Leak
04	Drift
05	No fault found
etc.	

Action taken

Examples are:

Code	Meaning
01	Item replaced
02	Adjusted
03	Item repaired
etc.	

Discipline

Where more than one type of maintenance skill is used, as is often the case on big sites, it is desirable to record the maintenance discipline involved. These are useful data for future maintenance planning and costing. Thus:

Code	Meaning
01	Electrical
02	Instrument
03	Mechanical
etc.	

Free text

In addition to the coded report there needs to be some provision for free text in order to amplify the data.

Each of the above fields may run to several dozen codes, which would be issued to the field maintenance personnel as a handbook. Two suitable types of package for analysis of the data are *spreadsheets* and *databases*. If the data can be inputted directly into one of these packages, so much the better. In some cases the data are resident in a more wide-ranging, field-specific, computerized maintenance system. In those cases it will be worth writing a download program to copy the defect data into one of the above types of package.

Spreadsheets such as Lotus 1-2-3 and Excel allow the data, including text, to be placed in cells arranged in rows and columns. Sorting is available as well as mathematical manipulation of the data.

In some cases the quantity of data may be such that spreadsheet manipulation becomes slow and cumbersome, or is limited by the extent of the PC memory. The use of *database* packages permits more data to be handled and more flexible and fast sorting. Sorting is far more flexible than with spreadsheets since words within text, within headings or even 'sound-alike' words can be sorted.

13.5 Best Practice and Recommendations

The following list summarizes the best practice together with recommended enhancements for both manual and computer based field failure recording.

Recorded field information is frequently inadequate and it is necessary to emphasize that failure data must contain sufficient information to enable precise failures to be identified and failure distributions to be identified. They must, therefore, include:

(a) Adequate information about the symptoms and causes of failure. This is important because predictions are only meaningful when a system level failure is precisely defined. Thus component failures that contribute to a defined system failure can only be identified if the failure modes are accurately recorded. There needs to be a distinction between failures (which cause loss of system function) and defects (which may only cause degradation of function).

(b) Detailed and accurate equipment inventories enabling each component item to be separately identified. This is essential in providing cumulative operating times for the calculation of assumed constant failure rates and also for obtaining individual calendar times (or operating times or cycles) to each mode of failure and for each component item. These individual times to failure are necessary if failure distributions are to be analyzed by the Weibull method dealt with in Chapter 6.

(c) Identification of common cause failures by requiring the inspection of redundant units to ascertain if failures have occurred in both (or all) units. This will provide data to enhance models such as the one developed in Chapter 8.2. In order to achieve this it is necessary to be able to identify that two or more failures are related to specific field items in a redundant configuration. It is therefore important that each recorded failure also identifies which specific item (i.e. tag number) it refers to.

(d) Intervals between common cause failures. Because common cause failures do not necessarily occur at precisely the same instant it is desirable to be able to identify the time elapsed between them.

(e) The effect that a 'component part' level failure has on failure at the system level. This will vary according to the type of system, the level of redundancy (which may postpone system level failure), etc.

(f) Costs of failure, such as the penalty cost of system outage (e.g. loss of production) and the cost of corrective repair effort and associated spares and other maintenance costs.

(g) The consequences in the case of safety-related failures (e.g. death, injury, environmental damage), which are not so easily quantified.

(h) Consideration of whether a failure is intrinsic to the item in question or was caused by an external factor. External factors might include:

process operator error induced failure

maintenance error induced failure

failure caused by a diagnostic replacement attempt

modification induced failure.

(i) Effective data screening to identify and correct errors and to ensure consistency. There is a cost issue here in that effective data screening requires significant man-hours to study the field failure returns. In the author's experience an average of as much as one hour per field return can be needed to enquire into the nature of a given failure and to discuss and establish the underlying cause. Both codification and narrative are helpful to the analyst and, whilst each has its own merits, a combination is required in practice. Modern computerized maintenance management systems offer possibilities for classification and codification of failure modes and causes. However, this relies on motivated and trained field technicians to input accurate and complete data. The option to add narrative should always be available.

(j) Adequate information about the environment (e.g. weather in the case of unprotected equipment) and operating conditions (e.g. unusual production throughput loadings).

13.6 Analysis and Presentation of Results

Once collected, data must be analyzed and put to use or the system of collection will lose credibility and, in any case, the cost will have been wasted. A Pareto analysis of defects is a powerful method of focusing attention on the major problems. If the frequency of each defect type is totaled and the types then ranked in descending order of frequency it will usually be seen that a high percentage of the defects are spread across only a few types. A still more useful approach, if cost information is available, is to multiply each defect type frequency by its cost and then to rerank the categories in descending order of cost. Thus the most expensive group of defects, rather than the most frequent, heads the list, as can be seen in Figure 13.1.

Note the emphasis on cost and that the total has been shown as a percentage of sales. It is clear that engineering effort could profitably be directed at the first two items, which together account for 38% of the failure cost. The first item is a mechanical design problem and the second a question of circuit tolerancing.

1. *Summary of Data*

Number of machines in field	50	
Operating hours (this period)	5320	
Number of corrective calls	39	
Total cost of calls	£4250	labor, travel and spares
Total cost as % of sales	4%	

2. *Incident Analysis*

Repetitive Failures	Frequency	Cost £	% of total
a) Mechanical transporter assembly — belt adjustment	4	935	22
b) Receiver carrier detector drift	9	680	16
c) Electromechanical relays	4	340	8
d) Gear meshing	3	340	8
e) Printed board 182c output VT2	2	300	7
f) Lamps	2	170	4
Non Repetitive Faults			
g) Printed board 424a IC5 ⎫			
h) Printed board 111e R2 ⎬	15	1485	35
etc. ⎭			
	39	4250	100

Figure 13.1: Quarterly incident report summary – product Y.

It is also useful to know whether the failure rate of a particular failure type is increasing, decreasing or constant. This will influence the engineering response. A decreasing failure rate indicates the need for further action in tests to eliminate the early failures. Increasing failure rate shows wearout, requiring either a design solution or preventive replacement. Constant failure rate suggests a reliability level that is inherent to that design configuration. Chapter 6 explains how failure data can be analyzed to quantify these trends. The report in Figure 13.1 might well contain other sections showing reliability growth, analysis of wearout, progress on engineering actions since the previous report, etc.

13.7 Manufacturers' data

There is a rapidly increasing trend to quote failure rates offered by equipment manufacturers. This dangerous pratcice was covered in Section 4.5 of Chapter 4 but is mentioned again here for completness.

13.8 Anecdotal Data

Although not as formal as data based on written maintenance records, this important source should not be overlooked. Quantities of failures quoted by long-serving site personnel are likely to be fairly accurate and might even, in some cases, be more valuable than records-based written data. The latter pass from maintainer to record keeper to analyst and may lose accuracy due to interpretation through the chain of analysis. Anecdotal data, on the other hand, can be challenged and interpreted first hand.

13.9 Examples of Failure Report Forms

Although very old, Figure 13.2 shows an example of a well-designed and thorough failure recording form as once used by the European companies of the International Telephone and Telegraph Corporation. This single form strikes a balance between the need for detailed failure information and the requirement for a simple reporting format. A feature of the ITT form is the use of four identical print-through forms. The information is therefore accurately recorded four times with minimum effort.

Figure 13.3 shows the author's recommended format taking into account the list of items in Section 13.5.

ITTE Failure Report and Action Form						
	System	Sub-system	Module/sub-assembly			

To be completed on site

	System	Sub-system	Module/sub-assembly
Type			
Serial number			
Location/Identification			
On-time (cumulative)			
Down time (this failure)			
Active repair time			

Report number:
Report date:
Report completed by:
Company:

System status:
- Field service
- Field trial
- Production prototype
- Model

Effect of failure on system:
- Complete system failure
- Major degradation
- Minor degradation
- None

On-site diagnosis:
- No defect found
- Part failure
- Installation defect
- Manufacturing defect
- Design defect
- Program defect
- On-site human error
- Other

Details of symptoms, diagnosis and failure:

Action taken:
- Replace module
- Repair
- Modification
- Program reload
- Other

Details of action taken.

To be completed at Designated Centre

Project engineering action:	Name	Company/Dept	Signature	Date completed
– Consolidate with filed data				
– For immediate analysis/action by				
Engineering				
Manufacturing				
Quality assurance				
Purchasing				
Other				

Analysis and action taken:

Engineering change no.:
Dated:
Follow-up report Ref. no.:
Dated:
Name:
Signature:
Date:

For information to:

Figure 13.2: ITT Europe failure report and action form

SERIAL NUMBER ....

DATE (and time) OF INCIDENT/EVENT/FAILURE

DATE ITEM INSTALLED (or replaced or refurbished)

MAINTENANCE TECHNICIAN (Provides traceability)

DISCIPLINE (e.g. Electrical, Mechanical, Instrumentation)

FAILED COMPONENT ITEM DESCRIPTION (e.g. Motor)

SUBSYSTEM (e.g. Support system)

DESCRIPTION OF FAULT/CAUSE (Failure mode, e.g. Windings open circuit)

'TAG', 'SERIAL NUMBER' (HENCE DATE OF INSTALLATION AND REFURB)
e.g. System xyz, Unit abc, Motor type zzz, serial no. def,

DOWN TIME [if known]/REPAIR TIME
e.g. 4 hrs repair, 24 hrs outage

TIME TO FAILURE (COMPUTED FROM DATE AND TAG NUMBER)
e.g. This date minus date of installation
e.g. This date minus date of last refurbishment

PARTS USED (in the repair)
e.g. New motor type zzz, serial no. efg

ACTION TAKEN (e.g. Replace motor)

HOW CAUSED
Intrinsic (e.g. RANDOM HARDWARE FAILURE) versus extrinsic (GIVE CAUSE IF EVIDENT)

HOW FOUND/DIAGNOSED
e.g. Customer report, technician discovered open circuit windings

RESULT OF FAILURE ON SYSTEM
e.g. Support system un-usable, process trip, no effect

COMMON CAUSE FAILURE e.g. redundancy defeated
time between CCFs ....
attributable to SEPARATION/DIVERSITY/COMPLEXITY/HUMAN FACTOR/ENVIRONMENT

ENVIRONMENT/OPERATING CONDITION
e.g. temp, humidity, 50% throughput, equipment unattended

NARRATIVE ....

Figure 13.3: Recommended failure data recording form

CHAPTER 14

Factors Influencing Down Time

The two main factors governing down time are equipment design and maintenance philosophy. In general, it is the active repair elements that are determined by the design and the passive elements which are governed by the maintenance philosophy. Designers must be aware of the maintenance strategy and of the possible equipment failure modes. They must understand that production difficulties can often become field problems since, if assembly is difficult, maintenance will be well-nigh impossible. Achieving acceptable repair times involves simplifying diagnosis and repair.

14.1 Key Design Areas

14.1.1 Access

Low-reliability parts should be the most accessible and must be easily removable with the minimum of disturbance. There must be enough room to withdraw such devices without touching or damaging other parts. On the other hand, the technician must be discouraged from removing and checking easily exchanged items as a substitute for the correct diagnostic procedure. The use of captive screws and fasteners is highly desirable as they are faster to use and eliminate the risk of losing screws in the equipment. Standard fasteners and covers become familiar and hence easier to use. The use of outriggers, which enables printed boards to be tested while still electrically connected to the system, can help to reduce diagnosis time. On the other hand, this type of online diagnosis can induce faults and is sometimes discouraged. In general, it is a good thing to minimize online testing by employing easily interchanged units together with alarms and displays providing diagnostic information and easy identification of the faulty unit.

Every LRA (least replaceable assembly) should be capable of removal without removing any other LRA or part. The size of the LRA affects the speed of access. The overall aim is for speedy access consistent with minimum risk of accidental damage.

14.1.2 Adjustment

The amount of adjustment required during normal system operation, and after LRA replacement, can be minimized (or eliminated) by generous tolerancing in the design, aimed at low sensitivity to drift.

Reliability, Maintainability and Risk. DOI: 10.1016/B978-0-08-096902-2.00014-3

Where adjustment is by a screwdriver or other tool, care should be taken to ensure that damage cannot be done to the equipment. Guide holes, for example, can prevent a screwdriver from slipping.

Where adjustment requires that measurements are made, or indicators observed, then the displays or meters should be easily visible while the adjustment is made.

It is usually necessary for adjustments and alignments to be carried out in a sequence and this must be specified in the maintenance instructions. The designer should understand that where drift in a particular component can be compensated for by the adjustment of some other item then, if that adjustment is difficult or critical, the service engineer will often change the drifting item, regardless of its cost.

14.1.3 Built-In Test Equipment

As with any test equipment, Built-In Test Equipment (BITE) should be an order of magnitude more reliable than the system of which it is part, in order to minimize the incidence of false alarms or incorrect diagnosis. Poor-reliability BITE will probably reduce the system availability.

The number of connections between the system and the built-in test equipment should be minimized to reduce the probability of system faults induced by the BITE. It carries the disadvantages of being costly, inflexible (designed around the system; it is difficult to modify) and of requiring some means of self-checking. In addition, it carries a weight, volume and power supply penalty but, on the other hand, greatly reduces the time required for realization diagnosis and checkout.

14.1.4 Circuit Layout and Hardware Partitioning

It is advisable to consider maintainability when designing and laying out circuitry. In some cases it is possible to identify a logical sequence of events or signal flow through a circuit, and fault diagnosis is helped by a component layout that reflects this logic. Components should not be so close together as to make damage likely when removing and replacing a faulty item.

The use of integrated circuits introduces difficulties. Their small size and large number of leads make it necessary for connections to be small and close together, which increases the possibility of damage during maintenance. In any case, field maintenance at circuit level is almost impossible owing to the high function density involved. Because of the high maintenance cost of removing and resoldering these devices, the question of plug-in ICs arises. Another point of view emphasizes that IC sockets increase both cost and the possibility of connector failure. The decision for or against is made on economic grounds and must be taken on the basis of field failure rate, socket cost and repair time. The IC is a functional unit in itself and therefore circuit layout is less capable of representing the circuit function.

In general, the cost of microelectronics hardware continues to fall and thus the printed circuit board is more and more considered as a throwaway unit.

14.1.5 Connections

Connections present a classic trade-off between reliability and maintainability. The following types of connection are ranked in order of reliability, starting with the most reliable. A comparison of failure rates is made by means of the following:

wrapped joint	0.00003 per 10^6 hrs
welded connection	0.002 per 10^6 hrs
machine-soldered joint	0.0003 per 10^6 hrs
crimped joint	0.0003 per 10^6 hrs
hand-soldered joint	0.0002 per 10^6 hrs
edge connector (per pin)	0.001 per 10^6 hrs.

Since edge connectors are less reliable than soldered joints, there needs to be a balance between having a few large plug-in units and a larger number of smaller throwaway units with the associated reliability problem of additional edge connectors. Boards terminated with wrapped joints rather than with edge connectors are two orders more reliable from the point of view of the connections, but the maintainability penalty can easily outweigh the reliability advantage. Bear in mind the time taken to make ten or twenty wrapped joints compared with that taken to plug in a board equipped with edge connectors.

The following are approximate times for making the different types of connection assuming that appropriate tools are available:

edge connector (multi-contact)	10 s
solder joint (single-wire)	20 s
wrapped joint	50 s.

As can be seen, maintainability ranks in the opposite order to reliability. In general, a high-reliability connection is required within the LRA, where maintainability is a secondary consideration. The interface between the LRA and the system requires a high degree of maintainability and the plug-in or edge connector is justified. If the LRA is highly reliable, and therefore unlikely to require frequent replacement, termination by the reliable wrapped joints could be justified. On the other hand a medium- or low-reliability unit would require plug and socket connection for quick interchange.

The reliability of a solder joint, hand or flow, is extremely sensitive to the quality control of the manufacturing process. Where cable connectors are used it should be ensured, by labeling or polarizing, that plugs will not be wrongly inserted in sockets or inserted in wrong sockets. Mechanical design should prevent insertion of plugs in the wrong configuration and also prevent damage to pins by clumsy insertion.

Where several connections are to be made within or between units, the complex of wiring is often provided by means of a cableform (loom) and the terminations (plug, solder or wrap) made according to an appropriate document. The cableform should be regarded as an LRA and local repairs should not be attempted. A faulty wire may be cut back, but left in place, and a single wire added to replace the link, provided that this does not involve the possibility of electrical pickup or misphasing.

14.1.6 Displays and Indicators

Displays and indicators are effective in reducing the diagnostic, checkout and alignment contributions to active repair time. Simplicity should be the keynote and a 'go, no go' type of meter or display will require only a glance. The use of stark color changes, or other obvious means, to divide a scale into areas of 'satisfactory operation' and 'alarm' should be used. Sometimes a meter, together with a multiway switch, is used to monitor several parameters in a system. It is desirable that the anticipated (normal) indication be the same for all the applications of the meter so that the correct condition is shown by little or no movement as the instrument is switched to the various test points. Displays should never be positioned where it is difficult, dangerous or uncomfortable to read them.

For an alarm condition an audible signal, as well as visual displays, is needed to draw attention to the fault. Displays in general, and those relating to alarm conditions in particular, must be more reliable than the parent system since a failure to indicate an alarm condition is potentially dangerous.

If equipment is unattended then some alarms and displays may have to be extended to another location and the reliability of the communications link then becomes important to the availability of the system.

The following points concerning meters are worth noting:

1. False readings can result from parallax effects owing to scale and pointer being in different planes. A mirror behind the pointer helps to overcome this difficulty.
2. Where a range exists outside which some parameter is unacceptable, then either the acceptable or the unacceptable range should be colored or otherwise made readily distinguishable from the rest of the scale (Figure 14.1(a)).
3. Where a meter displays a parameter that should normally have a single value, then a center-zero instrument can be used to advantage and the circuitry configured such that the normal acceptable range of values falls within the mid-zone of the scale (Figure 14.1(b)).
4. Linear scales are easier to read and less ambiguous than logarithmic scales, and consistency in the choice of scales and ranges minimizes the possibility of misreading (Figure 14.1). On the other hand, there are occasions when the use of a non-linear response or false-zero meter is desirable.

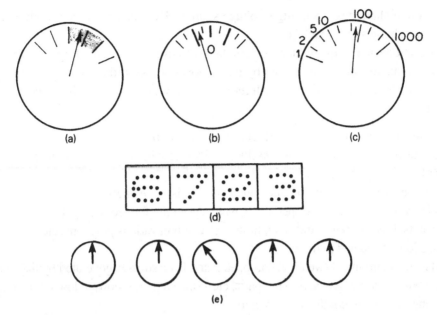

Figure 14.1: Meter displays. (a) Scale with shaded range; (b) scale with limits; (c) logarithmic scale; (d) digital display; (e) alignment of norms

5. Digital displays are now widely used and are superior to the analogue pointer-type of instrument where a reading has to be recorded (Figure 14.1(d)). The analogue type of display is preferable when a check or adjustment within a range is required.
6. When a number of meters are grouped together it is desirable that the pointer positions for the *normal* condition are alike. Figure 14.1(e) shows how easily an incorrect reading is noticed.

Consistency in the use of color codes, symbols and labels associated with displays is highly desirable. Filament lamps are not particularly reliable and should be derated. More reliable LEDs and liquid crystal displays are now widely used.

All displays should be positioned as near as possible to the location of the function or parameter to which they refer and mounted in an order relating to the sequence of adjustment. Unnecessary displays merely complicate the maintenance task and do more harm than good. Meters need be no more accurate than the measurement requirement of the parameter involved.

14.1.7 Handling, Human and Ergonomic Factors

Major handling points to watch are:
*   Weight, size and shape of removable modules. The LRA should not be capable of self-damage owing to its own instability, as in the case of a thin lamina construction.

- Protection of sharp edges and high-voltage sources. Even an unplugged module may hold dangerous charges on capacitors.
- Correct handles and grips reduce the temptation to use components for the purpose.
- When an inductive circuit is broken by the removal of a unit, then the earth return should not be via the frame. A separate earth return via a pin or connection from the unit should be used.

The following ergonomic factors also influence active repair time:
- Design for minimum maintenance skills considering what type of personnel are actually available.
- Beware of over-miniaturization – incidental damage is more likely.
- Consider comfort and safety of personnel when designing for access; e.g. body position, movements, limits of reach and span, limit of strength in various positions, etc.
- Illumination – fixed and portable.
- Shield from environment (weather, damp, etc.) and from stresses generated by the equipment (heat, vibration, noise, gases, moving parts, etc.) since repair is slowed down if the mainte-nance engineer has to combat these factors.

14.1.8 Identification

Identification of components, test points, terminals, leads, connectors and modules is helped by standardization of appearance. Color codes should not be complex since over 5% of the male population suffer from some form of color blindness. Simple, unambiguous numbers and symbols help in the identification of particular functional modules. The physical grouping of functions simplifies the signs required to identify a particular circuit or LRA.

In many cases programable hardware devices contain software (code). It is important to be able to identify the version of code resident in the device and this is often only possible by way of the component labeling.

14.1.9 Interchangeability

Where LRAs are interchangeable this simplifies diagnosis, replacement and checkout, owing to the element of standardization involved. Spares provisioning then becomes slightly less critical in view of the possibility of using a non-essential, redundant unit to effect a repair in some other part of the system. Cannibalization of several failed LRAs to yield a working module also becomes possible although this should never become standard field practice.

The smaller and less complex the LRA, the greater the possibility of standardization and hence interchangeability. The penalty lies in the number of interconnections between LRAs and the system (less reliability) and the fact that the diagnosis is referred to a lower level (greater skill and more equipment).

Interchange of non-identical boards or units should be made mechanically impossible. At least, pin conventions should be such that insertion of an incorrect board cannot cause damage either to that board or to other parts of the equipment. Each value of power supply must always occupy the same pin number.

14.1.10 Least Replaceable Assembly

The LRA is that replaceable module at which local fault diagnosis ceases and direct replacement occurs. Failures are traced only to the LRA, which should be easily removable (see Section 14.1.5), replacement LRAs being the spares holding. It should rarely be necessary to remove an LRA in order to prove that it is faulty, and no LRA should require the removal of any other LRA for diagnosis or for replacement.

The choice of level of the LRA is one of the most powerful factors in determining maintainability. The larger the LRA, the faster the diagnosis. Maintainability, however, is not the only factor in the choice of LRA. As the size of the LRA increases, so does its cost and the cost of spares holding. The more expensive the LRA, the less likely is a throwaway policy to be applicable. Also, a larger LRA is less likely to be interchangeable with any other. The following compares various factors as the size of LRA increases:

System maintainability	Improves
LRA reliability	Decreases
Cost of system testing (equipment and manpower)	Decreases
Cost of individual spares	Increases
Number of types of spares	Decreases

14.1.11 Mounting

If components are mounted so as to be self-locating then replacement is made easier. Mechanical design and layout of mounting pins and brackets can be made to prevent transposition where this is undesirable as in the case of a transformer, which must not be connected the wrong way round. Fragile components should be mounted as far as possible from handles and grips.

14.1.12 Component Part Selection

Main factors affecting repair times are:

availability of spares – delivery
reliability/deterioration under storage conditions
ease of recognition
ease of handling
cost of parts
physical strength and ease of adjustment.

14.1.13 Redundancy

Circuit redundancy within the LRA (usually unmonitored) increases the reliability of the module, and this technique can be used in order to make it sufficiently reliable to be regarded as a throwaway unit. Redundancy at the LRA level permits redundant units to be removed for preventive maintenance while the system remains in service.

Although improving both reliability and maintainability, redundant units require more space and weight. Capital cost is increased and the additional units need more spares and generate more maintenance. System availability is thus improved but both preventive and corrective maintenance costs increase with the number of units.

14.1.14 Safety

Apart from legal and ethical considerations, safety-related hazards increase active repair time by requiring greater care and attention. An unsafe design will encourage short cuts or the omission of essential activities. Accidents add, very substantially, to the repair time.

Where redundancy exists, routine maintenance can be carried out after isolation of the unit from high voltage and other hazards. In some cases routine maintenance is performed under power, in which case appropriate safeguards must be incorporated into the design. The following practices should be the norm:

- Isolate high voltages under the control of microswitches that are automatically operated during access. The use of a positive interlock should bar access unless the condition is safe.
- Weights should not have to be lifted or supported.
- Use appropriate handles.
- Provide physical shielding from high voltage, high temperature, etc.
- Eliminate sharp points and edges.
- Install alarm arrangements. The exposure of a distinguishing color when safety covers have been removed is good practice.
- Ensure adequate lighting.

14.1.15 Software

The availability of programable LSI (large-scale integration) devices has revolutionized the approach to circuit design. More and more electronic circuitry is being replaced by a standard microprocessor architecture with the individual circuit requirements achieved within the software (program) that is held in the memory section of the hardware. Under these conditions diagnosis can no longer be supported by circuit descriptions and measurement information. Complex sequences of digital processing make diagnosis impossible with traditional test equipment.

Production testing of this type of printed-board assembly is possible only with sophisticated computer-driven Automatic Test Equipment (ATE) and, as a result, field diagnosis can be only to board level. Where printed boards are interconnected by data highways carrying dynamic digital information, even this level of fault isolation may require field test equipment consisting of a microprocessor loaded with appropriate software for the unit under repair.

14.1.16 Standardization

Standardization leads to improved familiarization and hence shorter repair times. The number of different tools and test equipment is reduced, as is the possibility of delay due to having incorrect test gear. Fewer types of spares are required, reducing the probability of exhausting the stock.

14.1.17 Test Points

Test points are the interface between test equipment and the system, and are needed for diagnosis, adjustment, checkout, calibration and monitoring for drift. Their provision is largely governed by the level of LRA chosen and they will usually not extend beyond what is necessary to establish that an LRA is faulty. Test points within the LRA will be dictated by the type of board test carried out in production or in second-line repair.

In order to minimize faults caused during maintenance, test points should be accessible without the removal of covers and should be electrically buffered to protect the system from misuse of test equipment. Standard positioning also reduces the probability of incorrect diagnosis resulting from wrong connections. Test points should be grouped in such a way as to facilitate sequential checks. The total number should be kept to a minimum consistent with the diagnosis requirements. Unnecessary test points are likely to reduce rather than increase maintainability.

The above 17 design parameters relate to the equipment itself and not to the maintenance philosophy. Their main influence is on the active repair elements such as diagnosis, replacement, checkout, access and alignment. Maintenance philosophy and design are, nevertheless, interdependent. Most of the foregoing have some influence on the choice of test equipment. Skill requirements are influenced by the choice of LRA, by displays and by standardization. Maintenance procedures are affected by the size of modules and the number of types of spares. The following section will examine the ways in which maintenance philosophy and design act together to influence down times.

14.2 Maintenance Strategies and Handbooks

Both active and passive repair times are influenced by factors other than equipment design. Consideration of maintenance procedures, personnel and spares provisioning

is known as maintenance philosophy and plays an important part in determining overall availability. The costs involved in these activities are considerable and it is therefore important to strike a balance between over- and under-emphasizing each factor.
They can be grouped under seven headings:

organization of maintenance resources
maintenance procedures
tools and test equipment
personnel – selection, training and motivation
maintenance instructions and manuals
spares provisioning
logistics.

14.2.1 Organization of Maintenance Resources

It is usual to divide the maintenance tasks into three groups in order first, to concentrate the higher skills and more important test equipment in one place and second, to provide optimum replacement times in the field. These groups, which are known by a variety of names, are as follows.

14.2.1.1 First-line maintenance – Corrective maintenance – Call – Field maintenance

This will entail diagnosis only to the level of the LRA, and repair is by LRA replacement. The technician either carries spare LRAs or has rapid access to them. Diagnosis may be aided by a portable intelligent terminal, especially in the case of microprocessor-based equipment. This group may involve two grades of technician, the first answering calls and the second being a small group of specialists who can provide backup in the more difficult cases.

14.2.1.2 Preventive maintenance – Routine maintenance

This will entail scheduled replacement/discard (see Chapter 16) of defined modules and some degree of cleaning and adjustment. Parametric checks to locate dormant faults and drift conditions may be included.

14.2.1.3 Second-line maintenance – Workshop – Overhaul shop – Repair depot

This is for the purpose of:
1. Scheduled overhaul and refurbishing of units returned from preventive maintenance.
2. Unscheduled repair and/or overhaul of modules that have failed or become degraded.

Deeper diagnostic capability is needed and therefore the larger, more complex, test equipment will be found at the workshop together with full system information.

14.2.2 Maintenance Procedures

For any of the above groups of staff it has been shown that fast, effective and error-free maintenance is best achieved if a logical and formal procedure is followed on each occasion. A haphazard approach based on the subjective opinion of the maintenance technician, although occasionally resulting in spectacular short cuts, is unlikely to prove the better method in the long run. A formal procedure also ensures that calibration and essential checks are not omitted, that diagnosis always follows a logical sequence designed to prevent incorrect or incomplete fault detection, that correct test equipment is used for each task (damage is likely if incorrect test gear is used) and that dangerous practices are avoided. Correct maintenance procedure is ensured only by accurate and complete manuals and thorough training. A maintenance procedure must consist of the following:

> making and interpreting test readings;
> isolating the cause of a fault;
> part (LRA) replacement;
> adjusting for optimum performance (where applicable).

The extent of the diagnosis is determined by the level of fault identification and hence by the least replaceable assembly. A number of procedures are used:

1. Stimuli–response where the response to changes of one or more parameters is observed and compared with the expected response.
2. Parametric checks where parameters are observed at displays and test points and are compared with expected values.
3. Signal injection where a given pulse, or frequency, is applied to a particular point in the system and the signal observed at various points, in order to detect where it is lost, or incorrectly processed.
4. Functional isolation wherein signals and parameters are checked at various points in a sequence designed to eliminate the existence of faults before or after each point. In this way, the location of the fault is narrowed down.
5. Robot test methods where automatic test equipment is used to fully 'flood' the unit with a simulated load, in order to allow the fault to be observed.

Having isolated the fault, a number of repair methods present themselves:

1. Direct replacement of the LRA.
2. Component replacement or rebuilding, using simple construction techniques.
3. Cannibalization from non-essential parts.

In practice, direct replacement of the LRA is the usual solution owing to the high cost of field repair and the need for short down times in order to achieve the required equipment availability.

Depending upon circumstances, and the location of a system, repair may be carried out either immediately a fault is signaled or only at defined times, with redundancy being relied upon to maintain service between visits. In the former case, system reliability depends on the mean repair time and in the latter, upon the interval between visits and the amount of redundancy provided.

14.2.3 Tools and Test Equipment

The following are the main considerations when specifying tools and test equipment.

1. *Simplicity*: test gear should be easy to use and require no elaborate set-up procedure.
2. *Standardization*: the minimum number of types of test gear reduces the training and skill requirements and minimizes test equipment spares holdings. Standardization should include types of displays and connections.
3. *Reliability*: test gear should be an order of magnitude more reliable than the system for which it is designed, since a test equipment failure can extend down time or even result in a system failure.
4. *Maintainability*: ease of repair and calibration will affect the non-availability of test gear. Ultimately it reduces the amount of duplicate equipment required.
5. *Replacement*: suppliers should be chosen bearing in mind the delivery time for replacements and for how many years they will be available.

There is a trade-off between the complexity of test equipment and the skill and training of maintenance personnel. This extends to BITE, which, although introducing some disadvantages, speeds and simplifies maintenance.

BITE forms an integral part of the system and requires no setting-up procedure in order to initiate a test. Since it is part of the system, weight, volume and power consumption are important. A customer may specify these constraints in the system specification (e.g. power requirements of BITE not to exceed 2% of mean power consumption). Simple BITE can be in the form of displays of various parameters. At the other end of the scale, it may consist of a programed sequence of stimuli and tests, which culminate in a 'print-out' of diagnosis and repair instructions. There is no simple formula, however, for determining the optimum combination of equipment complexity and human skill. The whole situation, with the variables mentioned, has to be considered and a trade-off technique found that takes account of the design parameters together with the maintenance philosophy.

There is also the possibility of ATE being used for field maintenance. In this case, the test equipment is quite separate from the system and is capable of monitoring several parameters simultaneously and on a repetitive basis. Control is generally by software and the maintenance task is simplified.

When choosing simple portable test gear, there is a choice of commercially available general-purpose equipment, as against specially designed equipment. Cost and ease of replacement favor the general-purpose equipment whereas special-purpose equipment can be made simpler to use and more directly compatible with test points.

In general, the choice between the various test equipment options involves a trade-off of complexity, weight, cost, skill levels, time scales and design, all of which involve cost, with the advantages of faster and simpler maintenance.

14.2.4 Personnel Considerations

Four staffing considerations influence the maintainability of equipment:

> training given
> skill level employed
> motivation
> quantity and distribution of personnel.

More complex designs involve a wider range of maintenance and hence more training is required. Proficiency in carrying out corrective maintenance is achieved by a combination of knowledge and diagnostic skill. Whereas knowledge can be acquired by direct teaching methods, skill can be gained only from experience, in either a simulated or a real environment. Training must, therefore, include experience of practical fault finding on actual equipment. Sufficient theory, in order to understand the reasons for certain actions and to permit logical reasoning, is required, but an excess of theoretical teaching is both unnecessary and confusing. A balance must be achieved between the confusion of too much theory and the motivating interest created by such knowledge.

A problem with very-high-reliability equipment is that some failure modes occur so infrequently that the technicians have little or no field experience of their diagnosis and repair. Refresher training with simulated faults will be essential to ensure effective maintenance, should it be required. Training maintenance staff in a variety of skills (e.g. electronic as well as electromechanical work) provides a flexible workforce and reduces the probability of a technician being unable to deal with a particular failure unaided. Less time is wasted during a repair and transport costs are also reduced.

Training of customer maintenance staff is often given by the contractor, in which case an objective test of staff suitability may be required. Well-structured training that provides flexibility and proficiency improves motivation, since confidence, and the ability to perform a number of tasks, brings job satisfaction in demonstrating both speed and accuracy. In order to achieve a given performance, specified training and a stated level of ability are assumed. Skill levels must be described in objective terms of knowledge, dexterity, memory, visual acuity, physical strength, inductive reasoning and so on.

Staff scheduling requires a knowledge of the equipment failure rates. Different failure modes require different repair times and have different failure rates.

The MTTR may be reduced by increasing the effort from one to two technicians but any further increase in personnel may be counter-productive and not significantly reduce the repair time.

Personnel policies are usually under the control of the customer and, therefore, close liaison between contractor and customer is essential before design features relating to maintenance skills can be finalized. In other words, the design specification must reflect the personnel aspects of the maintenance philosophy.

14.2.5 Maintenance Manuals

14.2.5.1 Requirements

The main objective of a maintenance manual is to provide all the information required to carry out each maintenance task without reference to the base workshop, design authority or any other source of information. It may, therefore, include any of the following:

- specification of system performance and functions
- theory of operation and usage limitations
- method of operation
- range of operating conditions
- supply requirements
- corrective and preventive maintenance routines
- permitted modifications
- description of spares and alternatives
- list of test equipment and its check procedure
- disposal instructions for hazardous materials.

The actual manual might range from a simple card, which could hang on a wall, to a small library of information comprising many handbooks for different applications and users. Field reliability and maintainability are influenced, in no small way, by the maintenance instructions. The design team, or the maintainability engineer, has to supply information to the handbook writer and to collaborate if the instructions are to be effective.

Consider the provision of maintenance information for a complex system operated by a well-managed organization. The system will be maintained by a permanent team (A) based on site. This team of technicians, at a fair level of competence, service a range of systems and, therefore, are not expert in any one particular type of equipment. Assume that the system incorporates some internal monitoring equipment and that specialized portable test gear is available for both fault diagnosis and for routine checks. This local team carries out all the routine checks and repairs most faults by means of module replacement. There is a limited local stock of some modules (LRAs), which is replenished from a central depot which serves several sites. The depot also stocks those replacement items not normally held on-site.

Based at the central depot is a small staff of highly skilled specialist technicians (B) who are available to the individual sites. Available to them is further specialized test gear and also basic instruments capable of the full range of measurements and tests likely to be made. These technicians are called upon when the first-line (on-site) procedures are inadequate for diagnosis or replacement. This team also visits the sites in order to carry out the more complex or critical periodic checks.

Also at the central depot is a workshop staffed with a team of craftsmen and technicians (C) who carry out the routine repairs and the checkout of modules returned from the field. The specialist team (B) is available for diagnosis and checkout whenever the (C) group is unable to repair modules.

A maintenance planning group (D) is responsible for the management of the total service operation, including cost control, coordination of reliability and maintainability statistics, system modifications, service manual updating, spares provisioning, stock control and, in some cases, a post-design service.

A preventive maintenance team (E), also based at the depot, carries out the regular replacements and adjustments to a strict schedule.

Group A will require detailed and precise instructions for the corrective tasks that it carries out. A brief description of overall system operation is desirable to the extent of stimulating interest but it should not be so detailed as to permit unorthodox departures from the maintenance instructions. There is little scope for initiative in this type of maintenance since speedy module diagnosis and replacement is required. Instructions for incident reporting should be included and a set format used.

Group B requires a more detailed set of data since it has to carry out fault diagnosis in the presence of intermittent, marginal or multiple faults not necessarily anticipated when the handbooks were prepared. Diagnosis should nevertheless still be to LRA level since the philosophy of first-line replacement holds.

Group C will require information similar to that of Group A but will be concerned with the diagnosis and repair of modules. It may well be that certain repairs require the fabrication of piece parts, in which case the drawings and process instructions must be available.

Group D requires considerable design detail and a record of all changes. This will be essential after some years of service when the original design team may not be available to give advice. Detailed spares requirements are essential so that adequate, safe substitutions can be made in the event of a spares source or component type becoming unavailable. Consider a large population item that may have been originally subject to stringent screening for high reliability. Obtaining a further supply in a small quantity but to the same standard may be impossible, and their replacement with less-assured items may have to be considered. Consider also an item selected to meet a wide range of climatic conditions. A particular user may well select a cheaper replacement meeting his or her own conditions of environment.

Group E requires detailed instructions since, again, little initiative is required. Any departure from the instructions implies a need for Group A.

14.2.5.2 Types of manual

Preventive maintenance procedures will be listed in groups by service intervals, which can be by calendar time, switch-on time, hours flown, miles traveled, and so on, as appropriate. As with calibration intervals, the results and measurements at each maintenance should be used to lengthen or shorten the service interval as necessary. The maintenance procedure and reporting requirements must be very fully described so that little scope for initiative or interpretation is required. In general, all field maintenance should be as routine as possible and capable of being fully described in a manual. Any complicated diagnosis should be carried out at the workshop and module replacement on-site used to achieve this end. In the event of a routine maintenance check not yielding the desired result, the technician should either be referred to the corrective maintenance procedure or told to replace the suspect module.

In the case of *corrective maintenance* (callout for failure or incident) the documentation should first list all the possible indications such as printouts, alarms, displays, etc. Following this, routine functional checks and test point measurements can be specified. This may involve the use of a portable 'intelligent' terminal capable of injecting signals and making decisions based on the responses. A fault dictionary is a useful aid and should be continuously updated with data from the field and/or design and production areas. Full instructions should be included for isolating parts of the equipment or taking precautions where safety is involved. Precautions to prevent secondary failures being generated should be thought out by the designer and included in the maintenance procedure.

Having isolated the fault and taken any necessary precautions, the next consideration is the diagnostic procedure followed by repair and checkout. Diagnostic procedures are best described in a logical flow chart. Figure 14.2 shows a segment of a typical diagnostic algorithm involving simple Yes/No decisions with paths of action for each branch. Where such a simple process is not relevant and the technician has to use initiative, then the presentation of schematic diagrams and the system and circuit descriptions are important. Some faults, by their nature or symptoms, indicate the function that is faulty and the algorithm approach is most suitable. Other faults are best detected by observing the conditions existing at the interfaces between physical assemblies or functional stages. Here the location of the fault may be by a bracketing/elimination process. For example 'The required signal appears at point 12 but is not present at point 20. Does it appear at point 16? No, but it appears at point 14. Investigate unit between points 14 and 16'. The second part of Figure 14.2 is an example of this type of diagnosis presented in a flow diagram. In many cases a combination of the two approaches may be necessary.

14.2.6 Spares Provisioning

Figure 14.3 shows a simple model for a system having n of a particular item and a nominal spares stock of r. The stock is continually replenished either by repairing failed items or by ordering

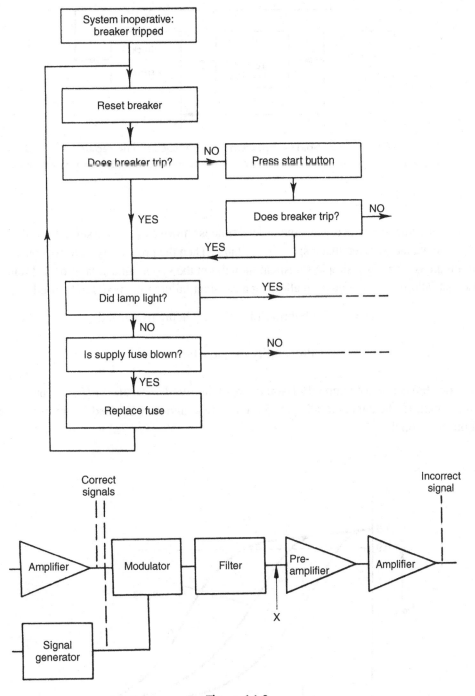

Figure 14.2

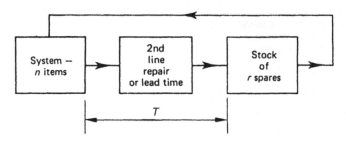

Figure 14.3: Spares replacement from second-line repair

new spares. In either case the repair time or lead time is shown as T. It is assumed that the system repair is instantaneous, given that a spare is available. Then the probability of a stockout causing system failure is given by a simple statistical model. Let the system unavailability be U and assume that failures occur at random allowing a constant failure rate model to be used.

$$U = 1 - \text{Probability of stock not being exhausted}$$

$$= 1 - \text{Probability of 0 to } r \text{ failures in } T.$$

Figure 14.4 shows a set of simple Poisson curves that give P_{0-r} against n/T for various values of spares stock, r. The curves in Chapter 5 are identical and may be used to obtain answers based on this model.

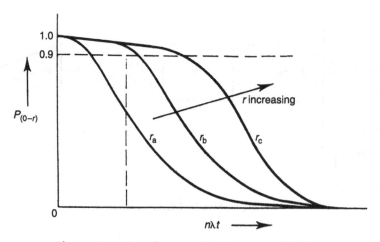

Figure 14.4: Set of curves for spares provisioning

A more realistic, and therefore more complex, unavailability model would take account of two additional parameters:

- the down time of the system while the spare (if available) is brought into use and the repair carried out;
- any redundancy. The simple model assumed that all *n* items were needed to operate. If some lesser number were adequate then a partial redundancy situation would apply and the unavailability would be less.

The simple Poisson model will not suffice for this situation and a more sophisticated technique, namely the Markov method described in Chapter 8, is needed for the calculations.

Figure 14.5 shows a typical state diagram for a situation involving four units and two spares. The lower left hand state represents four good items, with none failed and two spares. This is the 'start' state. A failure (having the rate 4λ) brings the system to the state, immediately to the right, where three are three operating with one failure but still two spares. The transition diagonally upwards to the left represents a repair (i.e. replacement by a spare). The subsequent transition downwards represents a procurement of a new spare and brings the system back to the 'start' state. The other states and transitions model the various possibilities of failure and spares states for the system.

If no redundancy exists then the availability (1 − unavailability) is obtained by evaluating the probability of being in any of the three states shown in the left hand column of the state diagram. 'three out of four' redundancy would imply that the availability is obtained from considering the probability of being in any of the states in the first two left hand columns, and so on.

Numerical evaluation of these states is obtained from the computer package COMPARE for each case of number of items, procurement time and repair time. Values of unavailability can be obtained for a number of failure rates and curves are then drawn for each case to be assessed.

The appropriate failure rate for each item can then be used to assess the unavailability associated with each of various spares levels.

Figure 14.6 gives an example of unavailability curves for specific values of MDT, turnaround time and redundancy.

The curves show the unavailability against failure rate for zero, one, and two spares. The curve for infinite spares gives the unavailability based only on the 12 hrs down time. It can only be seen in Figure 14.6 by understanding that for all values greater than two spares the line cannot be distinguished from the 2+ line. In other words, for two spares and greater, the unavailability is dominated by the repair time. For that particular example the following observations might be made when planning spares:

- For failure rates greater than about 25×10^{-6} per hour the unavailability is still significant even with large numbers of spares. Attention should be given to reducing the down time.

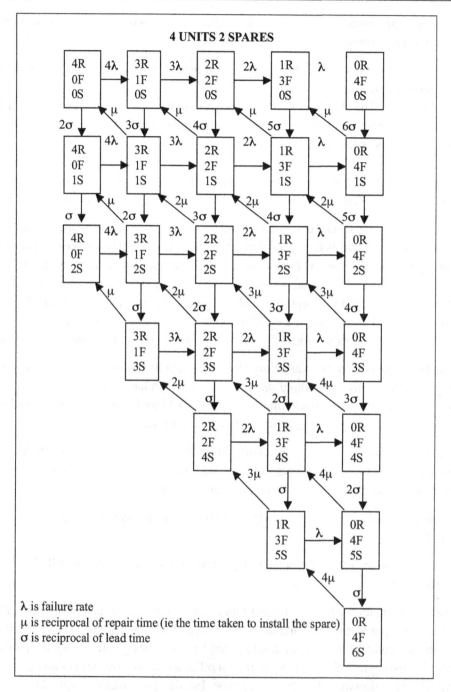

Figure 14.5: Markov state diagram-four units, two spares

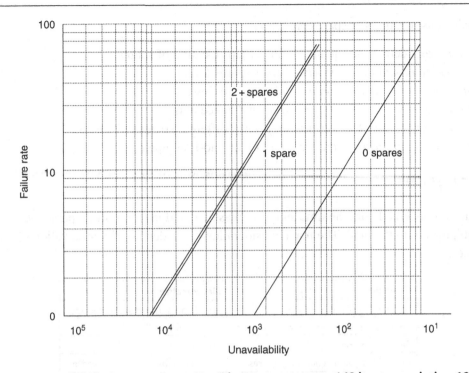

Figure 14.6: Unavailability/spares curves– N = 8 items; procurement 168 hours; repair time 12 hours

- For failure rates less than about 3×10^{-6} per hour, one spare is probably adequate and no further analysis is required.

It must be stressed that this is only one specific example and that the values will change considerably as the different parameters are altered.

The question arises as to whether spares that have been repaired should be returned to a central stock or retain their identity for return to the parent system. Returning a part to its original position is costly and requires a procedure so that initial replacement is only temporary. This may be necessary where servicing is carried out on equipment belonging to different customers – indeed some countries impose a legal requirement to this end. Another reason for retaining a separate identity for each unit occurs owing to wearout, when it is necessary to know the expired life of each item.

Stock control is necessary when holding spares and inputs are therefore required from:

preventive and corrective maintenance in the field
second-line maintenance
warranty items supplied.

The main considerations of spares provisioning are:

1. *Failure rate*: determines quantity and perhaps location of spares.
2. *Acceptable probability of stockout*: fixes spares level.
3. *Turnaround of second-line repair*: affects lead time.
4. *Cost of each spare*: affects spares level and hence item 2.
5. *Standardization and LRA*: affects number of different spares to be held.
6. *Lead time on ordering*: effectively part of second-line repair time.

14.2.7 Logistics

Logistics is concerned with the time and resources involved in transporting personnel, spares and equipment into the field. The main consideration is the degree of centralization of these resources.

Centralize	Decentralize
Specialized test equipment	Small tools and standard items
Low utilization of skills and test gear	Where small MTTR is vital
Second-line repair	Fragile test gear
Infrequent (high-reliability) spares	Frequent (low-reliability) spares

A combination will be found where a minimum of onsite facilities, that ensures repair within the specified MTTR is provided. The remainder of the spares backup and low utilization test gear can then be centralized. If availability is to be kept high by means of a low MTTR then spares depots have to be established at a sufficient number of points to permit access to spares within a specified time.

14.2.8 The User and the Designer

The considerations discussed in this chapter are very much the user's concern. It is necessary, however, to decide upon them at the design stage since they influence, and are influenced by, the engineering of the product. The following table shows a few of the relationships between maintenance philosophy and design.

Skill level of maintenance technician	Amount of built-in test equipment required Level of LRA replacement in the field
Tools and test equipment	LRA fixings, connections and access Test points and equipment standardization Ergonomics and environment
Maintenance procedure	Built-in test equipment diagnostics Displays Interchangeability

The importance of user involvement at the very earliest stages of design cannot be over-emphasized. Maintainability objectives cannot be satisfied merely by placing requirements on the designer and neither can they be considered without recognizing that there is a strong link between repair time and cost. The maintenance philosophy has therefore to be agreed while the design specification is being prepared.

14.2.9 Computer Aids to Maintenance

The availability of computer packages makes it possible to set up a complete preventive maintenance and spare-part provisioning scheme using computer facilities. The system is described to the computer by delineating all the parts and their respective failure rates, and routine maintenance schedules and the times to replenish each spare. The operator will then receive daily schedules of maintenance tasks with a list of spares and consumables required for each. There is automatic indication when stocks of any particular spare fall below the minimum level.

These minimum spares levels can be calculated from a knowledge of the part failure rate and ordering time if a given risk of spares stockout is specified.

Packages exist for optimum maintenance times and spares levels. The COMPARE package offers the type of reliability centered maintenance calculations described in Chapter 16.

CHAPTER 15

Predicting and Demonstrating Repair Times

15.1 Prediction Methods

The best-known methods for maintainability prediction are described in US Military Handbook 472. The methods described in this handbook, although applicable to a range of equipment developed at that time, have much to recommend them and are still worth attention. Unfortunately, the quantity of data required to develop these methods of prediction is so great that, with increasing costs and shorter design lives, it is unlikely that models will continue to be developed. On the other hand, calculations requiring the statistical analysis of large quantities of data lend themselves to computer methods and the rapid increase of these facilities makes such a calculation feasible if the necessary repair-time data for a very large sample of repairs (say, 10 000) are available.

Any realistic maintainability prediction procedure must meet the following essential requirements:

1. The prediction must be fully documented and described and subject to recorded modification as a result of experience.
2. All assumptions must be recorded and their validity checked where possible.
3. The prediction must be carried out by engineers who are not part of the design group and therefore not biased by the objectives.

Prediction, valuable as it is, should be followed by demonstration as soon as possible in the design program. Maintainability is related to reliability in that the frequency of each repair action is determined by failure rates. Maintainability prediction therefore requires a knowledge of failure rates in order to select the appropriate, weighted, sample of tasks. The prediction results can therefore be no more reliable than the accuracy of the failure rate data. Prediction is applicable only to the active elements of repair time since it is those that are influenced by the design.

There are two approaches to the prediction task. The first is a work study method that analyzes each task in the sample by breaking it into definable work elements. This requires an extensive databank of average times for a wide range of tasks on the equipment type in question. The second approach is empirical and involves rating a number of maintainability

Reliability, Maintainability and Risk. DOI: 10.1016/B978-0-08-096902-2.00015-5

factors against a checklist. The resulting 'scores' are converted to an MTTR by means of a nomograph that was obtained by regression analysis of the data.

The methods (called procedures) in US Military Handbook 472 are over twenty years old and it is unlikely that the databases are totally relevant to modern equipment. In the absence of alternative methods, however, procedure 3 is recommended because the prediction will still give a fair indication of the repair time and also because the checklist approach focuses attention on the practical features affecting repair time. Procedure 3 is therefore described here in some detail.

15.1.1 US Military Handbook 472 – Procedure 3

Procedure 3 was developed by RCA for the US Air Force and was intended for ground systems. It requires a fair knowledge of the design detail and maintenance procedures for the system being analyzed. The method is based on the principle of predicting a sample of the maintenance tasks. It is entirely empirical since it was developed to agree with known repair times for specific systems, including search radar, data processors and a digital data transmitter with r.f. elements. The sample of repair tasks is selected on the basis of failure rates and it is assumed that the time to diagnose and correct a failure of a given component is the same as for any other of that component type. This is not always true, as field data can show.

Where repair of the system is achieved by replacement of sizeable modules (that is, a large LRA) the sample is based on the failure rate of these high-level units.

The predicted repair time for each sample task is arrived at by considering a checklist of maintainability features and by scoring points for each feature. The score for each feature increases with the degree of conformity with a stated 'ideal'. The items in the checklist are grouped under three headings: design, maintenance support and personnel requirements. The points scored under each heading are appropriately weighted and related to the predicted repair time by means of a regression equation, which is presented in the form of an easily used nomograph.

Figure 15.1 shows the score sheet for use with the checklist and Figure 15.2 presents the regression equation nomograph. I deduce the regression equation to be:

$$\log_{10}\text{MTTR} = 3.544 - 0.0123C - 0.023(1.0638A + 1.29B)$$

where A, B and C are the respective checklist scores.

Looking at the checklist it will be noted that additional weight is given to some features of design or maintenance support by the fact that more than one score is influenced by a particular feature.

The checklist is reproduced, in part, in the following section but the reader wishing to carry out a prediction will need a copy of US Military Handbook 472 for the full list. The application of the checklist to typical tasks is, in the author's opinion, justified as an aid to maintainability design even if repair time prediction is not specifically required.

15.1.2 Checklist – Mil 472 – Procedure 3

The headings of each of the checklists are as follows:

Checklist A:

1.   Access (external)
2.   Latches and fasteners (external)

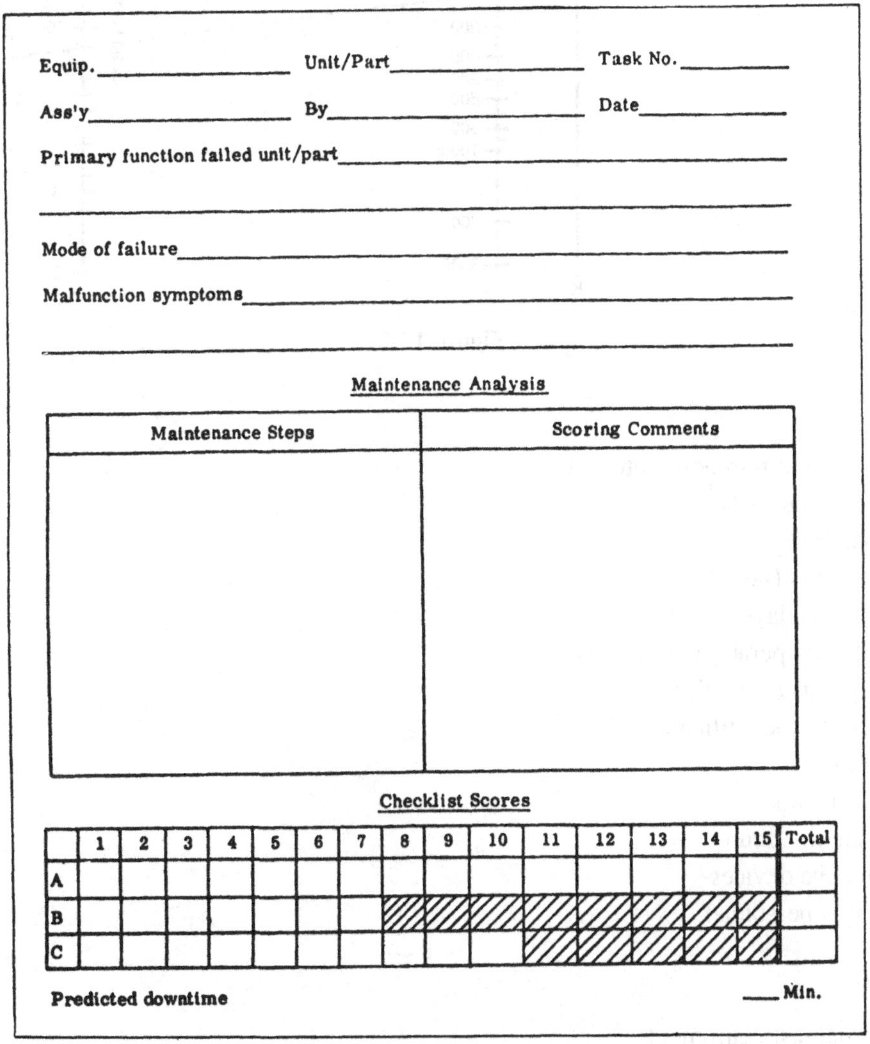

Figure 15.1

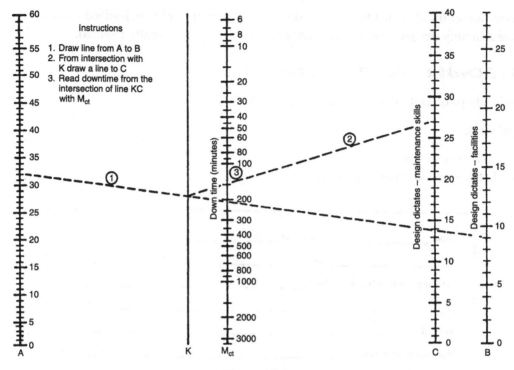

Figure 15.2

3. Latches and fasteners (internal)
4. Access (internal)
5. Packaging
6. Units/parts (failed)
7. Visual displays
8. Fault and operation indicators
9. Test points availability
10. Test points identification
11. Labeling
12. Adjustments
13. Testing in circuit
14. Protective devices
15. Safety – personnel.

Checklist B:

1. External test equipment
2. Connectors

3. Jigs and fixtures
4. Visual contact
5. Assistance operations
6. Assistance technical
7. Assistance supervisory.

Checklist C:

1. Arm, leg, and back strength
2. Endurance and energy
3. Eye – hand
4. Visual
5. Logic
6. Memory
7. Planning
8. Precision
9. Patience
10. Initiative.

Three items from each of checklists A and B and the scoring criteria for all of checklist C are reproduced as follows.

15.1.2.1 Checklist A – Scoring Physical Design Factors

1. *Access (external)*: determines if the external access is adequate for visual inspection and manipulative actions. Scoring will apply to external packaging as related to maintainability design concepts for ease of maintenance. This item is concerned with the design for external visual and manipulative actions that would precede internal maintenance actions. The following scores and scoring criteria will apply:

Scores

(a) Access adequate both for visual and manipulative tasks (electrical and mechanical)	4
(b) Access adequate for visual, but not manipulative, tasks	2
(c) Access adequate for manipulative, but not visual, tasks	2
(d) Access not adequate for visual or manipulative tasks	0

Scoring criteria

An explanation of the factors pertaining to the above scores is consecutively shown. This procedure is followed throughout for other scores and scoring criteria.
   (a) To be scored when the external access, while visual and manipulative actions are being performed on the exterior of the subassembly, does not present difficulties because of obstructions (cables, panels, supports, etc.).

(b) To be scored when the external access is adequate (no delay) for visual inspection, but not for manipulative actions. External screws, covers, panels, etc., can be located visually; however, external packaging or obstructions hinders manipulative actions (removal, tightening, replacement, etc.).

(c) To be scored when the external access is adequate (no delay) for manipulative actions, but not for visual inspections. This applies to the removal of external covers, panels, screws, cables, etc., which present no difficulties; however, their location does not easily permit visual inspection.

(d) To be scored when the external access is inadequate for both visual and manipulative tasks. External covers, panels, screws, cables, etc., cannot be easily removed nor visually inspected because of external packaging or location.

2. *Latches and fasteners (external)*: determines if the screws, clips, latches, or fasteners outside the assembly require special tools, or if significant time was consumed in the removal of such items. Scoring will relate external equipment packaging and hardware to maintainability design concepts. Time consumed with preliminary external disassembly will be proportional to the type of hardware and tools needed to release them and will be evaluated accordingly.

Scores

(a)	External latches and/or fasteners are captive, need no special tools, and require only a fraction of a turn for release	4
(b)	External latches and/or fasteners meet two of the above three criteria	2
(c)	External latches and/or fasteners meet one or none of the above three criteria	0

Scoring criteria

(a) To be scored when external screws, latches, and fasteners are:
1. captive
2. do not require special tools
3. can be released with a fraction of a turn.
Releasing a 'DZUS' fastener which requires a 90-degree turn using a standard screwdriver is an example of all three conditions.

(b) To be scored when external screws, latches, and fasteners meet two of the three conditions stated in (a) above. An action requiring an Allen wrench and several full turns for release shall be considered as meeting only one of the above requirements.

(c) To be scored when external screws, latches, and fasteners meet only one or none of the three conditions stated in (a) above.

3. *Latches and fasteners (internal)*: determines if the internal screws, clips, fasteners or latches within the unit require special tools, or if significant time was consumed in the removal of such items. Scoring will relate internal equipment hardware to maintainability

design concepts. The types of latches and fasteners in the equipment, and standardization of these throughout the equipment will tend to affect the task by reducing or increasing required time to remove and replace them. Consider 'internal' latches and fasteners to be within the interior of the assembly.

Scores

(a) Internal latches and/or fasteners are captive, need no special tools, and require only a fraction of a turn for release	4
(b) Internal latches and/or fasteners meet two of the above three criteria	2
(c) Internal latches and/or fasteners meet one or none of the above three criteria	0

Scoring Criteria

(a) To be scored when internal screws, latches and fasteners are:
    1. captive
    2. do not require special tools
    3. can be released with a fraction of a turn.
    Releasing a 'DZUS' fastener which requires a 90-degree turn using a standard screwdriver would be an example of all three conditions.

(b) To be scored when internal screws, latches, and fasteners meet two of the three conditions stated in (a) above. A screw that is captive can be removed with a standard or Phillips screwdriver, but requires several full turns for release.

(c) To be scored when internal screws, latches, and fasteners meet one of three conditions stated in (a) above. An action requiring an Allen wrench and several full turns for release shall be considered as meeting only one of the above requirements.

15.1.2.2 Checklist B – Scoring Design Dictates – Facilities

The intent of this questionnaire is to determine the need for external facilities. Facilities, as used here, include material such as test equipment, connectors, etc., and technical assistance from other maintenance personnel, supervisor, etc.

1. *External test equipment:* determines if external test equipment is required to complete the maintenance action. The type of repair considered maintainably ideal would be one which did not require the use of external test equipment. It follows, then, that a maintenance task requiring test equipment would involve more task time for set-up and adjustment and should receive a lower maintenance evaluation score.

Scores

(a) Task accomplishment does not require the use of external test equipment	4
(b) One piece of test equipment is needed	2
(c) Several pieces (two or three) of test equipment are needed	1
(d) Four or more items are required	0

Scoring criteria

(a) To be scored when the maintenance action does not require the use of external test equipment. Applicable when the cause of malfunction is easily detected by inspection or built-in test equipment.

(b) To be scored when one piece of test equipment is required to complete the maintenance action. Sufficient information is available through the use of one piece of external test equipment for adequate repair of the malfunction.

(c) To be scored when two or three pieces of external test equipment are required to complete the maintenance action. This type of malfunction would be complex enough to require testing in a number of areas with different test equipment.

(d) To be scored when four or more pieces of test equipment are required to complete the maintenance action. Involves an extensive testing requirement to locate the malfunction. This would indicate that a least maintainable condition exists.

2. *Connectors*: determines if supplementary test equipment requires special fittings, special tools, or adaptors to adequately perform tests on the electronic system or subsystem. During troubleshooting of electronic systems, the minimum need for test equipment adaptors or connectors indicates that a better maintainable condition exists.

Scores

(a) Connectors to test equipment require no special tools, fittings, or adaptors	4
(b) Connectors to test equipment require some special tools, fittings, or adaptors (less than two)	2
(c) Connectors to test equipment require special tools, fittings, and adaptors (more than one)	0

Scoring criteria

(a) To be scored when special fittings or adaptors and special tools are not required for testing. This would apply to tests requiring regular test leads (probes or alligator clips) that can be plugged into or otherwise secured to the test equipment binding post.

(b) Applies when one special fitting, adaptor or tool is required for testing. An example would be if testing had to be accomplished using a 10 dB attenuator pad in series with the test set.

(c) To be scored when more than one special fitting, adaptor, or tool is required for testing. An example would be when testing requires the use of an adaptor and an r.f. attenuator.

3. *Jigs or fixtures*: determines if supplementary materials such as block and tackle, braces, dollies, ladder, etc., are required to complete the maintenance action. The use of such items during maintenance would indicate the expenditure of a major maintenance time and pinpoint specific deficiencies in the design for maintainability.

Scores

(a) No supplementary materials are needed to perform task	4
(b) No more than one piece of supplementary material is needed to perform task	2
(c) Two or more pieces of supplementary material are needed	0

Scoring criteria

(a) To be scored when no supplementary materials (block and tackle, braces, dollies, ladder, etc.) are required to complete maintenance. Applies when the maintenance action consists of normal testings and the removal or replacement of parts or components can be accomplished by hand, using standard tools.

(b) To be scored when one supplementary material is required to complete maintenance. Applies when testing or when the removal and replacement of parts requires a stepladder for access or a dolly for transportation.

(c) To be scored when more than one supplementary material is required to complete maintenance. Concerns the maintenance action requiring a step ladder and dolly adequately to test and remove the replaced parts.

15.1.2.3 Checklist C – Scoring Design Dictates – Maintenance Skills

This checklist evaluates the personnel requirements relating to physical, mental, and attitude characteristics, as imposed by the maintenance task.

Evaluation procedure for this checklist can best be explained by way of several examples. Consider the first question, which deals with arm, leg and back strength. Should a particular task require the removal of an equipment drawer weighing 100 pounds (45 kg), this would impose a severe requirement on this characteristic. Hence, in this case the question would be given a low score (0–1). Assume another task that, owing to small size and delicate construction, required extremely careful handling. Here question 1 would be given a high

Scores

	Score
1. Arm, leg, and back strength	
2. Endurance and energy	___
3. Eye-hand coordination, manual dexterity and neatness	___
4. Visual acuity	___
5. Logical analysis	___
6. Memory – things and ideas	___
7. Planfulness and resourcefulness	___
8. Alertness, cautiousness and accuracy	___
9. Concentration, persistence and patience	___
10. Initiative and incisiveness	___

score (4), but the question dealing with eye-hand coordination and dexterity would be given a low score. Other questions in the checklist relate to various personnel characteristics important to maintenance task accomplishment. In completing the checklist, the task requires that each of these characteristics should be viewed with respect to average technician capabilities.

Scoring criteria

Quantitative evaluations of these items range from 0 to 4 and are defined in the following manner:

1. The maintenance action requires a *maximum* effort on his part.
2. The maintenance action requires an *above average* effort on his part.
3. The maintenance action requires an *average* effort on the part of the technician.
4. The maintenance action requires a *below average* effort on the part of the technician.
5. The maintenance action requires a *minimum* effort on the part of the technician.

15.1.3 Using a Weighted Sample

Clearly the MTTR of a complex piece of equipment cannot be assessed on the basis of one mode of failure. There will be a range of repair times depending on the symptoms and ease of replacement of the item in question. In practice a sample size will be chosen and a representative range of failures chosen. The size of the sample will usually be determined by the resources available for the study. Having agreed a sample size it is sensible to spread the choice over the range of technologies involved (i.e. electromechanical items, electronics, pneumatic items). Furthermore, choosing the higher failure rate components will effectively maximize the sample size by representing a higher proportion of the total failure of the equipment.

Having carried out the assessments the average MTTR is obtained by the weighted average of the individual MTTRs as shown in the following table.

Item Failure Rate	Assessed MTTR	
λ_1 Transformer s/c	$MTTR_1$	$\lambda_1 \times MTTR_1$
λ_2 Actuator jams	$MTTR_2$	$\lambda_2 \times MTTR_2$
λ_3 C23 s/c	$MTTR_3$	$\lambda_3 \times MTTR_3$
λ_4 Relay o/c	$MTTR_4$	$\lambda_4 \times MTTR_4$
$\Sigma\lambda$ (Total failure rate)		$\Sigma \lambda$ (MTTR)
	AVERAGE MTTR $= \dfrac{\Sigma(\lambda \times MTTR)}{\Sigma\lambda}$	

15.2 Demonstration Plans

15.2.1 Demonstration Risks

Where demonstration of maintainability is contractual, it is essential that the test method, and the conditions under which it is to be carried out, are fully described. If this is not observed then disagreements are likely to arise during the demonstration. Both supplier and customer wish to achieve the specified mean time to repair at minimum cost and yet a precise demonstration having acceptable risks to all parties is extremely expensive. A true assessment of maintainability can only be made at the end of the equipment life and anything less will represent a sample.

Figure 15.3 shows a typical test plan for observing the mean time to repair of a given item. Just as in Chapter 5, the curve shows the relationship of the probability of passing the test against the batch failure rate, then Figure 15.3 relates that probability to the actual MTTR.

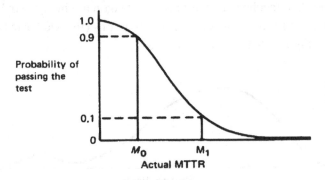

Figure 15.3: MTTR demonstration test plan

For a MTTR of $M0$ the probability of passing the test is 90% and for a value of $M1$ it falls to 10%. In other words, if $M0$ and $M1$ are within 2:1 of each other then the test has a good discrimination.

A fully documented procedure is essential and the only reference document available is US Military Standard 471A – *Maintainability Verification/Demonstration/Evaluation* – 27 March 1973. This document may be used as the basis for a contractual agreement, in which case both parties should carefully assess the risks involved. Statistical methods are usually dependent on assumptions concerning the practical world and it is important to establish their relevance to a particular test situation. In any maintainability demonstration test it is absolutely essential to fix the following:

* method of test demonstration task selection
* tools and test equipment available
* maintenance documentation
* skill level and training of test subject
* environment during test
* preventive maintenance given to test system.

15.2.2 US Military Standard 471A (1973)

This document replaces US Military Standard 471 (1971) and MIL 473 (1971) – Maintainability Demonstration. It contains a number of sampling plans for demonstrating maintenance times for various assumptions of repair time distribution. A task sampling plan is also included and describes how the sample of simulated failures should be chosen. Test plans choose either the log normal assumption or make no assumption of distribution. The log normal distribution frequently applies to systems using consistent technologies such as computer and data systems, telecommunications equipment, control systems and consumer electronics, but equipment with mixed technologies such as aircraft flight controls, microprocessor-controlled mechanical equipment and so on are likely to exhibit bimodal distributions. This results from two repair time distributions (for two basic types of defect) being superimposed. Figure 15.4 illustrates this case.

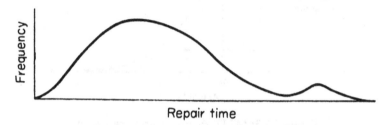

Figure 15.4: Distribution of repair times

The method of task sample selection involves stratified sampling. This involves dividing the equipment into functional units and, by ascribing failure rates to each unit, determining the relative frequency of each maintenance action. Taking into account the quantity of each unit the sample of tasks is spread according to the anticipated distribution of field failures. Random sampling is used to select specific tasks within each unit once the appropriate number of tasks has been assigned to each. The seven test plans are described as follows:

Test Method 1

The method tests for the mean repair time (MTTR). A minimum sample size of 30 is required and an equation is given for computing its value. Equations for the producer's and consumer's risks, α and β, and their associated repair times are also given. Two test plans are given. Plan A assumes a log normal distribution of repair times while plan B is distribution free, that is, it applies in all cases.

Test Method 2

The method tests for a percentile repair time. This means a repair time associated with a given probability of not being exceeded. For example, a 90 percentile repair time of one hour means that 90% of repairs are effected in one hour or less and that only 10% exceed this value. This test assumes a log normal distribution of repair times. Equations are given for calculating the sample size, the risks and their associated repair times.

Test Method 3

The method tests the percentile value of a specified repair time. It is distribution free and therefore applies in all cases. For a given repair time, values of sample size and pass criterion are calculated for given risks and stated pass and fail percentiles. For example, if a median MTTR of 30 min is acceptable, and if 30 min as the 25th percentile (75% of values are greater) is unacceptable, the test is established as follows. Producer's risk is the probability of rejection although 30 min is the median, and consumer's risk is the probability of acceptance although 30 min is only the 25th percentile. Let these both equal 10%. Equations then give the value of sample size as 23 and the criterion as 14. Hence if more than 14 of the observed values exceed 30 min the test is failed.

Test Method 4

The method tests the median time. The median is the value, in any distribution, such that 50% of values exceed it and 50% do not. Only in the normal distribution does the median equal the mean. A log normal distribution is assumed in this test, which has a fixed sample size of 20. The test involves comparing log MTTR in the test with log of the median value required in a given equation.

Test Method 5

The method tests the 'chargeable down time per flight'. This means the down time attributable to failures as opposed to passive maintenance activities, test-induced failures, modifications, etc. It is distribution free with a minimum sample size of 50 and can be used, indirectly, to demonstrate availability.

Test Method 6

The method is applicable to aeronautical systems and tests the 'man-hour rate'. This is defined as

$$\frac{\text{Total chargeable maintenance man-hours}}{\text{Total demonstration flight hours}}$$

Actual data are used and no consumer or producer risks apply.

Test Method 7

This is similar to test method 6 and tests the man-hour rate for simulated faults. There is a minimum sample size of 30.

Test methods 1–4 are of a general nature whereas methods 5–7 have been developed with aeronautical systems in mind. In applying any test the risks must be carefully evaluated.

There is a danger, however, of attaching an importance to results in proportion to the degree of care given to the calculations. It should therefore be emphasized that attention to the items listed in Section 15.2.1 in order to ensure that they reflect the agreed maintenance environment is of equal if not greater importance.

15.2.3 Data Collection

It would be wasteful to regard the demonstration test as no more than a means of determining compliance with a specification. Each repair is a source of maintainability design evaluation and a potential input to the manual. Diagnostic instructions should not be regarded as static but be updated as failure information accrues. If the feedback is to be of use, it is necessary to record each repair with the same detail as is called for in field reporting. The different repair elements of diagnosis, replacement, access, etc. should be listed separately, together with details of tools and equipment used. Demonstration repairs are easier to control than field maintenance and should therefore be better documented.

In any maintainability (or reliability) test the details should be fully described in order to minimize the possibilities of disagreement. Both parties should understand fully the quantitative and qualitative risks involved.

CHAPTER 16

Quantified Reliability Centered Maintenance

16.1 What is QRCM?

Quantitative Reliability Centered Maintenance (QRCM) involves calculations to balance the cost of excessive maintenance against that of the unavailability arising from insufficient maintenance. The following simple example illustrates one of the techniques that will be dealt with in this chapter.

Doubling the proof-test interval of a shutdown system on an off-shore production platform might lead to an annual saving of 2 man-days (say £2000). The cost in increased production unavailability might typically be calculated as 8×10^{-7}, in which case the annual loss would be $8 \times 10^{-7} \times$ say £50 (per barrel) $\times$ say 50 000 (barrels) $\times$ 365 (days) = £730. In this case the reduction in maintenance is justified as far as cost is concerned.

QRCM is therefore the use of reliability techniques to optimize:

- replacement (discard) intervals
- spares holdings
- proof-test intervals
- condition monitoring.

The first step in planning any QRCM strategy is to identify the critical items affecting plant unavailability since the greater an item's contribution to unavailability (or hazard) the more potential savings are to be made from reducing its failure rate.

Reliability modeling techniques lend themselves to this task in that they allow comparative availabilities to be calculated for a number of maintenance regimes. In this way the costs associated with changes in maintenance intervals, spares holdings and preventive replacement (discard) times can be compared with the savings achieved.

An important second step is to obtain site-specific failure data. Although QRCM techniques can be applied using GENERIC failure rates and down times, there is better precision from site-specific data. These are not, however, always available and published data sources (such as FARADIP.THREE) may have to be used. These are described in Chapter 4.

Reliability, Maintainability and Risk. DOI: 10.1016/B978-0-08-096902-2.00016-7

Because of the wide range of generic failure rates, plant-specific data are preferred and an accurate plant register goes hand in hand with this requirement. Plant registers are often out of date and should be revised at the beginning of a new QRCM initiative. Thought should be given to a rational, hierarchical numbering for plant items, which will assist in sorting like items, related items and items with like replacement times for purposes of maintenance and spares scheduling.

Good data are essential because, in applying QRCM, it is vital to take account of the way in which failures are distributed with time. We need to know if the failure rate is constant or whether it is increasing or decreasing. Preventive replacement (discard), for example, is only justified if there is an increasing failure rate.

16.2 The QRCM Decision Process

The use of these techniques depends upon the failure distribution, the degree of redundancy and whether the cost of the maintenance action is justified by the saving in operating costs, safety or environmental impact. Figure 16.1 is a QRCM decision algorithm. As each equipment item is considered the QRCM algorithm provides the logic which leads to the use of each of the techniques.

Using Figure 16.1 consider an unrevealed failure that, if it coincides with some other failure, leads to significant consequences such as the shutdown of a chemical plant. Assume that there is no measurable check whereby the failure can be pre-empted. Condition monitoring is not therefore appropriate. Assume, also, that the failure rate is not increasing, therefore preventive discard cannot be considered. There is, however, an optimum proof-test interval whereby the cost of proof test can be balanced against the penalty cost of the coincident failures.

16.3 Optimum Replacement (Discard)

Specific failure data are essential for this technique to be applied sensibly. There are no generic failure data describing wearout parameters that would be adequate for making discard decisions. Times to failure must be obtained for the plant items in question and the Weibull techniques described in Chapter 6 applied. Note that units of time may be hours, cycles, operations or any other suitable base.

Only a significant departure of the shape parameter from ($\beta = 1$) justifies considering discard.

If $\beta \leq 1$ then there is no justification for replacement or even routine maintenance. If, on the other hand, $\beta > 1$ then there may be some justification for considering a preventive replacement before the item has actually failed. This will only be justified if the costs associated with an unplanned replacement (due to failure) are greater than those of a planned discard/replacement.

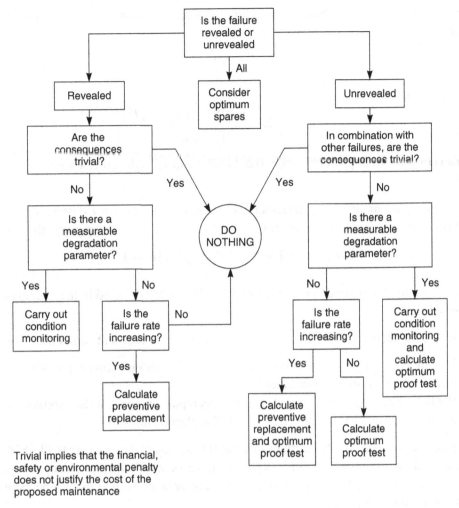

Figure 16.1: The QRCM decision algorithm

If this is the case then it is necessary to calculate:

- (a) The likelihood of a failure (i.e. $1 - \exp(-t/\eta)\beta$) in a particular interval times the cost of the unplanned failure;
- (b) The cost of planned replacements during that interval.

The optimum replacement interval that minimizes the sum of the above two costs can then be found. Two maintenance philosophies are possible:

- Age replacement
- Block replacement.

For the age replacement case, an interval starts at time $t = 0$ and ends either with a failure or with a replacement at time $t = T$, whichever occurs first. The probability of surviving until time $t = T$ is $R(T)$, thus the probability of failing is $1 - R(T)$. The average duration of all intervals is given by:

$$\int_0^T R(t)\,dt$$

Thus the cost per unit time is:

$$\frac{[\pounds_u \times (1 - R(T)) + \pounds_p \times R(T)]}{\int_0^T R(t)\,dt}$$

where £u is the cost of unplanned outage (i.e. failure) and £p is the cost of a planned replacement.

For the block replacement case, replacement always occurs at time $t = T$ despite the possibility of failures occurring before time $t = T$. For this case the cost per unit time is:

$$(\pounds_u \times T)/\text{MTBF} \times T + \pounds_p/T = \pounds_u/\text{MTBF} + \pounds_p/T$$

Note that, since the failure rate is not constant ($\beta > 1$), the MTBF used in the formula varies as a function of T.

There are two maintenance strategies involving preventive replacement (discard):

- (a) If a failure occurs, replace it and then wait the full interval before replacing again. This is known as AGE replacement.
- (b) If a failure occurs, replace it and nevertheless replace it again at the expiration of the existing interval. This is known as BLOCK replacement.

AGE replacement would clearly be more suitable for expensive items whereas BLOCK replacement might be appropriate for inexpensive items of which there are many to replace. Furthermore, BLOCK replacement is easier to administer since routine replacements then occur at regular intervals.

The COMPARE software package calculates the replacement interval for both cases and such that the sum of the following two costs is minimized:

- The cost of unplanned replacement taking account of the likelihood that it will occur PLUS
- The cost of the scheduled replacement.

The program requests the unplanned and planned maintenance costs as well as the SHAPE and SCALE parameters.

Clearly the calculation is not relevant unless:
- SHAPE parameter $\beta > 1$
  AND
- Unplanned cost > planned cost.

COMPARE provides a table of total costs (for the two strategies) against potential replacement times as can be seen in the following table, where 1600 hours (nearly 10 weeks) is the optimum. It can be seen that the age and block replacement cases do not yield quite the same cost per unit time and that block replacement is slightly less efficient. The difference may, however, be more than compensated for by the savings in the convenience of replacing similar items at the same time. Chapter 6 has already dealt with the issue of significance and of mixed failure modes.

- Shape parameter (Beta) = 2.500
- Scale parameter (Eta) = 4000 hrs
- Cost of unscheduled replacement = £4000
- Cost of planned replacement = £500.

Replacement Interval	Cost per unit Time	
	Age Replace	Block Replace
1 000	0.6131	0.6234
1 200	0.5648	0.5777
1 400	0.5429	0.5582
1 600	**0.5381**	**0.5554**
1 800	0.5451	0.5637
2 000	0.5605	0.5796
2 200	0.5820	0.6006
2 400	0.6080	0.6250
2 600	0.6372	0.6515
2 800	0.6688	0.6789
3 000	0.7018	0.7064

16.4 Optimum Spares

There is a cost associated with carrying spares, namely capital depreciation, space, maintenance, etc. In order to assess an optimum spares level it is necessary to calculate the unavailability that will occur at each level of spares holding. This will depend on the following variables:

- Number of spares held
- Failure rate of the item
- Number of identical items in service
- Degree of redundancy within those items
- Lead time of procurement of spares
- Replacement time (unit down time) when an item fails.

This relationship can be modeled by means of Markov state diagram analysis and was fully described in Chapter 14 (Section 14.2.6).

It should be noted that as the number of spares increases, there is a diminishing return in terms of improved unavailability until the so-called infinite spares case is reached. This is where the unavailability is dominated by the repair time and thus increased spares holding becomes ineffectual. At this point, only an improvement in repair time or in failure rate can increase the availability.

The cost of unavailability can be calculated for, say, zero spares. The cost saving in reduced unavailability can then be compared with the cost of carrying one spare and the process repeated until the optimum spares level is assessed.

The COMPARE package automatically creates successive runs for different spares levels and displays them in tabular form. Figure 14.5 shows the Markov state diagram for four units with up to two spares.

16.5 Optimum Proof Test

In the case of redundant systems where failed redundant units are not revealed then the option of periodic proof test arises. Although the failure rate of each item is constant, the system failure rate actually increases.

The unavailability of a system can be calculated using the methods described in Chapter 8. It is clearly dependent partly on the proof-test interval, which determines the down time of a failed (dormant) redundant item.

The technique involves calculating an optimum proof-test interval for revealing dormant failures. It seeks to trade off the cost of the proof test (i.e. preventive maintenance) against the reduction in unavailability.

It applies where coincident dormant failures cause unavailability. An example would be the failure to respond of both a 'high' alarm and a 'high high' signal.

The unavailability is a function of the instrument failure rates and the time for which dormant failures persist. The more frequent the proof test, which seeks to identify the dormant failures, then the shorter is the down time of the failed items.

Assume that the 'high' alarm and 'high high' signal represent a duplicated redundant arrangement. Thus, one instrument may fail without causing plant failure (shutdown).

It has already been shown that the reliability of the system is given by:

$$R(t) = 2e^{-\lambda t} - e^{-2\lambda t}$$

Thus the probability of failure is $1 - R(t)$

$$= 1 - 2e^{-\lambda t} + e^{-2\lambda t}$$

If the cost of an outage (i.e. lost production) is $£_u$ then the expected cost, due to outage, is:

$$= (1 - 2e^{-\lambda t} + e^{-\lambda t}) \times £_u$$

Now consider the proof test, which costs $£_p$ per visit. If the proof-test interval is T then the expected cost, due to preventive maintenance, is:

$$= (2e^{-\lambda t} - e^{-2\lambda t}) \times £_p$$

The total cost per time interval is thus:

$$= [(1 - 2e^{-\lambda t} + e^{-\lambda t}) \times £_u] + [(2e^{-\lambda t} - e^{-2\lambda t}) \times £_p]$$

The average length of each interval is $\int_0^T R(t)dt$

$$= 3/2\lambda - 2/\lambda e^{-\lambda T} + 1/2\lambda e^{-2\lambda T}$$

The total cost per unit time can therefore be obtained by dividing the above expression into the preceding one.

The minimum cost can be found by tabulating the cost against the proof-test interval (T). In the general case the total cost per unit time is:

$$\frac{(1 - R(T)) \times £_u] + [R(T) \times £_p]}{\int_0^T R(t)\,dt}$$

Again, the COMPARE package performs this calculation and provides an optimum interval (approximately three years) as can be seen in the following example.

- Total number of units = 2
- number of units required = 1
- MTBF of a single unit = 10. years
- cost of unscheduled outage = £2000
- cost of a planned visit = £100.

Proof-Test Interval	Cost per Unit Time
1.000	117.6
1.700	86.88
2.400	78.98
3.100	**77.79**
3.800	79.18
4.500	81.65
5.200	84.56
5.900	87.60
6.600	90.61
7.300	93.51
8.000	96.28

16.6 Condition Monitoring

Many failures do not actually occur spontaneously but develop over a period of time. It follows, therefore, that if this gradual 'degradation' can be identified it may well be possible to pre-empt the failure. Overhaul and replacement are then both realistic options. During the failure mode analysis it may be possible to determine parameters that, although not themselves causing a hazard or equipment outage, are indicators of the degradation process.

In other words, the degradation parameter can be monitored and action taken to prevent failure by observing trends. Trend analysis would be carried out on each of the measurements in order to determine the optimum point for remedial action.

It is necessary for there to be a reasonable time period between the onset of the measurable degradation condition and the actual failure. The length (and consistency) of this period will determine the optimum inspection interval.

There are a number of approaches to determining the inspection interval. Methods involving a gradual increase in interval run the risk of suffering the failure. This may be expensive or hazardous. Establishing the interval by testing, although safer, is expensive, may take time and relies on simulated operating environments. However, in practice, a sensible mixture of experience and data can lead to realistic intervals being chosen. By concentrating on a specific failure mode (say valve diaphragm leakage) and by seeking out those with real operating experience it is possible to establish realistic times. Even limited field and test data will enhance the decision.

The following list provides some examples of effects that can be monitored:

- regular gas and liquid emission leak checks
- critical instrumentation parameter measurements (gas, fire, temp, level, etc.)
- insulation resistivity
- vibration measurement and analysis of characteristics
- proximity analysis
- shock pulse monitoring
- acoustic emission
- corrosive states (electro-chemical monitoring)
- dye penetration
- spectrometric oil analysis
- electrical insulation
- hot spots
- surface deterioration
- state of lubrication and lubricant
- plastic deformation
- balance and alignment.

CHAPTER 17

Systematic Failures, Especially Software

The techniques dealt with so far in this book apply largely to catastrophic hardware failures. The term 'random hardware failures' covers the constant failure rate part of the bathtub curve and the term 'dependent failures' embraces the common causes dealt with earlier.

It was mentioned in Section 1.3 that, due to the complexity of modern engineering products, system failure does not always involve a single component part failure. More subtle factors, such as the following, often dominate the system failure rate:

- failure resulting from software elements
- failure due to environmental factors
- failure due to ambiguity in the specification
- failure due to timing constraints within the design
- failure due to combinations of component parameter tolerance.

Systematic failures, therefore, are additional to those which we quantify by means of failure rate (or even Weibull) methods. Since they do not relate to past 'failure' data, it follows that it is very difficult to imagine their being predicted by the modeling techniques we have considered so far.

Therefore, qualitative measures (**often referred to as life cycle activities**) are called for in the hope that they will minimize these systematic failures. In Chapter 22, which deals specifically with safety-related systems, the concept of integrity levels will be introduced. This involves prescribing varying levels of rigor for these qualitative measures, according to the degree of integrity required.

The following sections summarize these defenses with particular reference to software-related failure.

17.1 Programable Devices

For the last thirty years programable devices have made a significant impact on electronic circuit design. The main effect has been to reduce the number of different circuit types by virtue of there being a more or less standard computer architecture. The hardware, in combination with software programming, provides the individual circuit functions previously

Reliability, Maintainability and Risk. DOI: 10.1016/B978-0-08-096902-2.00017-9

achieved by differences in hardware. The term 'software' refers to the instructions needed to enable a programmable device to function, including the associated hierarchy of documents required to produce the code. This use of programming at the circuit level, now common with most industrial and consumer products, brings with it some associated quality and reliability problems. When applied to microprocessors at the circuit level, the programming, which is semi-permanent and usually contained in ROM (Read Only Memory), is known as Firmware. The necessary increase in function density of devices in order to provide the large quantities of memory in small packages has matched this trend.

Computing and its associated software was once seen in the three broad categories of *Mainframe computing*, *Minicomputing* and *Microprocessing*. The microprocessor has now largely replaced both the minicomputer and the mainframe for normal industrial applications.

We now tend to group them as:

1. *PCs*
    Physical size has become largely irrelevant. These devices carry out computational and data handling tasks using proprietary software packages. In some cases real-time control is implemented by interfacing with instrumentation and with field devices.
2. Programmable Logic Controllers (PLCs)
    Although similar in construction to the above devices, there is a more consistent architecture with the accent on providing input and output ports which allow large amounts of 'data in' and 'command out' capability in analogue, digital and loop modes. Languages are usually proprietary to the device in question and are frequently much more simple than the well-known branching programming languages. Thus, programming is simplified and the opportunity for errors in code design is significantly reduced.
3. Embedded microcomputing
    In this case the computer architecture is provided by processor and memory in the form of LSI chips. Individual designs are carried out at printed board level and programming can be at assembler (machine code) level as well as by means of high-level programming languages. On one hand there is total flexibility of design but, on the other, opportunities for timing-related circuit problems, as well as code design errors, are much greater.

From the safety-integrity and from the reliability point of view, there are both advantages and disadvantages arising from programmable design solutions. The relative magnitude of the advantages and disadvantages will vary according to the type of device (as described earlier):

Reliability/integrity Advantages	Reliability/integrity Disadvantages
• Fewer different types of hardware	• Difficult to 'inspect' software for errors
• Consistent architectures (CPU/bus/I-O)	• Test involves a limited sample of execution possibilities
• Common approach to hardware design	• Control of changes is more complicated and more critical
• Easier to support several versions in the field	• Difficult to impose programming 'standard approaches'
• Simpler to implement modifications	• Very difficult to predict meaningful 'failure rates' of software-related systematic failures

17.2 Software-related Failures

The question arises as to how a software failure (or for that matter a systematic failure) is defined. Unlike hardware, there is no physical change associated with a unit that is 'functional' at one moment and 'failed' at the next. Software failures are in fact errors that, owing to the complexity of a computer program, do not become evident until the combination of conditions brings the error to light. The effect is nevertheless the same as any other failure. Unlike the hardware bathtub curve, there is no wearout characteristic but only a continuing burn-in. However, each time that a change to the software is made the error rate may well rise, as shown in Figure 17.1. As a result of software errors there has been, for some time, an interest in developing methods of controling the activities of programmers and of reducing software complexity by attempts at standardization.

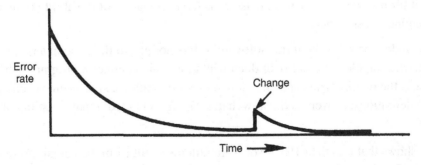

Figure 17.1: Software error curve

Figure 17.2 illustrates the idea of software failures in programable systems. It introduces the concept of fault/error/failure. Faults may occur in both hardware and software. Software faults, often known as bugs, will appear as a result of particular portions of code being used for the first time under a particular set of circumstances.

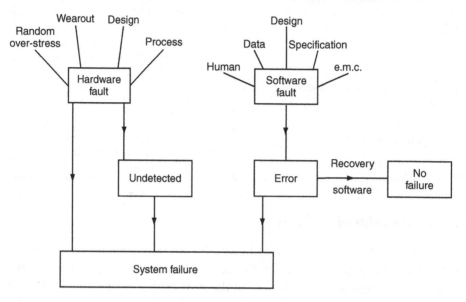

Figure 17.2: Fault/error/failure

A fault in a programed system does not necessarily result in either an error or a failure. A long time may elapse before that code is used under the circumstances that lead to failure.

A fault (bug) may lead to an error that occurs when the system reaches an incorrect state. That is, a bit, or bits, takes an incorrect value in some memory location.

An error may propagate to become a failure if the system does not contain error-recovery software capable of detecting and eliminating the error.

Failure, be it for hardware or software reasons, is the termination of the ability of an item to perform the function specified.

It should be understood that the term 'software' refers not just to the code in a programable device but to the complete hierarchy of documentation which defines a programmable system. This embraces the requirements specification, data specifications, subsystem specifications and module definitions, as well as the flowcharts, listings and media that constitute the entire software.

Experience shows that less than 1% of software failures result from the actual 'production' of the firmware. This is hardly surprising since the act of inputting code is often self-checking and errors are fairly easy to detect. This leaves the design and coding activities as the source of failures. Within these, fewer less than 50% of errors are attributed to the coding activity. Software reliability is therefore inherent in the design process of breaking down the requirements into successive levels of specification.

17.3 Software Failure Modeling

Numerous attempts have been made to define models that enable software failure rates to be predicted from the initial failures observed during integration and test or from parameters such as the length and nature of the code. The latter suffers from the difficulty that, in software, there are no elements (as with hardware components) with failure characteristics that can be taken from experience and used for predictive purposes. This type of prediction is therefore unlikely to prove successful. The former method (i.e. modeling based on the early failures) suffers from a difficulty which is illustrated by this simple example. Consider the following failure pattern based on four days of testing:

Day one   10 failures
Day two   9 failures
Day three 8 failures
Day four   7 failures

To predict, from these data, when we might observe 6 failures per day is not too difficult, but what we need to know is when the failure rate will be 10^{-4} or perhaps 10^{-5}. It is not likely that the information required is in fact contained within the data available at this point.
Figure 17.3 illustrates the coarseness of the data and the fact that the tail of the distribution is not well defined and by no means determined by the shape of the left-hand end.
Extrapolation, in these circumstances, is very dangerous.

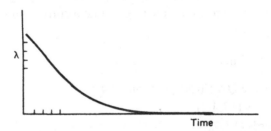

Figure 17.3: Decreasing software failures

A number of models have been developed. They rely on various assumptions concerning the nature of the failure process, such as the idea that failure rate is determined by the number of potential failures remaining in the program. These are by no means revealed solely by the passage of calendar time, since repeated executions of the same code will not usually reveal further failures.

Present opinion is that no one model is better than any other, and it must be said that, in any case, an accurate prediction only provides a tool for scheduling rather than a long-term field reliability assessment. The models include:

- *Jelinski Moranda*: this assumes that failure rate is proportional to the remaining fault content. Remaining faults are assumed to be equally likely to occur.
- *Musa*: program execution rather than calendar time is taken as the variable.
- *Littlewood Verall*: assumes successive execution time between failures to be an exponentially distributed random variable.
- *Structured models*: these attempt to break software into subunits. Rules for switching between units and for the failure rate of each unit are developed.
- *Seeding and tagging*: this relies on the injection of known faults into the software. The success rate of debugging of the known faults is used to predict the total population of failures by applying the ratio of success to the revealed non-seeded failures. For this method to be successful one has to assume that the seeded failures are of the same type as the unknown failures.

Clearly, the number of variables involved is large and their relationship to failure rate far from precise. It is the author's view that actually implementing life-cycle activities in software Quality Assurance are more effective than attempts at prediction.

17.4 Software Quality Assurance (Life Cycle Activities)

Software QA, like hardware QA, is aimed at preventing failures. It is based on the observation that software failures are predominantly determined by the design. Experience in testing real-time software controlled systems shows that 50% of software 'bugs' result from unforeseen combinations of real-time operating events which the program instructions cannot accommodate. As a result, the algorithm fails to generate a correct output or instruction and the system fails.

Software QA is concerned with:

Organization of Software QA Effort (Section 17.4.1)
Documentation Controls (17.4.2)
Programming Standards (17.4.3)
Design Features (17.4.4)
Code Inspections and Walkthroughs (17.4.5)
Integration and Test (17.4.6)

The following sections outline these areas and this chapter concludes with a number of checklist questions suitable for use as an aide-memoire for audit or even as design guidelines.

17.4.1 Organization of Software QA

There needs to be an identifiable organizational responsibility for software QA. The important point is that the function can be identified. In a small organization, individuals often carry out a number of tasks. It should be possible to identify written statements of responsibility for software QA, the maintenance of standards and the control of changes.

There should be a quality manual, quality plans and specific test documents controlled by QA independently of the project management. They need not be called by those names and may be contained in other documents. It is the intent that is important. Main activities should include:

> configuration control
> library of media and documentation
> design review
> auditing
> test planning.

17.4.2 Documentation Controls

There must be an integrated hierarchy of specification/documents that translate the functional requirements of the product through successive levels of detail to the actual source code. In the simplest case this could be satisfied by:

> a functional description; and
> a flowchart or set of high-level statements; and
> a program listing.

In more complex systems there should be a documentation hierarchy because it is the only way to view to the software and without it there is nothing to review or audit. The design must focus onto a user requirements specification, which is the starting point in a top-down approach.

In auditing software it is important to look for such a hierarchy and to establish a diagram similar to Figure 17.4, which reflects the product, its specifications and numbering system. Failure to obtain this information is a sure indicator that software is being produced with less than adequate controls. Important documents are:

- *User requirements specification*: describes the functions required of the system. It should be unambiguous and complete and should describe what is required and not how it is to be achieved. It should be quantitative, where possible, to facilitate test planning. It states *what* is required and must not pre-empt and hence constrain the design.

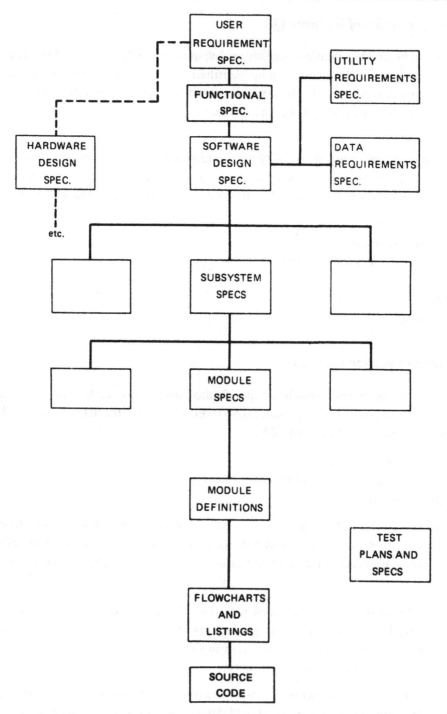

Figure 17.4: The document hierarchy

- *Functional specification*: whereas the user requirements specification states *what* is required, the functional specification outlines *how* it will be achieved. It is usually prepared by the developer in response to the requirements.
- *Software design specification*: takes the above requirements and, with regard to the hardware configuration, describes the functions of processing that are required and addresses such items as language, memory requirements, partitioning of the program into accessible subsystems, inputs, outputs, memory organization, data flow, etc.
- *Subsystem specification*: this should commence with a brief description of the subsystem function. Interfaces to other subsystems may be described by means of flow diagrams.
- *Module specification*: treating the module as a black box, it describes the interfaces with the rest of the system and the functional performance as perceived by the rest of the software.
- *Module definition*: describes the working of the software in each module. It should include the module test specification, stipulating details of input values and the combinations that are to be tested.
- *Charts and diagrams*: a number of techniques are used for charting or describing a module. The most commonly known is the flowchart, shown in Figure 17.5. There are, however, alternatives, particularly in the use of high-level languages. These involve diagrams and pseudo-code.
- *Utilities specification*: this should contain a description of the hardware requirements, including the operator interface and operating system, the memory requirements, processor hardware, data communications and software support packages.
- *Development notebooks*: an excellent feature is the use of a formal development notebook. Each designer opens a loose-leaf file in which are kept all specifications, listings, notes, change documentation and correspondence pertaining to that project.

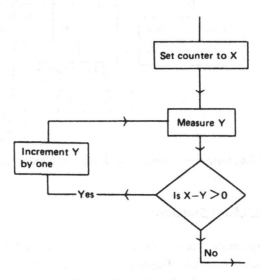

Figure 17.5: Flowchart

17.4.2.1 Change Control

As with hardware, the need to ensure that changes are documented and correctly applied to all media and program documents is vital. All programs and their associated documents should therefore carry issue numbers. A formal document and software change procedure is required (see Figure 17.6) so that all change proposals are reviewed for their effect on the total system.

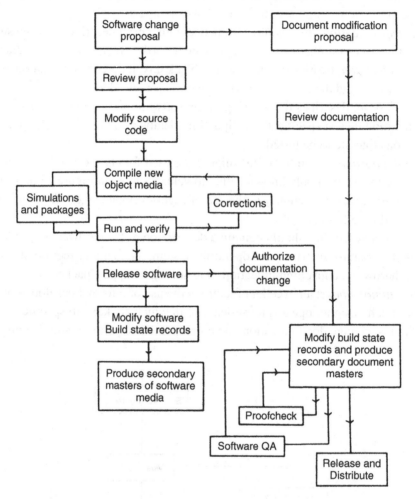

Figure 17.6: Software change and documentation procedure

17.4.3 Programming (Coding) Standards

The aim of structured programming is to reduce program complexity by using a library of defined structures wherever possible. The human brain is not well adapted to retaining random information and sets of standard rules and concepts substantially reduce the likelihood of error. A standard approach to creating files, polling output devices, handling interrupt routines, etc. constrains the programmer to use the proven methods. The use of specific subroutines is a further step in this direction. Once a particular sequence of program steps has been developed in order to execute a specific calculation, then it should be used as a library subroutine by the rest of the team. Re-inventing the wheel is both a waste of time and an unnecessary source of failure if an error-free program has already been developed.

A good guide is thirty to sixty lines of coding plus twenty lines of comment. Since the real criterion is that the module shall be no larger than to permit a total grasp of its function (that is, it is perceivable), it is likely that the optimum size is a line print page (three at most).

The use of standard sources of information is of immense value. Examples are:

Standard values for constants
Code templates (standard pieces of code for given flowchart elements)
Compilers

The objective is to write clear, structured software, employing well-defined modules whose functions are readily understood. There is no prize for complexity.

There are several methods of developing the module on paper. They include:

flow diagrams
hierarchical diagrams
structured box diagrams
pseudo-code.

17.4.4 Fault-Tolerant Design Features

Fault tolerance can be enhanced by attention to a number of design areas. These features include:

- Use of redundancy, which is expensive. The two options are dual processing and alternate path (recovery blocks).
- Use of error-checking software involving parity bits or checksums together with routines for correcting the processing.
- Timely display of fault and error codes.
- Generous tolerancing of timing requirements.
- Ability to operate in degraded modes.
- Error confinement. Programming to avoid error proliferation or, failing that, some form of recovery.
- Watchdog timer techniques involve taking a feedback from the microprocessor and, using that clocked rate, examining outputs to verify that they are dynamic and not stuck in one state. The timer itself should be periodically reset.

- Faults in one microprocessor should not be capable of affecting another. Protection by means of buffers at inputs and outputs is desirable so that a faulty part cannot pull another part into an incorrect state. Software routines for regular checking of the state (high or low) of each part may also be used.
- Where parts of a system are replicated the use of separate power supplies can be considered, especially since the power supply is likely to be less reliable than the replicated processor.

17.4.5 Reviews

There are two approaches to review of code:

1. *Code inspection*: where the designer describes the overall situation and the module functions to the inspection team. The team study the documentation and, with the aid of previous fault histories, attempt to code the module. Errors are sought and the designer then carries out any rework, which is then re-inspected by the team.
2. *The structured walkthrough*: in which the designer explains and justifies each element of code until the inspection team is satisfied that they agree and understand each module.

17.4.6 Integration and Test

There are various types of testing that can be applied to software:

- *Dynamic testing*: this involves executing the code with real data and I/O. At the lowest level this can be performed on development systems as is usually the case with module testing. As integration and test proceeds, the dynamic tests involve more of the actual equipment until the functional tests on the total equipment are reached. Aids to dynamic testing include automatic test beds and simulators, which are now readily available.
- *Path testing*: this involves testing each path of the software. In the case of flowcharted design there are techniques for 'walking through' each path and determining a test. It is difficult in a complex program to be sure that all combinations have been checked. In fact the number of combinations may be too high to permit all paths to be tested.
- *Software proving by emulation*: an 'intelligent' communications analyzer or other simulator having programable stimulus and response facilities is used to emulate parts of the system not yet developed. In this way the software can be made to interact with the emulator, which appears as if it were the surrounding hardware and software. Software testing can thus proceed before the total system is complete.
- *Functional testing*: the ultimate empirical test is to assemble the system and to test every possible function. This is described by a complex test procedure and should attempt to cover the full range of environmental conditions specified.

- *Load testing*: the situation may exist where a computer controls a number of smaller microprocessors, data channels or even hard-wired equipment. The full quantity of these peripheral devices may not be available during test, particularly if the system is designed for expansion. In these cases, it is necessary to simulate the full number of inputs by means of a simulator. A further micro- or minicomputer may well be used for this purpose. Test software will then have to be written which emulates the total number of devices and sends and receives data from the processor under test.

Be most suspicious of repeated slips in a test program. This is usually a symptom that the test procedure is only a cover for debug. Ideally, a complete error-free run of the test procedure is needed after debug, although this is seldom achieved in practice with large systems.

The practice of pouring in additional personnel to meet the project schedule is ineffective. The division of labor, below module level, actually slows down the project.

17.5 Modern/Formal Methods

The traditional software QA methods, described in the previous section, are essentially open-ended checklist techniques. They have been developed over the last fifteen years but would be greatly enhanced by the application of more formal and automated methods. The main problem with the existing open-ended techniques is that they provide no formal measures as to how many of the hidden errors have been revealed.

The term 'formal methods' is much used and much abused. It covers a number of methodologies and techniques for specifying and designing systems, both non-programable and programmable. They can be applied throughout the life cycle including the specification stage and the software coding itself.

The term is used here to describe a range of mathematical notations and techniques applied to the rigorous definition of system requirements that can then be propagated into the subsequent design stages. The strength of formal methods is that they address the requirements at the beginning of the design cycle. One of the main benefits of this is that formalism applied at this early stage may lead to the prevention, or at least early detection, of incipient errors. The cost of errors revealed at this stage is dramatically less than if they are allowed to persist until commissioning or even field use. This is because the longer they remain undetected the more serious and far reaching are the changes required to correct them.

The three major quality problems with software are illustrated in Figure 17.7. First, the statement of requirements is in free language and thus the opportunity for ambiguity, error and omission is at a maximum. The very free language nature of the requirements makes it impossible to apply any formal or mathematical review process at this stage. It is well known that the majority of serious software failures originate in this part of the design cycle.

Second, the source code, once produced, can only be reviewed by open-ended techniques as described in Section 17.4.4. Again, the discovery of ten faults gives no clue as whether one, ten or one hundreed remain. Third, the use of the software (implying actual execution of the code) is effectively a very small sample of the execution paths and input/output combinations that are possible in a typical piece of real-time software. Functional test is, thus, only a small contribution to the validation of a software system.

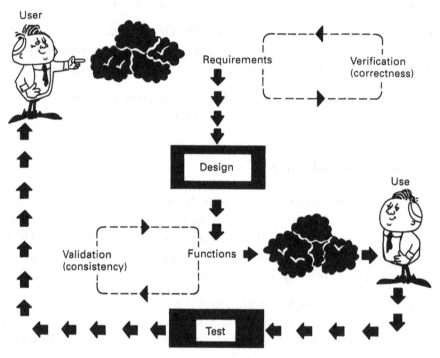

Figure 17.7: The quality problem

In these three areas of the design cycle, there are specific developments, described in the following sections:

17.5.1 Requirements Specification and Design

There is emerging a group of design languages involving formal graphical and algebraic methods of expression. For requirements, such tools as VDM (Vienna Development Method), OBJ (Object Oriented Code) and Z (a method developed at Oxford University) are now in use. They require formal language statements and, to some extent, the use of Boolean expressions. The advantage of these methods is that they substantially reduce the opportunity for ambiguity and omission and provide a more formal framework against which to validate the requirements.

Especial interest in these methods has been generated in the area of safety-related systems in view of their potential contribution to the safety-integrity of systems in whose design they are used.

The potential benefits are considerable but they cannot be realized without properly trained people and appropriate tools. Formal methods are not easy to use. As with all languages, it is easier to read a piece of specification than it is to write it. A further complication is the choice of method for a particular application. Unfortunately, there is not a universally suitable method for all situations.

Formal methods are equally applicable to the design of hardware and software. In fact they have been successfully used in the design of large-scale integration electronic devices such as, for example, the Viper chip produced by RSRE in Malvern, UK.

It should always be borne in mind that establishing the correctness of software, or even hardware, alone is no guarantee of correct system performance. Hardware and software interact to produce a system effect and it is the specification, design and validation of the system that matters. This system-wide view should also include the effects of human beings and the environment.

The potential for creating faults in the specification stage arises largely from the fact that it is carried out mainly in natural language. On one hand this permits freedom of expression and comprehensive description but, on the other, leads to ambiguity, lack of clarity and little protection against omission. The user communicates freely in this language, which is not readily compatible with the formalism being suggested here.

17.5.2 Static Analysis

This involves the algebraic examination of source code (not its execution). Packages are available (such as MALPAS from W.S. Atkins, Farnham, Surrey) that examine the code statements for such features as:

> the graph structure of the paths
> unreachable code
> use of variables
> dependency of variables upon each other
> actual semantic relationship of variables.

Consider the following piece of code:

```
BEGIN
INTEGER A, B, C, D, E
A:=0
NEXT: INPUT C:
IF C < 0 THEN GOTO EXIT:
B:=B+C
D:=B/A
GOTO NEXT:
PRINT B, D;
EXIT: END;
```

Static analysis will detect that:

(i)   B is not initialized before use.
(ii) E is never used
(iii) A is zero and is used as a divisor
(iv) The PRINT B, D; command is never used because of the preceding statement.

Static analysis is extremely powerful in that it enables the outputs of the various analyzers to be compared with the specification in order to provide a formal review loop between code and specification. A further advantage is that static analysis forces the production of proper specifications since they become essential in order to make use of the analyzer outputs.

Figure 17.8 shows the packages of MALPAS (one such static analysis tool). It acts on the source code and control flow analysis identifies the possible entry and exit points to the module, pieces of unreachable code and any infinitely looping dynamic halts. It gives an initial feel for the structure and quality of the program. Data use analysis identifies all the inputs and outputs of the module and checks that data are being correctly handled. For example, it checks that each variable is initialized before being used. Information flow analysis deduces the information on which each output depends. The path assessor is used to provide a measure of the complexity in that the number of paths through the code is reported for each procedure. Semantic analysis identifies the actions taken on each feasible path through a procedure. In particular, it rewrites imperative, step-by-step procedures into a declarative, parallel assignment form. The analyst can use this to provide an alternative perspective on the function of the procedure. The result of the analyzer is to tell the analyst

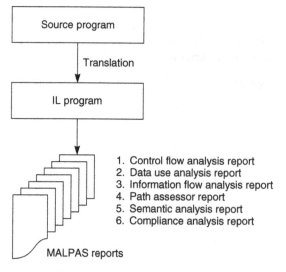

Figure 17.8: The MALPAS suite

the actual relationship of the variables to each other. Compliance analysis attempts to prove that a procedure satisfies a specified condition. For example, it could be used to check that the result of the procedure 'sort' is a sequence of items where each item is bigger than the preceding one. The report from the compliance analysis identifies those input values for which the procedure will fail.

17.5.3 Test Beds

During dynamic testing (involving actual execution of the code), automated 'test beds' and 'animators' enable testing to proceed with the values of variables being displayed alongside the portions of code under test. Numerous test 'tools' and so-called environments are commercially available and continue to be developed.

17.6 Software Checklists

17.6.1 Organization of Software QA

1. Is there a senior person with responsibility for software QA and does he or she have adequate competence and authority to resolve all software matters?
2. Is there evidence of regular reviews of software standards?
3. Is there a written company requirement for the planning of a software development?
4. Is there evidence of software training?
5. Is there a quality manual or equivalent documents?
6. Is there a system for labeling all software media?
7. Is there a quality plan for each development including:

> organization of the team
> milestones
> codes of practice
> QC procedures, including release
> purchased software
> documentation management
> support utilities
> installation
> test strategy.

8. Is there evidence of documented design reviews? The timing is important. So-called reviews that are at the completion of test are hardly design reviews.
9. Is there evidence of defect reporting and corrective action?
10. Are the vendor's quality activities carried out by people not involved in the design of the product that they are auditing?
11. Is there a fireproof media and file store?
12. Are media duplicated and separately stored?

17.6.2 Documentation Controls

1. Is there an adequate structure of documentation for the type of product being designed?
2. Do all the documents exist?
3. Do specifications define what must not happen as well as what must?
4. Is there a standard or guide for flowcharts, diagrams or pseudo-code in the design of modules?
5. Are there written conventions for file naming and module labeling?
6. Is there a person with specific responsibility for documentation control?
7. Is there a person with specific responsibility for change control?
8. Is there a distribution list for each document?
9. Are there established rules for the holding of originals?
10. Are all issues of program media accurately recorded?
11. Is there a system for the removal and destruction of obsolete documents from all work areas?
12. Are media containing non-conforming software segregated and erased?

17.6.3 Programming Standards

1. Is there a library of common program modules?
2. Is the 'top-down' approach to software design in evidence?
3. Is high-level or low-level language used? Has there been a conscious justification?
4. Is there a document defining program standards?
5. Is there reference to structured programming?
6. Is each of the following covered:

> block lengths
> size of codable units (module size)
> use of globals
> use of GOTO statements
> file, operator error, and unauthorized use security
> recovery conventions
> data organization and structures
> memory organization and backup
> error-correction software
> automatic fault diagnosis
> range checking of arrays
> use of PROM, EPROM, RAM, DISC, etc.
> structured techniques
> treatment of variables (that is, access)
> coding formats

      code layout
      comments (REM statements)
      rules for module identification.

17.6.4 Design Features

1. Is there evidence that the following are taken into consideration:

      electrical protection (mains, airborne)
      power supplies and filters
      opto isolation, buffers
      earthing
      battery backup
      choice of processors
      use of language
      rating of I/O devices
      redundancy (dual programming)
      data communications
      human/machine interface
      layout of hardware
      hardware configuration (e.g. multidrops)
      watchdog timers
      RAM checks
      error confinement
      error detection
      error recovery.

2. Are there syntax- and protocol-checking algorithms?
3. Are interfaces defined such that illegal actions do not corrupt the system or lock up the interface?
4. Are all data files listed (there should be a separate list)?
5. Were estimates of size and timing carried out?
6. Are the timing criteria of the system defined where possible?
7. Will it reconstruct any records that may be lost?
8. Are there facilities for recording system state in the event of failure?
9. Have acceptable degraded facilities been defined?
10. Is there a capability to recover from random jumps resulting from interference?
11. Are the following adequate:

      electrical protection (mains and e.m.i.)
      power suppliers and filters
      earthing.

12. Is memory storage adequate for foreseeable expansion requirements?
13. Are data link lengths likely to cause timing problems?
14. Are the following suitable for the application in hand:

> processor
> peripherals
> operating system
> packaging.

15. Is there evidence of a hardware/software trade-off study?
16. Is use made of watchdog timers to monitor processors?

> coding formats
> code layout
> comments (REM statements)
> rules for module identification.

17.6.5 Code Inspections and Walkthroughs

1. Are all constants defined?
2. Are all unique values explicitly tested on input parameters?
3. Are values stored after they are calculated?
4. Are all defaults explicitly tested on input parameters?
5. If character strings are created are they complete? Are all delimiters shown?
6. If a parameter has many unique values, are they all checked?
7. Are registers restored on exits from interrupts?
8. Should any register's contents be retained when re-using that register?
9. Are all incremental counts properly initialized (0 or 1)?
10. Are absolute addresses avoided where there should be symbolics?
11. Are internal variable names unique or confusing if concatenated?
12. Are all blocks of code necessary or are they extraneous (e.g. test code)?
13. Are there combinations of input parameters that could cause a malfunction?
14. Can interrupts cause data corruption?
15. Is there adequate commentary (REM statements) in the listing?
16. Are there time or cycle limitations placed on infinite loops?

17.6.6 Integration and Test

1. Are there written requirements for testing subcontracted or proprietary software?
2. Is there evidence of test reporting and remedial action?
3. Is there evidence of thorough environmental testing?
4. Is there a defect-recording procedure in active use?

5. Is there an independent test manager appointed for the test phase of each development program?
6. Is there a comprehensive system of test documentation (e.g. test plans, specifications, schedules) for each product?
7. Is there an effective system of calibration and control of test equipment?
8. Do test plans indicate a build-up of testing (e.g. module test followed by subsystem test followed by system test)?
9. Do test schedules permit adequate time for testing?
10. Is there evidence of repeated slip in the test program?
11. To what extent are all the paths in the program checked?
12. Does the overall design of the tests attempt to prove that the system behaves correctly for improbable real-time events (e.g. misuse tests)?

These checklists provide a brief overview of ways to limit systematic failures. In the case of safety-related applications, such lists have been extensively developed as a result of IEC 61508 (Functional Safety). This is dealt with in Chapter 22 and more fully in *The Safety Cirtical Systems Handbook* (*A straightforward guide to functional Safety IEC 61508*) 3rd edition, 2010, Smith D. J. and Simpson K. G. L., Butterworth-Heinemann, ISBN 9780080967813.

Legal, Management and Safety Considerations

Project Management and Competence

18.1 Setting Objectives and Making Specifications

Realistic reliability, maintainability and safety (RAMS) objectives need to be set with due regard to the customer's design and operating requirements and cost constraints. In the case of contract development or plant engineering, these are likely to be outlined in a tender document or a requirements specification. Liaison with the customer may be required to establish optimum economic reliability values that sensibly meet his or her requirements and which are achievable within the proposed technology at the costs allowed for. Over-specifying the requirement may delay the project when tests eventually show that objectives cannot be met and it is realized that budgets will be exceeded. It is not uncommon for safety targets to be overstated due to ill-informed targetting methods. Chapter 22 will address safety-integrity targets.

When specifying a failure rate (or MTBF) it is a common mistake to state a confidence level. However, the requirement stands alone and the addition of a confidence level implies a statistical demonstration and supposes that the failure rate or MTBF would be established by a single demonstration at the stated confidence. On the contrary, a design objective is a target and must be stated without statistical limitations.

Vague statements such as 'high reliability' and 'the highest quality' should be avoided at all costs. They are totally subjective and cannot be measured. Therefore they cannot be demonstrated or proved.

Consideration of the equipment type and the use to which it is put will influence the parameters chosen. Remember the advice given in Chapter 2 about the meaning and applicability of failure rate, MTBF, availability, MTTR, etc.

A major contribution to the problems associated with reliability and quality comes from the lack of (or inadequacy of) the engineering design specification. It should specify the engineering requirements in full, including reliability and MTTR parameters. These factors should include:

1. *Functional description*: speeds, functions (including safety-related functions), human interfaces and operating periods.

Reliability, Maintainability and Risk. DOI: 10.1016/B978-0-0809-6902-2.00018-0

2. *Environment*: temperature, humidity, etc.
3. *Design life*: related to wearout and replacement policy.
4. *Physical parameters*: size and weight restrictions, power supply limits.
5. *Standards*: BS, US MIL, Def Con, etc., standards for materials, components and tests.
6. *Finishes*: appearance and materials.
7. *Ergonomics*: human limitations and safety considerations.
8. *Reliability, availability, maintainability and safety integrity*: module reliability and MTTR objectives. Equipment R and M related to module levels.
9. *Manufacturing quantity*: projected manufacturing levels – first off, batch, flow.
10. *Maintenance philosophy*: type and frequency of preventive maintenance. Repair level, method of diagnosis, method of second-line repair.

18.2 Planning, Feasibility and Allocation

The design and assurance activities described in this book simply will not take place unless there is real management understanding and commitment to a reliability and maintainability program with specific resources having been allocated. Responsibilities have to be placed on individuals for each of the activities and a reliability program manager appointed with sufficient authority and the absence of conflicting priorities (e.g. program dates) to control the RAM activities. Milestones, with dates, will be required, against which progress can be measured as, for example:

> completion of feasibility study (including RAM calculations)
> reliability objectives for modules and for bought-out items allocated
> test specification prepared and agreed
> prototype tests completed
> modifications arising from tests completed
> demonstrations of reliability and maintainability
> design review dates.

The purpose of a feasibility study is to establish if the performance specification can be met within the constraints of cost, technology, time and so on. This involves a brief (high level) reliability prediction, based perhaps on a block diagram approach, in order to decide if the design proposal has a reasonable chance of being engineered to meet the requirements. Allocation of objectives has been emphasized in Chapter 11 and is important if the objectives are not to be met by a mixture of over- and under-design.

It is useful to remember that there are three levels of RAM measurement:

> *Prediction*: a modeling exercise that relies on the validity of historical failure rates to the design in question. This provides the lowest level of confidence. (See Chaper 4.)

Statistical demonstration test: this provides sample failure information (perhaps even zero failures in a given amount of time). It is usually in a test rather than field environment. Whilst providng more confidence than paper *prediction* it is still subject to statistical risk and the limitations of a test environment. (See Chaper 5.)

Field data: except in the case of very-high-reliability systems (e.g. submerged cable and repeater), realistic numbers of failures are obtained and can be used in a reliability growth program as well as for comparison with the original targets. (See Chaper 13.)

18.3 Program Activities

The extent of the reliability and maintainability activities in a project will depend upon:

The severity of the requirement.
The complexity of the product.
Time and cost constraints.
Consequences of failure (death, production loss, environment etc).
Safety-integrity targets.
The number of items to be produced.

A safety and reliability plan must be produced for each project or development. Without this there is nothing against which to audit progress and, therefore, no formal measure of progress towards the targets. Figure 18.1 re-iterates the RAMS Design Cycle, which provides a model against which to review the activities. Figure 1.2, in Section 1.5, gave more detail.

These have all been covered in the book and include:

* *Feasibility study*: an initial 'prediction' to ascertain if the targets are realistic or impossible.
* *Setting objectives*: discussed above with allocation and feasibility.
* *Contract requirements*: the formal agreement on the RAM targets, warranty, acceptance criteria, etc.
* *Design reviews*: these are intended to provide an evaluation of the design at defined milestones. The design review team should include a variety of skills and be chaired by a person independent of the design team. The following checklist is a guide to the factors which might be considered:
  1. Electrical factors involving critical features, component standards, circuit trade-offs, etc.
  2. Software reliability including configuration control, flowcharts, user documentation, etc.
  3. Mechanical features such as materials and finish, industrial design, ergonomics, equipment practice and so on.
  4. Quality and reliability covering environmental testing, RAM predictions and demonstrations, FMECA, test equipment and procedures, trade-offs, etc.

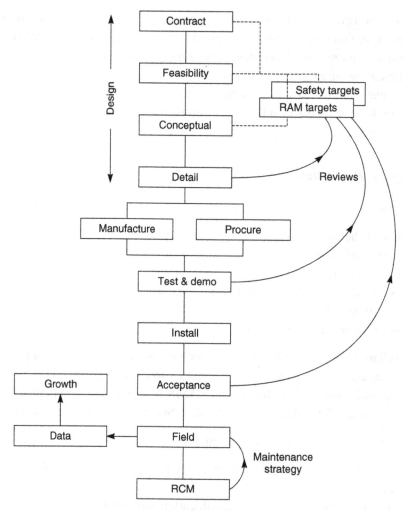

Figure 18.1: RAMS cycle

5. Maintenance philosophy including repair policy, MTTR prediction, maintenance resource forecasts, customer training and manuals.
6. Purchased items involving lead times, multiple sourcing, supplier evaluation and make/buy decisions.
7. Manufacturing and installation covering tolerances, burn-in, packaging and transport, costs, etc.
8. Other items include patents, value engineering, safety, documentation standards and product liability.

•   *RAMS predictions*: this focuses attention on the critical failure areas, highlights failures that are difficult to diagnose and provides a measure of the design reliability against the objectives. FMEA, FTA and other modeling exercises are used, in the design reviews, to measure conformance to the RAMS targets.

- *Design trade-offs*: these may be between R and M and may involve sacrificing one for the other as, for example, between the reliability of the wrapped joint and the easy replaceability of a connector. Major trade-offs will involve the design review whereas others will be made by the designer.
- *Prototype tests*: these cover marginal, functional, parametric, environmental and reliability tests. It is the first opportunity to observe reliability in practice and to make some comparison against the predictions.
- *Parts selection and approval*: involves field tests or seeking field information from other users. The continued availability of each part is important and may influence the choice of supplier.
- *Demonstrations*: since these involve statistical sampling, test plans have to be calculated at an early stage so that the risks can be evaluated.
- *Spares provisioning*: this affects reliability and maintainability and has to be calculated during design.
- *Data collection and failure analysis*: failure data, with the associated stress information, are essential to reliability growth programs and also to future predictions. A formal failure-reporting scheme should be set up at an early stage so that tests on the earliest prototype modules contribute towards the analysis.
- *Reliability growth*: establishing reporting and analysis to confirm that field reliability growth meets targets.
- *Training*: design engineers should be trained to a level where they can work with the R and M specialist. Customer training of maintenance staff is another aspect which may arise.

18.4 Responsibilities and Competence

RAMS are an integral part of the design process. In many cases mere lip service is given and this leads to little more than high-level predictions being carried out too late in the design. These have no effect whatever in bringing the design nearer to the targets. Reliability and maintainability are engineering parameters and the responsibility for their achievement is therefore primarily with the design team. Quality assurance techniques play a vital role in achieving the goals but cannot be used to 'test in' reliability to a design that has its own inherent level. Three distinct responsibilities therefore emerge that are complementary but do not replace each other. See Figure 18.2.

The need for adequate competency implies a number of factors including:

- responsibilities and level of supervision
- the link between severity of consequences and degree of competence
- the link between severity of the target and the degree of competence

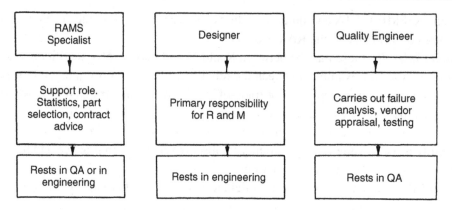

Figure 18.2: Responsibilities

- the link between design novelty and rigor of competence
- relevance of previous experience
- engineering application knowledge
- technology knowledge
- safety engineering knowledge
- legal/regulatory knowledge
- relevance of qualifications
- the need for training to be documented.

An early guidance document in this area was the IET/BCS *Competency Guidelines for Safety-related Systems Practitioners*, which listed twelve safety-related job functions (described as functions) broken down into specific tasks. Guidance was provided on setting up a review process and in assessing capability. The three levels of competence described in the document are:

- *The Supervised Practitioner* who can carry out one of the above jobs but requiring review of the work.
- *The Practitioner* who can work unsupervised and can manage and check the work of a Supervised Practitioner.
- *The Expert* who will be keeping abreast of the state of art and will be able to tackle novel scenarios.

More recently, the HSE document (2007) *Managing competence for safety-related systems* was produced in co-operation with the IET and the BCS. In outline its structure is:

- *Phase one*: plan (define purpose and scope).
- *Phase two*: design, competence criteria, processes and methods.
- *Phase three*: operate, select and recruit, assess competence, develop competence, assign responsibilities, monitor, deal with failure, manage assessors' and managers' competence, manage supplier competence, manage information, manage change.

- *Phase four*: audit & review.

In order to implement and control competency, a competency register is essential. Experience and training should be logged so that individuals can be assessed for the suitability to carry out tasks as defined in the company's procedure. Figure 18.3 shows a typical format for an assessment document for each person. These would form the competency register within an organization.

Name	Xxxxxxx	
Qualifications	BSc, MSc in Safety (xx University)	
Date of employment	Xxxxxxx	
Training	In-house appreciation course Technis certificate in R&FS (distinction)	May 2008 April 2010
Professional	Paper on QRA and maximum tolerable risk comparisons (SaRS Journal)	2009
Task in the life-cycle	**Experience**	**Level of expertise (as defined in company procedure)**
Risk analysis	Lead SIL determination team 5 processes (2009)	FS Manager
Requirements	Reviewed requ's specs for new instrumentation (ESD and HIPPs systems) and drafted FS requ's	FS Assessor
Design	No experience to date	N/A
Assessment	Introduced fault tree tool and carried out 6 assessments of ESD systems against SIL targets. Analyzed field data over a 3-year period and produced failure rate sheet for instruments and actuators	FS Manager
Regulatory	Attended 3 meetings with HSE representatives: a) Review of human factors elements of company safety submissions b) Review of SIL targets c) Review of life-cycle claims	FS Assessor
etc	etc	
etc	etc	
Training needs	Design of ESD architectures and choice of instrumentation to meet SIL targets Review of life-cycle techniques and measures	
Last review	31 May 2010 by xxx and yyy	

Figure 18.3: Competency register entry

18.5 *Functional Safety Capability*

IEC 61508 (Functional Safety) has led to the activities associated with achieving functional safety becoming an area of specific responsibility. Although largely achieved as a result of working to an ISO 9000-based management system, they nevertheless require an additional procedure(s) to provide the detail.

The main items can be summarized as follows:

Contract/project review: all contracts, prior to acceptance by a company, need to be examined to ascertain if they involve safety-related requirements. These requirements may be stated directly by the client or may be implicit by reference to some standard. Clients may not always use appropriate terms to refer to safety-related applications or safety-integrity requirements. Therefore, a competent person will need to decide if a contract is safety related or not. Similarly, all major modifications, new product designs or major projects have to be assessed in this way.

Control of modifications: change proposals need to be declared safety related or not. The change proposal document must therefore have a space dedicated to describing the impact of the change, which must be declared safety related or not by a competent person.

Project management (i.e. control of specific tasks): every project requires a 'safety authority' who is independent of the actual design tasks. Again, it has to be someone who is deemed competent to plan, instigate and review all the functional safety assessment activities associated with the project.

Competence: the references to competence (in the RAMS context) imply that all the safety-related tasks (e.g. carrying out assessments, doing audits, acting as project manager) have been defined and that a 'register' is maintained identifying those who are able to carry out each task. This needs to be reviewed on a regular basis and used to select individuals for each task. It will also identify training needs (or the need for consultancy assistance) where all the skills are not currently available in-house.

Remedial action: the defect recording documentation (used in test, commissioning and in the field) will also need to cater, in the same way as the change proposal, for identifying the safety-related implications of incidents and failures. Responsibilities for carrying out remedial action will also need to be defined.

Functional safety validation: this consists of a positive demonstration that all the safety-related requirements have been achieved. It is usually satisfied by the satisfactory completion of the final acceptance test together with a cross-reference to each of the requirements. A matrix is often generated to fulfil this need. In addition, a documented close out of all the outstanding remedial actions from reviews, tests, defects and changes can be appended to the matrix.

This topic is addressed in detail in *The Safety Critical Systems Handbook*, D. J. Smith and K. G. L. Simpson, 3rd edition, ISBN 9780080967813.

18.6 Standards and Guidance Documents

There are numerous standards that might be called for. The more frequently quoted, although by no means new, are:

- BS 5760: *Reliability of systems, equipment and components*: this is in a number of parts. Part 1 is Guide to Reliability Programme Management and outlines the reliability activities such as have been dealt with in this book. Other parts deal with prediction, data, practices and so on.
- UK Ministry of Defence Standard 00-40 *Reliability and maintainability*: this is in eight parts. Parts 1 and 2 are concerned with project requirements and the remainder with requirements documents, training, procurement and so on.
- US Military Standard 785A *Reliability Program for Systems and Equipment Development and Production*: specifies program plans, reviews, predictions and so on.
- US Military Standard 470 *Maintainability Program Requirements*: a document, from 1966, which covers the program plan and specifies activities for design criteria, design review, trade-offs, data collection, predictions and status reporting.

CHAPTER 19

Contract Clauses and Their Pitfalls

19.1 Essential Areas

Since the late 1950s in the USA, reliability and maintainability requirements have appeared in both military and civil engineering contracts. These contracts often carry penalties for failure to meet these objectives. For over thirty years in the UK, suppliers of military and commercial electronic and telecommunication equipment have also found that clauses specifying reliability and maintainability are included in invitations to tender and in the subsequent contracts. For at least ten years the focus on safety-related requirements means that, today, the vast majority of contracts for plant and equipment specify these parameters, for hazardous failure modes. The knock-on effect is that suppliers of instrumentation are called upon to demonstrate conformance to some failure rate (or PFD) target and to show that they meet the criteria described in Chapter 22.

Suppliers of highly reliable and maintainable equipment are often well able to satisfy these requests with little or no additional design or manufacturing effort, but nevertheless incur costs since a formal demonstration of these parameters may be required. Furthermore, failure-reporting procedures might not exist and therefore historical data as to a product's reliability or repair time may not be readily available.

Including reliability and safety-related requirements in a contract involves the suppliers of both good and poor equipment in additional activities. System effectiveness clauses in contracts range from a few words – specifying availability, failure rate or MTBF of all or part of the system – to many pages containing details of design and test procedures, methods of collecting failure data, methods of demonstrating reliability and repair time, limitations on component sources, limits to size and cost of test equipment, and so on. Two main types of pitfall arise from such contractual conditions:

1. Those due to the omission of essential conditions or definitions.
2. Those due to inadequately worded conditions that present ambiguities, concealed risks, eventualities unforeseen by both parties, etc.

Attention to the following headings is essential if reliability or maintainability is to be specified.

Reliability, Maintainability and Risk. DOI: 10.1016/B978-0-08-096902-2.00019-2

19.1.1 Definitions

If a mean time to repair or down time is specified, then the meaning of repair time must be defined in detail. Mean time to repair is often used when it is mean down time that is intended (see Figure 2.8 in Chapter 2).

Failure itself must also be thoroughly defined at system and module levels. It may be necessary to define more than one type of failure (for example, total system failure or degradation failure) or failures for different operating modes (for example, in flight or on ground) in order to describe all the requirements. MTBFs might then be ascribed to the different failure types. MTBFs and failure rates often require clarification as to the meaning of 'failure' and 'time'. The latter may refer to operating time, revenue time, clock time, etc. Types of failure that do not count for the purpose of proving the reliability (for example, maintenance induced or environment outside limits) have also to be defined and specifically excluded.

For process-related equipment it is usual to specify availability. Unless, however, some failure modes are defined, the figures can be of little value. In a safety system, failure can consist of a spurious alarm or of failure to respond to a genuine demand. Combining those two failure rates produces a totally misleading figure and the two modes must be evaluated separately. Figure 19.1 reminds us of the bathtub curve with early, random and wearout failures. Reliability parameters usually refer to random failures unless stated to the contrary, it being assumed that burn-in failures are removed by screening and wearout failures eliminated by preventive replacement.

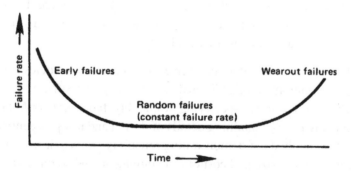

Figure 19.1: Bathtub curve

It should be remembered that the bathtub curve is a statistical picture and that, in practice, it is rarely possible to ascribe a particular failure to any of the three categories. It is therefore vital that, if reliability is being demonstrated by a test or in the field, these early and wearout failures are eliminated, as far as possible, by the measures already described. The specification should make clear which types of failure are being observed in a test.

Parameters should not be used without due regard to their meaning and applicability. Failure rate, for example, has no meaning except when describing random failures. Remember that in systems involving redundancy, constant failure rate may not apply except in the special cases outlined in Chapters 7–9. Availability, MTBF or reliability should then be specified in preference to failure rate.

Reliability and maintainability are often combined by specifying the useful parameter, availability. This can be defined in more than one way and should therefore be specifically defined. The usual form is the steady state availability, which is MTBF/(MTBF + MDT), where MDT is the mean down time.

19.1.2 Environment

A common mistake is not to specify the environmental conditions under which the product is to work. The specification is often confined to temperature range and maximum humidity, and this is not always adequate. Even these two parameters can create problems, as with temperature cycling under high-humidity conditions. Other stress parameters include pressure, vibration and shock, chemical and bacteriological attack, power supply variations and interference, radiation, human factors and many others. The combination or the cycling of any of these parameters can have significant results.

Where equipment is used in standby or held as spares, the environmental conditions will be different to those experienced by operating units. It is often assumed that because a unit is not powered, or in store, it will not fail. In fact the environment may be more conducive to failure under those circumstances. Self-generated heat and mechanical self-cleaning wiping actions are often important ingredients for reliability. If equipment is to be transported while the supplier is liable for failure, then the environmental conditions must be evaluated. On the other hand, over-specifying environmental conditions is a temptation that leads to over-design and higher costs. Environmental testing is expensive, particularly if large items of equipment are involved and if vibration tests are called for. These costs should be quantified by obtaining quotations from a number of test houses before any commitment is made to demonstrate equipment under environmental conditions.

Maintainability can also be influenced by environment. Conditions relating to safety, comfort, health and ergonomic efficiency will influence repair times since the use of protective clothing, remote-handling devices, safety precautions, etc. increases the active elements of repair time by slowing down the technician.

19.1.3 Maintenance Support

The provision of spares, test equipment, personnel, transport and the maintenance of both spares and test equipment is a responsibility that may be divided between supplier and customer or fall entirely on either. These responsibilities must be described in the contract and the supplier must be conscious of the risks involved in the customer not meeting his or her side of the agreement.

If the supplier is responsible for training the customer's maintenance staff then levels of skill and training details and competency have to be specified.

Maintenance philosophy, usually under customer control, plays a part in determining the achieved reliability. Periodic inspection of a non-attended system, during which failed redundant units are replaced, yields a different MTBF to the case of immediate repair of failed units, irrespective of whether they result in system failure. The maintenance philosophy must therefore be defined.

A contract may specify an MTTR supported by a statement such as 'identification of faulty modules will be automatic and will be achieved by automatic test means. No additional test equipment will be required for diagnosis.' This type of requirement involves considerable additional design effort in order to permit all necessary diagnostic signals to be made accessible and for measurements to be made. Additional hardware will be required in the form of either BITE or an 'intelligent' portable terminal with diagnostic capability. If such a requirement is overlooked when costing and planning the design the subsequent engineering delay and cost is likely to be considerable.

19.1.4 Demonstration and Prediction

The supplier might be called upon to give a statistical demonstration of either reliability or repair time. In the case of maintainability a number of corrective or preventive maintenance actions will be carried out and a given MTTR (or better) will have to be achieved for some proportion of the attempts. In this situation it is essential to define the tools and equipment to be used, the maintenance instructions, test environment and technician level. The method of task selection, the spares and the level of repair to be carried out also require stating. The probability of failing the test should be evaluated since some standard tests carry high supplier's risks. When reliability is being demonstrated then a given number of hours will be accumulated and a number of failures stated, above which the test is failed. Again, statistical risks apply and the supplier needs to calculate the probability of failing the test with good equipment and the customer that of passing inadequate goods.

Essential parameters to define here are environmental conditions, allowable failures (for example, maintenance induced), operating mode, preventive maintenance, burn-in, testing costs. It is often not possible to construct a reliability demonstration that combines sensible risks ($\leq 15\%$) for both parties with a reasonable length of test. Under these circumstances the acceptance of reliability may have to be on the basis of accumulated operating hours on previously installed similar systems.

An alternative to statistical or historical demonstrations of repair time and reliability is a guarantee period wherein all or part of the failure costs, and sometimes redesign costs, are borne by the supplier. In these cases great care must be taken to calculate the likely costs. It must be remembered that if 100 items of equipment meet their stated MTBF under random failure conditions, then after operating for a period equal to one MTBF, 63 of them, on average, will have failed.

From the point of view of producer's risk, a warranty period is a form of reliability demonstration since, having calculated the expected number of failures during the warranty, there is a probability that more will occur. Many profit margins have been absorbed by the unbudgeted penalty maintenance arising from this fact.

A reliability prediction is often called for as a type of demonstration. It is desirable that the data source is agreed between the two parties or else the 'numbers game' will ensue as specific component failure rates are 'negotiated' by each party seeking to turn the prediction to his or her favor.

19.1.5 Liability

The exact nature of the supplier's liability must be spelt out, including the maximum penalty that can be incurred. It is often the case that contracts will contain a clause stipulating that liquidated damages are to be payable upon non-performance or breach. In principle the amount of damages should be a genuine pre-estimate of the losses that would be experienced by the injured party upon such an event. The amount then is representative of genuine loss and thus the damages payable are deemed as compensatory and not punitive. The supplier should be aware of the amounts stated and how they are made up. This ensure that the claimed values are not disputed in the event of a failure to achieve the required performance, and resultant breach of contract.

If some qualifying or guarantee period is involved it is necessary to define when this commences and when the supplier is free of liability. The borders between delivery, installation, commissioning and operation are often blurred and therefore the beginning of the guarantee period will be unclear.

It is wise to establish a mutually acceptable means of Alternative Dispute Resolution (ADR) in case the interpretation of later events becomes the subject of a dispute. ADR can be achieved in a number of different ways but is usually done through mediation or arbitration. Whilst mediation is a less formal approach, whereby the parties in dispute are helped to achieve an amicable settlement by a mediator, arbitration is a formal process and is controlled In England and Wales by The Arbitration Act 1996.

If part of the liability for failure or repair is to fall on some other contractor, care must be taken in defining each party's area. The interface between equipment guaranteed by different suppliers may be physically easy to define but there exists the possibility of failures induced in one item of equipment owing to failure or degraded performance in another. This point should be considered where more than one supplier is involved.

19.2 Other Areas

The following items are often covered in a detailed invitation to tender.

19.2.1 Reliability and Maintainability Program

The detailed activities during design, manufacturing and installation are sometimes spelt out contractually. In a development contract this enables the customer to monitor the reliability and maintainability design activities and to measure progress against agreed milestones. Sometimes standard program requirements are used, for example:

US Military Standard 470, Maintainability Program Requirements
US Military Standard 785, Requirements for Reliability Program
BS 4200: Part 5 Reliability programmes for equipment
BS 5760 Reliability of constructed and manufactured products, systems, equipment and components
The life-cycle requirements of *IEC 61508 (2010) - Functional Safety.*

Typical activities specified are:

Prediction: data sources, mathematical models.
Testing: methods and scheduling of design, environmental and other tests.
Design review: details of participation in design reviews.
Failure mode and effect analysis: details of method and timing.
Failure reporting: failure reporting documents and reporting procedures.

19.2.2 Reliability and Maintainability Analysis

The supplier may be required to offer a detailed reliability or maintainability prediction together with an explanation of the techniques and data used. Alternatively, a prediction may be requested using defined data and methods of calculation. Insistence on optimistic data makes it more difficult to achieve the predicted values whereas pessimistic data leads to over-design.

19.2.3 Storage

The equipment may be received by the customer and stored for some time before it is used under conditions different to normal operation. If there is a guarantee period then the storage conditions and durations will have to be defined. The same applies to storage and transport of spares and test equipment.

19.2.4 Design Standards

Specific design standards are sometimes described or referenced in contracts or their associated specifications. These can cover many areas, including:

> printed-board assemblies – design and manufacture
> wiring and soldering
> nuts, bolts and threads
> finishes
> component ratings
> packaging.

A problem exists that these standards are very detailed and most manufacturers have their own version. Although differences exist in the fine detail they are usually overlooked until some formal acceptance inspection takes place, by which time retrospective action is difficult, time-consuming and costly.

19.2.5 Safety-Related Equipment

Clauses relating to the safety-integrity (i.e. the reliability in respect of hazardous failure modes) of equipment are now commonplace. Sometimes specific integrity requirements are made for defined failure modes. These may be expressed as qualitative 'probability of failure on demand' or 'failure rate' targets. Alternatively they may be expressed as a SIL requirement as defined in IEC 61508 (see Chapter 22). Examples might be:

- The probability of failure on demand of the shut down system in response to outlet overpressure shall not exceed 5×10^{-3}.
- The failure rate in respect of delivering unignited gas from the appliance shall not exceed 3×10^{-3} pa.
- The safety-integrity of the emergency shutdown system with respect to failure to respond to any valid input shall meet the requirements of SIL 2 of IEC 61508.

Specific failure modes are not always defined by the customer as, for example:

- The emergency shutdown system shall meet the requirements of SIL 2 of IEC 61508.

It then becomes necessary to establish, with the customer, what the hazardous failure modes are of the application for which he is obtaining the equipment.

Even less specific (and indeed meaningless) requirements are those such as:

- The emergency shutdown system shall meet the requirements of IEC 61508.

Here, it may become necessary for the vendor to supply the expertise needed to establish hazards (i.e. HAZOP) and to target the integrity levels needed to meet credible risk criteria.

19.3 Pitfalls

The previous sections have dealt with aspects of reliability and maintainability likely to be mentioned in an invitation to tender or in a contract. There are pitfalls associated with the omission or inadequate definition of these factors and some of the more serious are outlined below.

19.3.1 Definitions

The most likely area of dispute is the definition of what constitutes a failure and whether or not a particular incident ranks as one or not. There are levels of failure (system, unit, etc.), types of failure (catastrophic, degradation, etc.), causes of failure (random, systematic, over-stress, etc.) and there are effects of failure (dormant, hazardous, etc.). For various combinations of these, different MTBF and MTTR objectives with different penalties may be set. It is seldom sufficient, therefore, to define failure as not performing to specification since there are so many combinations covered by that statement. Careful definition of the failure types covered by the contract is therefore important.

19.3.2 Repair Time

It was shown in Chapter 2 that repair times could be divided into elements. Initially they can be grouped into active and passive elements and, broadly speaking, the active elements are dictated by system design and the passive by maintenance and operating arrangements. For this reason, the supplier should never guarantee any part of the repair time that is influenced by the user.

19.3.3 Statistical Risks

A statistical maintainability test is described by a number of repair actions and an objective MTTR that must not be exceeded on more than a given number of attempts. A reliability test involves a number of hours and a similar pass criterion of a given number of failures. In both cases producer and consumer risks apply, as explained in earlier chapters, and unless these risks are calculated they can prove to be unacceptable. Where published test plans are quoted, it is never a bad thing to recalculate the risks involved. It is not difficult to find a test that requires the supplier to achieve an MTBF 50 times the value which is to be proved in order to stand a reasonable chance of passing the test.

19.3.4 Quoted Specifications

Sometimes a reliability or maintainability program or test plan is specified by calling up a published standard. Definitions are also sometimes dealt with in this way. The danger with blanket definitions lies in the possibility that not all the quoted terms are suitable and that the standards will not be studied in every detail.

19.3.5 Environment

Environmental conditions affect both reliability and repair times. Temperature and humidity are the most usual factors to be specified and the problems associated with cycling have already been pointed out. If other factors are likely to be present in field use then they must either be specifically excluded from the range of environments for which the product is guaranteed or included, and therefore allowed for in the design and in the price. It is not desirable to specify every possible parameter, since this leads to over-design.

19.3.6 Liability

When stating the supplier's liability it is important to establish its limit in terms of both cost and time. Suppliers must ensure that they know when they are finally free of liability.

19.3.7 In Summary

The biggest pitfall of all is to assume that either party wins any advantage from ambiguity or looseness in the conditions of a contract. In practice, the hours of investigation and negotiation which ensue from a dispute far outweigh any advantage that might have been secured, to say nothing of the loss of goodwill and reputation. If every effort is made to cover all the areas discussed as clearly and simply as possible, then both parties will gain.

19.4 Penalties

There are various ways in which a penalty may be imposed on the basis of maintenance costs or the cost of system outage. In the case of liquidated damages it must be remembered that any cash penalty must be a genuine and reasonable pre-estimate of the damages thought to result. Some alternatives are briefly outlined.

19.4.1 Apportionment of Costs During Guarantee

Figure 19.2(a) illustrates the method where the supplier pays the total cost of corrective maintenance during the guarantee period. He or she may also be liable for the cost of redesign made necessary by systematic failures. In some cases the guarantee period recommences for those parts of the equipment affected by modifications. Whilst it must be borne in mind that the injured party has a duty in law to mitigate and limit any losses incurred due to a breach of contract, in practice a disadvantage of this arrangement is that it gives the customer no great incentive to minimize maintenance costs until the guarantee has expired. If the maintenance

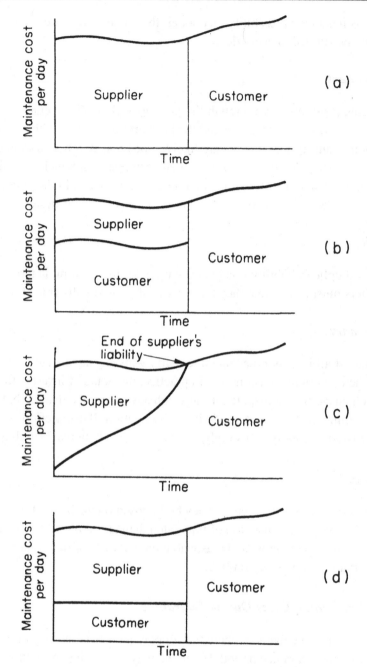

Figure 19.2: Methods of planning penalties

is carried out by the customer and paid for by the supplier then the latter's control over the preventive maintenance effectiveness is minimal. The customer should never be permitted to benefit from poor maintenance, for which reason this method is not very desirable.

An improvement of this is obtained by Figure 19.2(b), whereby the supplier pays a proportion of the costs during the guarantee and both parties therefore have an incentive to minimize costs. In Figure 19.2(c) the supplier's proportion of the costs decreases over the liability period. In Figure 19.2(d) the customer's share of the maintenance costs remains constant and the supplier pays the excess. The arrangements in (b) and (c) both provide mutual incentives. Arrangement (d), however, provides a mixed incentive. The customer has, initially, a very high incentive to reduce maintenance costs but once the ceiling has been reached this disappears. On the other hand, (d) recognizes the fact that for a specified MTBF the customer should anticipate a given amount of repair. Above this amount the supplier pays for the difference between the achieved and contracted values.

19.4.2 Payment According to Down Time

The above arrangements involve penalties related to the cost of repair. Some contracts, however, demand a payment of some fixed percentage of the contract price during the down time. Provided that the actual sum paid is less than the cost of the repair this method is similar to Figure 19.2(b), although in practice it is not likely to be so generous. In any case, an arrangement of this type must be subject to an upper limit.

19.4.3 In Summary

The general position in contract law is that, in the event of breach, the injured party should be placed in the position that they would have been in had the contract been properly discharged. The rationale underlying this is that the injured party has 'committed no wrong' and should not then be penalized as a result. The practice, however, is that the amount of loss genuinely experienced and fully mitigated is often difficult to evaluate without some level of forensic analysis. It is therefore desirable to agree mutually acceptable quantums and/ or rectification requirements within the original contract documents. Except in case (a) it would not be practicable for the supplier to carry out the maintenance. Usually the customer carries out the repairs and the supplier pays according to some agreed rate. In this case the supplier must require some control over the recording of repair effort and a right to inspect the customer's maintenance records and facilities from time to time. It should be remembered that achievement of reliability and repair time objectives does not imply zero maintenance costs. If a desired MTBF of 20 000h is achieved for each of ten items of equipment, then in one year (8760h) about four failures can be expected. On this basis (d) is fairer than (a). When part of a system is subcontracted to another supplier, then the prime contractor must ensure that he or she passes on an appropriate allocation of the reliability commitments in order to be protected.

19.5 Subcontracted Reliability Assessments

It is common in the development of large systems for either the designer or the customer to subcontract the task of carrying out failure mode analysis and reliability/safety assessments. It may be that the customer requires the designer to place such a contract with a firm of consultants approved by the customer. It is desirable for such work to be covered by a contract that outlines the scope of work and the general agreement between the two parties. Topics to be covered include:

- databank sources to be used
- traceability where non-published data are used
- target reliability, availability or MTBF
- specific duty cycles and environmental profiles
- extent of the failure mode analysis required
- types of recommendation required in the event of the prediction indicating that the design will not meet the objectives
- requirement for ranking of major contributors to system failure
- if the prediction indicates that the design more than meets the objective, a requirement to identify the areas of over-design
- identification of critical single-point or common cause failures
- identification of safety hazards
- recommendations for maintenance (e.g. replacement strategy, periodic inspection time)
- calculations of spares-holding levels for defined probabilities of stockout
- aspects of human error required in the analysis
- arrangements for control and review of the assessment work, including reporting (e.g. conceptual design report, interim prediction and report, detailed failure mode analysis, final design qualification report, etc.)
- schedules, costs, invoicing.

■ Examples

(a) The following requirements might well be placed in an invitation to tender for a piece of measuring equipment. They are by no means intended as a model contract and, in fact, contain a number of deliberately inappropriate statements. The reader might care to critique them from both the designer's and customer's points of view. The case study in Chapter 23 also contains some contract clauses for critique and discussion.

  1. Loss of measurement shall include the total loss of temperature recording as well as a loss of recording accuracy exceeding 20%:
     Mode 1: The loss of two or more consecutive measurements.

Mode 2: The loss of recording accuracy of temperature within the range (>1% to 20%).

2. Bidders shall satisfy 'XYZ' that the equipment will meet the following:
    MTBF (Mode 1) $\geq$ 5 years
    MTBF (Mode 2) $\geq$ 10 years.

3. The MTBF shall be achieved without the use of redundancy but by the use of appropriate component quality and stress levels. It shall be demonstrated by means of a failure mode analysis of the component parts. FARADIP.THREE shall be used as the failure rate data source except where alternative sources are approved by 'XYZ'.

4. The above specification takes no account of the infant mortality failures usually characterized by a decreasing failure rate in the early life of the equipment. The supplier shall determine a suitable burn-in period and arrange for the removal of these failures by an appropriate soak test.

5. No wearout failure mechanisms, characterized by an increasing failure rate, shall be evident in the life of the equipment. Any components requiring preventive replacement in order to achieve this requirement shall be highlighted to 'XYZ' for consideration and approval.

6. In the event of the MTBFs not being demonstrated, at 80% confidence, after 10 device years of operation have been accumulated then the supplier will carry out any necessary redesign and modification in order to achieve the MTBF objectives.

7. During the life of the equipment any systematic failures shall be dealt with by the supplier, who will carry out any necessary redesign and modification. A systematic failure is one that occurs three or more times for the same root cause.

(b) The following requirement is not uncommon:
    The failure mode 'downstream overpressure resulting from loss of control of the pressure reduction equipment' is safety-related. It will meet safety-integrity level SIL 2 of IEC 61508.

(*Note*: this implies a failure rate, for the failure mode described, of less than 10^{-2} pa. It also implies various qualitative requirements. This is dealt with in Chapter 22.) It is also not uncommon to call for a SIL to be met with no mention of the failure mode. This is, of course, meaningless.

CHAPTER 20

Product Liability and Safety Legislation

Product liability is the liability of a supplier, designer or manufacturer to the customer for injury or loss resulting from a defect in that product. This has been the focus of attention for some time. The first reason was the publication, in July 1985, of a directive by the European Community, and the second was the wave of actions under United States law that resulted in spectacular awards for claims involving death or injury. By 1984, sums awarded resulting from court proceedings often reached $1 million. Changes in the United Kingdom became inevitable and the Consumer Protection Act reinforced the application of strict liability. It is necessary, therefore, to review the legal position.

20.1 The General Situation

20.1.1 Contract Law

Terms are implied by statute where government chooses to regulate certain types of agreements in order to protect weaker parties from differences in bargaining power. It follows that terms implied by statute are enforceable no matter what the wishes of the parties to the contract. In consumer contracts for the supply of goods this is largely governed by the Sale of Goods Act 1979, which requires under section 13 that the goods correspond to the description given to them and under section 14(2) that they are of satisfactory quality and are free from defects and suitably durable. Where the consumer is relying on the skill and judgement of the seller, section 14(3) of the Act also requires that goods are fit for any purpose stated by the buyer. Where the contract is for supply of goods and services similar provisions exist under the Supply of Goods and Services Act 1982 with regard to goods being of satisfactory quality and that services are provided with reasonable care and skill. With the exception of third parties, specifically identified under the Contracts (Rights of Third Parties Act) 1999, Privity of Contract exists between the buyer and seller. This means that only the buyer has any remedy for injury or loss and then only against the seller, although the cascade effect of each party suing, in turn, the other would offset this. However, exclusion clauses are void for consumer contracts. This means that a condition excluding the seller from liability would be void in law. Note that a contract does not have to be in writing and that a sale, in this context, implies the existence of a contract.

Reliability, Maintainability and Risk. DOI: 10.1016/B978-0-08-096902-2.00020-9
Copyright © 2011 Elsevier Ltd. All rights reserved.

20.1.2 Common Law

The relevant area is that relating to the Tort of Negligence, for which a claim for damages can be made. A duty of care is owed to your neighbor, in law, and failure to exercise reasonable precautions with regard to acts and omissions in relation to one's skill, knowledge and the circumstances involved may constitute a breach of that care. In law your neighbor is someone who is so closely and directly affected by an act that they ought to have reasonably been in your contemplation as being affected by any acts or omissions in question. The standard test on professionals is thus objective, i.e. 'a reasonable practitioner', and not subjective, i.e. it need not be in the consciousness of the party who has breached the duty. A claim for damages, for common law negligence, is therefore open to anyone and not restricted as in Privity of Contract. On the other hand, the onus is with the plaintiff to prove that a duty of care existed, that there has been a breach of that duty, that the breach was causal to the loss, and the damage not too 'remote', i.e. it was not too far from the breach of duty concerned and could thus be 'foreseeable'. It follows that for a tortuous liability to accrue in negligence proof is required to demonstrate:

> There was a duty of care owed
> That the product was defective or there was a forseeable misuse and thus the duty was breached
> That the defect was causal to the injury
> That this was foreseeable and that the plaintiff's loss was not, as a result, too remote.

20.1.3 Statute Law

The main Acts relevant to this area are:

> *Sale of Goods Act 1979*:
>> Goods must be of satisfactory quality
>> Goods must be fit for purpose.
> *Supply of Goods and Services Act 1982*:
>> Goods to be of satisafctory quality
>> Services to be provided with reasonable care and skill.
> *Contract (Rights of Third Parties) Act 1999*.
> *Unfair Contract Terms Act 1977*:
>> Exclusion of personal injury liability is void
>> Exclusion of damage liability only if reasonable.
> *Consumer Protection Act 1987*:
>> Imposes strict liability
>> Replaces the Consumer Safety Act 1978
> *Product Safety Regulations 1994*.

Health and Safety at Work Act 1974, Section 6:

> Involves the criminal law. Places a duty to construct and install items, processes and materials without health or safety risks. It applies to places of work. Responsibility involves everyone including management. The Consumer Protection Act extends Section 6 of the Health and Safety at Work Act to include all areas of use. European legislation will further extend this (see Section 20.4.5).

20.1.4 In Summary

The present situation involves a form of strict liability but:

> With the exception of those nominated under Contracts (Rights of Third Parties) Act 1999 Privity of Contract excludes third parties in contract claims.
> The onus is to prove negligence unless the loss results from a breach of contract.
> Exclusion clauses, involving death and personal injury, are void.

20.2 Strict Liability

20.2.1 Concept

The concept of strict liability hinges on the idea that liability exists for no other reason than the mere existence of a defect. No breach of contract or act of negligence is required in order to incur responsibility and manufacturers will be liable for compensation if their products cause injury.

The various recommendations that are summarized later involve slightly different interpretations of strict liability ranging from the extreme case of everyone in the chain of distribution and design being strictly liable, to the manufacturers being liable unless they can prove that the defect did not exist when the product left them. The Consumer Protection Act was the UK's response to EU Directive 85/374 on product liability and makes manufacturers liable whether or not they were negligent, and is both civil and criminal in content. The Act has also been supplemented by the Product Safety Regulations 1994.

20.2.2 Defects

A defect, for the purposes of product liability, includes:

Manufacturing	Presence of impurities or foreign bodies
	Fault or failure due to manufacturing or installation
Design	Product not fit for the purpose stated
	Inherent safety hazard in the design
Documentation	Lack of necessary warnings
	Inadequate or incorrect operating and
	maintenance instructions resulting in a hazard

20.3 The Consumer Protection Act 1987

20.3.1 Background

In 1985, after nine years of discussion, the European Community adopted a directive on product liability and member states were required to put this into effect before the end of July 1988. The Consumer Protection Bill resulted in the Consumer Protection Act 1987, which establishes strict liability as described above.

20.3.2 Provisions of the Act

The Act provides that a producer (and this includes manufactuers, those who import from outside the EC and retailers of 'own brands') will be liable for damage caused wholly or partly by defective products, which includes goods, components and materials but excludes unprocessed agricultural produce. 'Defective' is defined as not providing such safety as people are generally entitled to expect, taking into account the manner of marketing, instructions for use, the likely uses and the time at which the product was supplied. Death, personal injury and damage (other than to the product) exceeding £275 are included.

The consumer must show that the defect caused the damage but no longer has the onus of proving negligence. Defences include:

- The state of scientific and technical knowledge at the time was such that the producer could not be expected to have discovered the defect. This is known as the 'development risks' defence.
- The defect resulted from the product complying with the law.
- The producer did not supply the product.
- The defect was not present when the product was supplied by the manufacturer.
- The product was not supplied in the course of business.
- The product was in fact a component part used in the manufacture of a further product and the defect was not due to this component.

In addition, the producer's liability may be reduced by the user's contributory negligence. Further, unlike the privity limitation imposed by contract law, any consumer is covered in addition to the original purchaser.

Notwithstanding the above there is still a requirement to demonstrate that the loss was causal to the defect and the requirements for standard of care are very similar to negligence. Furthermore there are very strict limitation periods of 3 years after becoming aware of the defect, or if the damage is latent the date of the knowledge of the damage provided that it is within a 10 year period.

The Act sets out a general safety requirement for consumer goods and applies it to anyone who supplies goods that are not reasonably safe having regard to the circumstances pertaining. These include published safety standards, the cost of making goods safe and whether or not the goods are new.

20.4 Health and Safety at Work Act 1974

20.4.1 Scope

Section 6 of this Act applies strict liability to articles produced for use at work, although the Consumer Protection Act extends this to all areas. It is very wide and embraces designers, manufacturers, suppliers, hirers and employers of industrial plant and equipment. We are now dealing with criminal law and failure to observe the duties laid down in the Act is punishable by fine or imprisonment. Claims for compensation are still dealt with in civil law.

20.4.2 Duties

The main items are:

to design and construct products without risk to health or safety
to provide adequate information to the user for safe operation
to carry out research to discover and eliminate risks
to make positive tests to evaluate risks and hazards
to carry out tests to ensure that the product is inherently safe
to use safe methods of installation
to use safe (proven) substances and materials.

20.4.3 Concessions

The main concessions are:

- It is a defence that a product has been used without regard to the relevant information supplied by the designer.
- It is a defence that the design was carried out on the basis of a written undertaking by the purchaser to take specified steps sufficient to ensure the safe use of the item.
- One's duty is restricted to matters within one's control.
- One is not required to repeat tests upon which it is reasonable to rely.

20.4.4 Responsibilities

Basically, everyone concerned in the design and provision of an article is responsible for it. Directors and managers are held responsible for the designs and manufactured articles of their companies and are expected to take steps to ensure safety in their products. Employees are also responsible. The 'buck' cannot be passed in either direction.

20.4.5 European Community Legislation

In 1989/1990 the EC agreed to a framework of directives involving health and safety. This legislation will eventually replace the Health and Safety at Work Act, being more prescriptive and detailed than the former. The directive mirrors the Health and Safety at Work Act by setting general duties on both employees and employers for all work activities.

In implementing this European legislation the Health and Safety Commission will attempt to avoid disrupting the framework that has been established by the Health and Safety at Work Act. The directive covers:

> the overall framework
> the workplace
> use of work equipment
> use of personal protective equipment
> manual handling
> display screen equipment.

20.4.6 Management of Health and Safety at Work Regulations 1992

These lay down broad general duties that apply to almost all Great Britain onshore and offshore activities. They are aimed at improving health and safety management and can be seen as a way of making more explicit what is called for by the H&SW Act 1974. They are designed to encourage a more systematic and better organized approach to dealing with health and safety, including the use of risk assessment.

20.5 Insurance and Product Recall

20.5.1 The Effect of Product Liability Trends

- An increase in the number of claims.
- Higher premiums.
- The creation of separate product liability policies.
- Involvement of insurance companies in defining quality and reliability standards and procedures.
- Contracts requiring the designer to insure the customer against genuine and frivolous consumer claims.

20.5.2 Some Critical Areas

- *All risks*: this means all risks specified in the policy. Check that your requirements are met by the policy.

- *Comprehensive*: essentially means the same as the above.
- *Disclosure*: the policy holder is bound to disclose any information relevant to the risk. Failure to do so, whether asked for or not, can invalidate a claim. The test of what should be disclosed is described as 'anything the prudent insurer should know'.
- *Exclusions*: the Unfair Contract Terms Act 1977 does not apply to insurance, so read and negotiate accordingly. For example, defects related to design could be excluded and this would considerably weaken a policy from the product liability standpoint.
- *Prompt notification of claims.*

20.5.3 Areas of Cover

Premiums are usually expressed as a percentage of turnover and cover is divided into three areas:

Product liability: cover against claims for personal injury or loss.
Product guarantee: cover against the expenses of warranty/repair.
Product recall: cover against the expenses of recall.

20.5.4 Product Recall

A design defect causing a potential hazard to life, health or safety may become evident when a number of products are already in use. It may then become necessary to recall, for replacement or modification, a batch of items, some of which may be spread throughout the chain of distribution and others in use. The recall may vary in the degree of urgency depending on whether the hazard is to life, health or merely reputation. A hazard that could reasonably be thought to endanger life or to create a serious health hazard should be treated by an emergency recall procedure. Where less critical risks involving minor health and safety hazards are discovered a slightly less urgent approach may suffice. A third category, operated at the vendor's discretion, applies to defects causing little or no personal hazard and where only reputation is at risk.

If it becomes necessary to implement a recall the extent will be determined by the nature of the defect. It might involve, in the worst case, every user or perhaps only a specific batch of items. In some cases the modification may be possible in the field and in others physical return of the item will be required. In any case, a full evaluation of the hazard must be made and a report prepared.

One person, usually the Quality Manager, must be responsible for the handling of the recall and must be directly answerable to the Managing Director or Chief Executive. The first task is to prepare, if appropriate, a 'hazard notice' in order to warn those likely to be exposed to the risk. Circulation may involve individual customers when traceable, field service staff, distributors, or even the news media. It will contain sufficient information to describe the nature of the hazard and the precautions to be taken. Instructions for returning the defective item can be included, preferably with a pre-paid return card. Small items can be returned with the card whereas large ones, or products to be modified in the field, will be retained while arrangements are made.

Where products are despatched to known customers a comparison of returns with output records will enable a 100% check to be made on the coverage. Where products have been despatched in batches to wholesalers or retail outlets the task is not so easy and the quantity of returns can only be compared with a known output, perhaps by area. Individual users cannot be traced with 100% certainty. Where customers have completed and returned record cards after purchase the effectiveness of the recall is improved.

After the recall exercise has been completed a major investigation into the causes of the defect must be made and the results progressed through the company's quality and reliability program. Causes could include:

> insufficient test hours
> insufficient test coverage
> insufficient information sought on materials
> insufficient industrial engineering of the product prior to manufacture
> insufficient production testing
> insufficient field/user trials
> insufficient user training.

CHAPTER 21

Major Incident Legislation

21.1 History of Major Incidents

Since the 1960s developments in the process industries have resulted in large quantities of noxious and flammable substances being stored and transmitted in locations that could, in the event of failure, affect the public. Society has become increasingly aware of these hazards as a result of major incidents that involved both process plant and public transport such as:

Aberfan (UK)	1966	144 deaths due to collapse of a coalmine waste tip
Flixborough (UK)	1974	28 deaths due to an explosion resulting from the stress failure of a temporary reactor by-pass, leading to an escape of cyclohexane
Beek (Netherlands)	1975	14 deaths due to propylene
Seveso (Italy)	1976	Unknown number of casualties due to a release of dioxin
San Carlos Holiday Camp (Spain)	1978	c. 150 deaths due to a propylene tanker accident
Three Mile Island (USA)	1979	0 immediate deaths. Incident due to a complex sequence of operator and physical events following a leaking valve allowing water into the instrument air. This led to eventual loss of cooling and reactor core damage
Bhopal (India)	1984	2000+ deaths following a release of methyl isocyanate due to some safety-related systems being out of service due to inadequate maintenance
Mexico City (Mexico)	1984	500+ deaths due to an LPG explosion at a refinery
Chernobyl (USSR)	1986	31 immediate deaths and unknown number of casualties following the meltdown of a nuclear reactor due to intrinsic reactor design and operating sequences
Herald of Free Enterprise (North sea)	1987	184 deaths due to capsize of Zeebrugge–Dover ferry
Piper Alpha (North Sea)	1988	167 deaths due to an explosion of leaking condensate following erroneous use of a condensate pump in a stream disabled for maintenance
Clapham (UK)	1988	34 deaths due to a rail crash resulting from a signalling failure
Kegworth (UK)	1989	47 deaths due to a Boeing 737 crash on landing involving erroneous shutdown of the remaining good engine
Cannon Street, London (UK)	1991	2 deaths and 248 injured due to a rail buffer-stop collision
Strasbourg (France)	1992	87 deaths due to A320 Airbus crash
Eastern Turkey	1992	400+ deaths due to methane explosion in a coal mine
Paddington (UK)	1999	31 deaths due to a rail crash (drawing attention to the debate over automatic train protection)
Paris (France)	2000	114 deaths due to the crash of a Concorde aircraft
Potters Bar (UK)	2002	7 deaths due to derailment of a train

(Cont.)

Reliability, Maintainability and Risk. DOI: 10.1016/B978-0-08-096902-2.00021-0

Near Reading (UK)	2004	6 deaths due to a rail crash (drawing attention to level crossings)
Glasgow (UK)	2004	9 deaths due to an explosion at the ICL plastics factory
Texas (USA)	2005	15 deaths and over 100 injured due to an explosion at a BP refinery following vessel maintenance
Buncefield (UK)	2005	Miraculously no deaths due to the timing (6 am Sunday morning) following the ignition of a vapor cloud from 250 000 litres of petrol leakage. Damage in excess of £750M
Qinghe Special Steel Corp (China)	2007	32 workers killed and 6 injured after spillage from a ladle holding molten steel
Georgia Sugar Refinary (USA)	2008	13 people killed and 42 injured in a dust explosion at a sugar factory
Connecticut Power Plant (USA)	2010	5 deaths and 27 casuatlies following a large explosion at a power plant at Middletown Conneticut
Deepwater Horizon, Gulf of Mexico (USA)	2010	11 deaths following a subsea oil pipe blow-out leading to rig explosion and large-scale environmental hydrocarbon release

It is important to note that in a very large number (if not all) of the above incidents human factors played a strong part. It has long been clear that major incidents seldom occur as a result of equipment failure alone but involve humans in the maintenance or operating features of the plant.

Media attention is frequently focused on the effects of such disasters and subsequent inquiries have brought the reasons behind them under increasingly closer scrutiny. The public is now very aware of the risks from major transport and process facilities and, in particular, those arising from nuclear installations. Debate concerning the comparative risks from nuclear and fossil-fuel power generation was once the province of the safety professionals. It is now frequently the subject of public debate. Plant-reliability assessment was, at one time, concerned largely with availability and throughput. Today it focuses equally on the hazardous failure modes.

21.2 Development of Major Incident Legislation

Following the Flixborough disaster in 1974, the Health and Safety Commission set up an Advisory Committee on Major Hazards (ACMH) in order to generate advice on how to handle major industrial hazards. It made recommendations concerning the compulsory notification of major hazards. Before these recommendations were fully implemented, the Seveso accident, in 1976 drew attention to the lack of formal controls throughout the EC. This prompted a draft European Directive in 1980, which was adopted as the so-called Seveso Directive (82/501/EEC) in 1982. Delays in obtaining agreement resulted in this not being implemented until September 1984. Its aim was:

> To prevent major chemical industrial accidents and to limit the consequences to people and the environment of any which do occur.

In the UK the HSC (Health and Safety Commission) introduced in January 1983 the Notification of Installations Handling Hazardous Substances (NIHHS) regulations. These required the notification of hazardous installations and that assessments be carried out of the risks and consequences.

The 1984 EC regulations were implemented in the UK as the CIMAH (Control of Industrial Major Accident Hazards regulations, 1984). They were concerned with people and the environment and cover processes and the storage of dangerous substances. A total of 178 substances were listed and the quantities of each that would render them notifiable. In these cases a safety case (nowadays called safety report) is required, that must contain a substantial hazard and operability study and a quantitative risk assessment. The purpose of the safety report is to demonstrate either that a particular consequence is relatively minor or that the probability of its occurrence is extremely small. It is also required to describe adequate emergency procedures in the event of an incident. The latest date for the submission of safety reports is three months prior to bringing hazardous materials on site.

As a result of lessons learnt from the Bhopal incident there were two subsequent amendments to the CIMAH regulations (1988 and 1990), which refined the requirements, added substances and revised some of the notifiable quantities. The first revision reduced the threshold quantities for some substances and the second revision was more comprehensive, concerning the storage of dangerous substances.

Following the offshore Piper Alpha incident in 1988, and the subsequent Cullen enquiry, the responsibility for UK offshore safety was transferred from the Department of Energy to a newly formed department of the HSE (Health and Safety Executive). Equivalent requirements to the CIMAH regulations are now applied to offshore installations and the latest date for submitting cases was November 1993.

Quantification of frequency, as well as consequences, in safety reports is now the norm and the role of human error in contributing to failures is attracting increasing interest. Emphasis is also being placed on threats to the environment.

The CIMAH regulations have now been replaced by a further directive on the Control of Major Accident Hazards (COMAH). The introduction of the COMAH regulations was as a result of the Seveso II directive and the COMAH regulations came into force on the 1st April 1999. The regulations were further amended in June 2005 to reflect further changes to the Seveso II directive. Although similar to CIMAH, the COMAH requirements are more stringent, with their aim being to treat risks to the environment as seriously as risks to people. The amendments to the COMAH regulations in 2005 include the addition of new named substances, modifications to exisiting named substances, changes to the aggregation rule involving the cumulative effect of several named substances and a broadening of the scope of regulations to sites such as mines, quarries, boreholes and landfill. The COMAH regulations include a number of requirements that place duties upon the operators including:

- provision of information to the public
- demonstration of management control systems

- identification of 'domino' effects
- details of worker participation.

The CIMAH requirements defined 'Top Tier' sites by virtue of the threshold quantities of substances. For example, 500 tonnes of bromine, 50 tonnes of acetylene or 100 tonnes of natural gas (methane) render a site 'Top Tier'.

To comply with the top tier regulations a plant operator was required to:

- prepare and submit to HSE a safety report
- draw up an onsite emergency plan
- provide information to local authorities for an offsite emergency plan
- provide information to the public
- report major accidents
- show, at any time, safe operation.

21.3 CIMAH Safety Reports

The safety report provides the HSE with a means of assessing the compliance with the CIMAH regulations. Second, and just as important, the exercise of producing the report increases awareness of the risks and focuses attention on providing adequate protection and mitigation measures. Therefore the safety report must:

- identify the scale and nature of potential hazards
- assess the likelihood and consequence of accidents
- describe the safeguards
- demonstrate management competence.

The contents of a safety report are addressed in Schedule 6 of the regulations and include:

- the nature of the dangerous substances, the hazards created by them and the means by which their presence is detected
- details of the geography, layout, staffing and processes on the site
- the procedures for controlling staff and processes (including emergency procedures) in order to provide safe operation
- a description of the potential accident scenarios and the events and pre-conditions that might lead to them.

QRA (Quantified Risk Assessment), whereby frequency as well as the consequences is quantified, is not a specific requirement for onshore safety reports. It is, however, becoming more and more the practice to provide such studies as safety report support material. For offshore installations QRA is required.

Reports are assessed by the HSE in two stages. The first is a screening process (completed within six weeks) which identifies reports clearly deficient in the schedule 6 requirements. Within 12 months a detailed assessment is carried out to reveal any issues that require follow-up action.

A typical safety report might consist of:

(a) *General plant information*:
   plant/process description (main features and operating conditions)
   personnel distribution on site
   local population distribution.

(b) *Hazard identification*:
   methodology used
   summary of HAZOP and recommendations
   comparative considerations
   conclusions from hazard identification.

(c) *Potential hazards and their consequences*:
   dangerous materials on site
   inventory of flammable/dangerous substances
   hazards created by the above
   analysis and detection of dangerous materials
   nature of hazards
   fire and explosion
   toxic hazards
   impact/dropped object
   unloading spillage
   natural hazards
   hazards and sources leading to a major accident.

(d) *Plant management*:
   structure and duties (including responsibilities)
   personnel qualification
   general manning arrangements
   operating policy and procedures
   shift system/transfer of information
   commissioning and start up of new plant
   training program
   interface between OM&S area
   support functions
   record keeping.

(e) *Plant safety features*:
   control instrumentation
   codes and standards
   integrity

   electrical distribution
   design
   protection
   changeover
   recovery
   emergency generator
   emergency procedure for power fail
   isolation for maintenance
   area classification
   safety systems
   ESD
   blowdown
   relief
   fire fighting
   design of system
   water supplies
   drenching systems
   foam
   halon
   rendezvous
   piping design
   material selection
   design code
   plant communications.

(f) *Emergency planning*:
   onsite emergency plans
   offsite emergency plan.

(g) *Other items*:
   site meteorological conditions
   plant and area maps
   meteorological reports
   health and safety policy
   location of dangerous substances
   site health and safety information sheets
   description of tools used in the analysis.

21.4 Offshore Safety Cases

The offshore safety case is assessed by the Offshore Safety Division of the HSE and assessment is in two stages:

- an initial screen to determine if the case is suitable for assessment and, if appropriate the preparation of an assessment work plan;
- detailed assessment leading to either acceptance or rejection.

The content of a safety case needs to cover sufficient detail to demonstrate that:

- the management system is adequate to ensure compliance with statutory health and safety requirements;
- adequate arrangements have been made for audit and the preparation of audit reports;
- all hazards with the potential to cause a major accident have been identified, their risks evaluated, and measures taken to reduce risks to persons to as low as reasonably practicable.

In general the list of contents shown for CIMAH site safety cases will be suitable. A QRA is obligatory for offshore cases and will include consequences and frequency. Additional items which are specific to offshore are:

- temporary refuge
- control of well pressure
- well and bore details
- seabed properties
- abandonment details.

There are three points at which a safety case must be submitted:

- *Design*: To be submitted early enough for the detailed design to take account of issues raised.

- *Pre-operational*: To be submitted six months before operation.

- *Abandonment*: To be submitted six months before commencement.

Particulars to be covered include:

- *Design safety case for fixed installation*:
    name and address
    safety management system
    scale plan of installation
    scale plan of location, conditions, etc.
    operation and activities
    number of persons on installation
    well operations
    pipelines

    detection equipment

    personnel protection (including performance standards)

    QRA

    design and construction codes of practice

    principal features of design and construction.

- *Operation safety case for fixed installation*:

    name and address

    scale plan of installation

    scale plan of location, conditions, etc.

    operation and activities

    number of persons on installation

    well operations

    pipelines

    detection equipment

    personnel protection (including performance standards)

    QRA

    limits of safe operation

    risks are lowest reasonably practicable

    remedial work particulars.

- *Safety case for a mobile installation*:

    name and address

    scale plan of installation

    operation and activities

    number of persons on installation

    well operations

    detection equipment

    personnel protection (including performance standards)

    QRA

    limits of safe operation

    environmental limits

    risks are lowest reasonably practicable

    remedial work particulars.

- *Safety case for abandonment of a fixed installation*:

    name and address

    scale plan of installation

    scale plan of location, conditions, etc.

    operation and activities

    number of persons on installation

    well operations

    pipelines

detection equipment
evacuation details
wells and pipelines present lowest reasonable risk.

21.5 Problem Areas

Reports must be site specific and the use of generic procedures and justifications is to be discouraged. Adopting the contents of procedures and documents from a similar site is quite valid provided care is taken to ensure that the end result is site specific. Initiating events as well as the impact on surroundings will vary according to the location so it cannot be assumed that procedures adequate for one site will necessarily translate satisfactorily to another. A pressure vessel directly in the flight path of a major airport or beneath an elevated section of motorway is more at risk from collision than one in a deserted location. A liquid natural gas site on a moor will have different impacts from one situated next to a factory.

The hazards from a dangerous substance may be various and it is necessary to consider secondary as well primary hazards. Natural gas, for example, can asphyxiate as well as cause fire and explosion. Furthermore the long-term exposure of ground to natural gas will result in the concentration of dangerous trace substances. Decommissioning of gas-holder sites therefore involves the removal of such impurities from the soil. Carbon disulfide is hazardous in that it is flammable. However, when burned it produces sulfur dioxide, which in turn is toxic.

The events that could lead to the major accident scenario have to be identified fully. In other words the fault tree approach (Chapter 8) needs to identify all the initiators of the tree. This is an open-ended problem in that it is a subjective judgement as to when they have ALL been listed. An obvious checklist would include, as well as hardware failures:

- earthquake
- human error
- software
- vandalism/terrorism
- external collision
- meteorology
- out of spec substances.

The HAZOP approach (Chapter 10) greatly assists in bringing varied views to bear on the problem.

Consequences must also be researched fully. There is a requirement to quantify the magnitude of outcome of the various hazards and the appropriate data and modelling tools are needed. The consequence of a pressurized pipeline rupture, for example, requires the appropriate mathematical treatment for which computer models are available. All eventualities need to be considered such as the meteorological conditions under which the ruptured pipeline will disgorge gas. Damage to the very features that provide protection and mitigation must also be considered when quantifying consequences.

21.6 The COMAH Directive (1999 and 2005 Amendment)

The COMAH directive, mentioned above, now replaces CIMAH. It places more emphasis on risk assessment and the main features are:

- The simplification that their application will be dependent on exceeding threshold quantities and the distinction between process and storage will no longer apply.
- The exclusion of explosive, chemical and waste disposal hazards at nuclear installations will be removed. The regulations do not, however, apply to offshore installations.
- Substances hazardous to the environment (as well as people) are introduced. In the first instance these will take account of the aquatic environment.
- More generic categories of substances are introduced. The 178 substances currently named will thus reduce to 37. A spin-off is that new substances are more easily catered for by virtue of their inclusion in a generic group.

More information than before will be publicly available, including off-site emergency plans.

- The competent authority in the UK will positively assess a safety report.
- The periodic update is five years instead of three years.
- More onus on demonstrating the effectiveness of proposed safety measures and on showing ALARP.

A key feature of the new regulations is that they cover both safety and the environment. The standard by which acceptable risk management will be judged is the As Low As Reasonably Practicable (ALARP) principle for human risks and Best Available Technology Not Entailing Excessive Cost (BATNEEC) for environmental risks, although it should be noted that the principles of proportionality remain fundamental to and underpin the intent of the regulations. They will be enforced by a competent authority comprising the HSE and the environment agency in England and Wales and the HSE and the Scottish Environmental Protection Agency in Scotland.

21.7 Rail

The importance of rail safety cases was reinforced by the Cullen Report following the public enquiry into the Paddington rail disaster of 1999. The Railway Safety (Miscellaneous Amendments) Regulations 2001 amend the Railway (Safety Case) Regulations 2000 to clarify that train operators' safety cases must cover escape arrangements following emergencies.

21.8 *Corporate Manslaughter and Corporate Homicide*

The Corporate Manslaughter and Corporate Homicide Act 2007, which came into force on the 6th April 2008, provides for the first time that companies and organizations can be found guilty of corporate manslaughter as a result of serious failures in the management of health and safety. This is a departure from the traditional view that the controlling mind of senior corporate officials must be held to be at the root of any failing. Serious failures of the management of a company or organization that result in death will be held as a gross breach of the duty of care owed by that company or organization to the deceased, this gross breach being held as a crime against the state and thus warranting punitive rather than mere compensatory penalties. It should be noted that whilst any prosecutions under this legislation will be against the corporate entity and not the individuals concerned, their duties and the ability to be prosecuted under existing health and safety and/or criminal law remain unaffected. The Act also removes the previous immunity held by the Crown.

In England, Wales and Northern Ireland the offence is called Corporate Manslaughter and in Scotland Corporate Homicide. Penalties upon prosecution include unlimited fines, remedial orders and publicity orders.

The corporation, department, police force, partnership, trade union, employers association etc. is guilty of the offence if by the way in which its activities are managed or organized a person's death is caused and this amounts to a gross breach of relevant duty of care owed by the organization to the deceased.

Where:

the way in which its activities are managed or organized by its senior management is a substantial element in the breach;

the conduct falls far below what can reasonably be expected of the organization in the circumstances;

senior management means persons who play a significant role in the making of decisions about the whole or a substantial part of the activities to be managed or organized, or the actual managing of the said activities.

The duty of care is owed to:

employees or other persons working for the organization or performing services for it

the occupier(s) of premises

supply or organization of goods and services

the carrying on of construction or maintenance

the carrying on of any other activity on a commercial basis

the keeping or organization of any plant, vehicle or other thing.

The first cases have now begun to be heard under the new Corporate Manslaughter and Corporate Homicide Act 2007 and it is likely that a number of cases will follow given the change in the approach taken by the legislature.

CHAPTER 22

Integrity of Safety-Related Systems

This chapter is a brief introduction to safety-related systems. *The Safety Critical Systems Handbook*, D. J. Smith and K. G. L. Simpson 3rd edition, ISBN 9780080967813, is a thorough treatment of this topic

22.1 Safety-Related or Safety-Critical?

As well as a focus of interest on major accident hazards there has long been awareness that many failures relate to the control and safety systems used for plant operation and protection. Examples of this type of equipment are fire detection systems, emergency shutdown systems, distributed control systems, rail signaling, automotive controls, medical electronics, nuclear control systems and aircraft flight controls.

Terms such as 'safety-related' and 'safety-critical' have become part of the engineering vocabulary. The distinction between them has become blurred and they have tended to be used synonymously.

'Safety-critical' has tended to be used where the hazard leads to fatality whereas 'safety-related' has been used in a broader context. There are many definitions, all of which differ slightly, for example:

- some distinguish between multiple and single deaths;
- some include injury, illness and incapacity without death;
- some include effects on the environment;
- some include system damage.

However, the current consensus distinguishes them as follows:

- Safety-related systems are those that, singly or together with other safety-related systems, achieve or maintain a safe state for equipment under their control.
- Safety-critical systems are those that, on their own, achieve or maintain a safe state for equipment under their control.

The difference involves the number of levels of protection. The term *safety-related application* implies a control or safety function where failure or failures could lead to death, injury or environmental damage.

Reliability, Maintainability and Risk. DOI: 10.1016/B978-0-08-096902-2.00022-2

The term *safety-related* applies to any hardwired or programable system where a failure, singly or in combination with other failures/errors, could lead to death, injury or environmental damage.

A piece of equipment, or software, cannot be excluded from this safety-related category merely by identifying that there are alternative means of protection. This would be to pre-judge the issue whereas a formal safety-integrity assessment would be required to determine the issue.

A distinction is made between control and protection systems. Control systems cause a process to perform in a particular manner whereas protection systems deal with fault conditions and their function is therefore to override the control system. Sometimes the equipment that provides these functions is combined and sometimes it is separate. Both can be safety-related and the relevant issue is whether or not the failure of a particular system can lead to a hazard, rather than whether or not it is called a safety system. The argument is often put forward (wrongly) that a system is not safety related because, in the event of its failure, another level of protection exists. An example might be a circuit for automatically closing a valve in the event of high pressure in a pipeline. This potentially dangerous pressure might also be mitigated by the additional protection afforded by a relief valve. This does not, however, mean that the valve-closing circuit ceases to be safety-related.

Until recently the design approach has generally been to ensure that, for each possible hazardous failure, there are at least two levels of protection. In other words two independent failures would be necessary in order for the hazard to occur. Using the approach described in the next section a single (simplex) arrangement could be deemed adequate although, usually, redundancy proves to be necessary in order to make the incident frequencies sufficiently low as to be acceptable.

22.2 Safety-Integrity Levels (SILs)

22.2.1 Targets

This section follows on from Section 10.2 in Chapter 10, which described how a maximum tolerable frequency of the risk of fatality is chosen. During the 1990s the concept of Safety-Integrity Levels (known as SILs) evolved and is used in the majority of guidance documents in this area. The concept is to divide the 'spectrum' of integrity into four discrete levels and then to lay down requirements for each level. Clearly, the higher the SIL then the more rigorous become the requirements. In IEC 61508 (and in most other documents) the four levels are defined as in Table 22.1.

Note that because the high-demand SIL bands are expressed as 'per annum' the tables appear to be numerically similar. However, failure rate and PFD, being different parameters, are NOT even the same dimensionally. The reason for there being two tables (high and low demand) is that there are two ways in which the integrity target may need to be described. The difference can best be understood by way of two examples.

Table 22.1: Safety-Integrity Levels

Safety-Integrity Level	High Demand Rate (dangerous failures/yr)	Low Demand Rate (probability of failure on demand)
4	$\geq 10^{-5}$ to $<10^{-4}$	$\geq 10^{-5}$ to $<10^{-4}$
3	$\geq 10^{-4}$ to $<10^{-3}$	$\geq 10^{-4}$ to $<10^{-3}$
2	$\geq 10^{-3}$ to $<10^{-2}$	$\geq 10^{-3}$ to $<10^{-2}$
1	$\geq 10^{-2}$ to $<10^{-1}$	$\geq 10^{-2}$ to $<10^{-1}$

Consider the motor car brakes. It is their rate of failure that is of concern because there is a high probability of suffering the hazard immediately each failure occurs. Hence we have the middle column of Table 22.1.

On the other hand, consider the motor car air bag. This is a low-demand protection system in the sense that demands on it are infrequent (years or even tens of years apart). Failure rate alone is of little use to describe the integrity since the hazard is not incurred immediately each failure occurs and we therefore have to take into consideration the test interval. In other words, since the demand is infrequent, failures may well be dormant and persist during the test interval. What is of interest is the combination of failure rate and down time and we therefore specify the probability of failure on demand (PFD): hence the right-hand column of Table 22.1.

Now look at the following examples.

Low demand

As a simple example of selecting an appropriate SIL, assume that the maximum tolerable frequency for an involuntary risk scenario (e.g. customer killed by explosion) is 10^{-5} pa (A) (see Table 2.1 in Chapter 2). Assume that 10^{-2} (B) of the hazardous events in question lead to fatality. Thus the maximum tolerable failure rate for the hazardous event will be $A/B = 10^{-3}$ pa (C). Assume that a fault tree analysis predicts that the unprotected process is only likely to achieve a failure rate of 2×10^{-1} pa (D) (i.e. 1/5 years). The maximum probablity of failure on demand of the safety system would need to be $E = C/D = 10^{-3}/2\ 10^{-1} = 5 \times 10^{-3}$. Consulting the right-hand column of Table 22.1, SIL 2 is applicable. This is an example of a low-demand safety-related system in that it is only called upon to operate at a frequency determined by the frequency of failure of the equipment under control (EUC) – in this case 2×10^{-1} pa. Note, also, that the target 'E' in the above paragraph is dimensionless by virtue of dividing a rate by a rate. Again, this is consistent with the right-hand column of Table 22.1.

High demand

Now consider a failure in a domestic appliance that leads to overheating and subsequent fire. Assume, again, that the target risk of fatality is said to be 10^{-5} pa. Assume that a study suggests that 1 in 400 incidents leads to fatality. It follows that the target maximum tolerable failure rate for the hazardous event can be calculated as $10^{-5} \times 400 = 4 \times 10^{-3}$ pa (i.e. 1/250 years).

Consulting the middle column of Table 22.1, SIL 2 is applicable. This is an example of a high-demand safety-related system in that it is 'at risk' continuously. Note, also, that the target in the above paragraph has the dimension of rate by virtue of multiplying a rate by a dimensionless number. Again, this is consistent with the middle column of Table 22.1.

More complex example

In the fault tree (Figure 22.1), Gate G1 describes the causes of some hazardous event. It would be quantified using the rate parameter. Dividing the target maximum tolerable failure rate associated with the top gate (GTOP) by the rate for Gate G1 provides a target PFD (probability of failure on demand) for the protection.

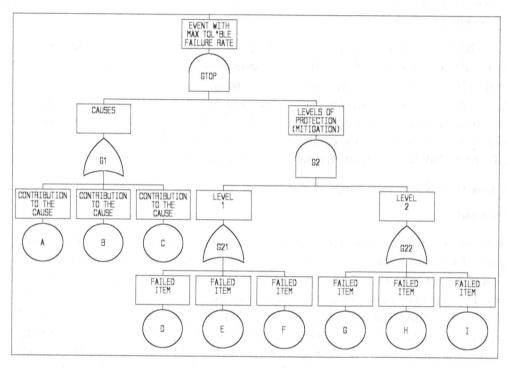

Figure 22.1: Fault tree

Independent levels of protection are then modeled as shown by gates G21 and G22 in Figure 22.1. It is important to remember that the use of an AND gate (e.g. Gate G2) implies that the events below that gate are totally independent of each other. A greater number of levels of protection (i.e. gates below G2) leads to larger PFDs being allocated for each and, thus, lower-integrity requirements will apply to each.

A maximum tolerable failure rate of 5.3×10^{-4} pa is taken as an example. Assume that the frequency of causes (i.e. Gate G1) is 10^{-1} pa. Thus the target PFD associated with Gate G2 becomes:

$$5.3 \times 10^{-4} \text{ pa} / 10^{-1} \text{ pa} = 5.3 \times 10^{-3} \text{ (Note that the result is dimensionally correct,}$$
i.e. a rate/rate becomes a PFD.)

A common mistake is to describe the scenario as 'a SIL 2 safety system'. This would ONLY be the case if the mitigation were to be a single element and not decomposed into separate independent layers. In Figure 22.1 there are two levels of protection for which the product of the two PFDs needs to be less than 5.3×10^{-3}.

Depending on the equipment in question this could involve a number of possibilities. Examples are shown in Table 22.2, that assume independent levels of protection. As can be seen, the safety integrity level is inferred only once the PFD associated with each level of protection has been assigned/assessed.

Table 22.2: Possible SIL Outcomes

	Level 1 PFD	Level 1 SIL	Level 2 PFD	Level 2 SIL
OPTION	$2\ 10^{-1}$	<1	$2.65\ 10^{-2}$	1
OPTION	$7.3\ 10^{-2}$	1	$7.3\ 10^{-2}$	1
OPTION	$7\ 10^{-1}$	<1	$7.57\ 10^{-3}$	2

It is worth noting that for a low-demand system the standard is being applied to an 'add-on' safety system that is separate from the normal control of the EUC (i.e. plant). On the other hand for a continuous system the standard is being applied to the actual control element because its failure will lead directly to the potential hazard even though the control element may require additional features to meet the required integrity. Note that the IEC 61508 standard requires that a safety-related system with a demand rate of greater than once per annum should be treated as 'high demand'. This topic is dealt with in far greater detail, with numerous examples, in *The Safety Critical Systems Handbook*.

One methodology, specifically mentioned in Part 3 of IEC 61511 (Annex F), is known as Layer of Protection Analysis (LOPA). LOPA provides a structured risk analysis that can follow on from a qualitative technique such as HAZOP. In general, formalized LOPA procedures tend to use order of magnitude estimates and are thus referred to as so called semi-quantitative methods. Also, they are tailored to low-demand safety functions. Nevertheless, many practitioners, despite using the term LOPA, actually carry out the analysis to a refinement level such as I have described in Section 10.2. This is commonly referred to as a quantitative approach.

In general the method described in Section 10.2 should be adopted in order to establish a maximum tolerable failure rate. The above examples (in this chapter) show how SIL targets may then be determined according to the type of safety-related system (high or low demand).

However, there is an alternative approach to establishing safety-integrity levels, known as the risk graph approach. This avoids quantifying the maximum tolerable risk of fatality by using qualitative judgements. Figure 22.2 gives an example of a risk graph.

The advantage is that the risk graph is easier and quicker to apply but, on the other hand, it is less precise. Order of magnitude decisions, with breakpoints, can lead to gross inaccuracies.

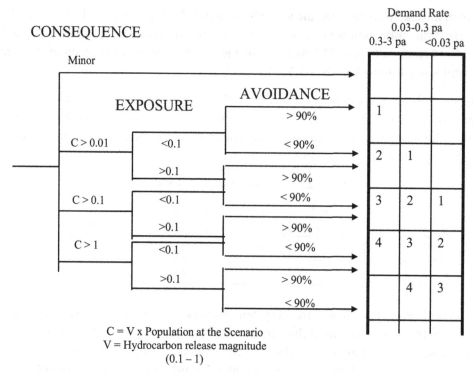

Figure 22.2: Risk graph

The author does not recommend this approach and the earlier quantitative risk-based methods are always to be preferred.

22.2.2 Assessing Equipment Against the Targets

22.2.2 .1 Quantitative versus qualitative features

It is important to take account of the fact that not all failures can be quantified and expressed by predicted failure rates. Random hardware failures are generally those for which failure rate data are available. On the other hand systematic failures, in particular software failures, cannot readily be expressed in that way since they are not random repeatable failures and the concept of a rate being used to predict future performance does not apply. Traditional reliability prediction is therefore not an option in this area.

As has already been explained, it is for this reason that the SIL concept is used. If there were only random hardware failures, we could talk in terms of the rate and there would be no need to establish 'bands' of targets. For systematic failures, however, since these can only be mitigated by qualitative life-cycle activities, it is necessary to define levels of rigor, in the life-cycle processes, appropriate to each level. The use of four levels is slightly artificial in that it can be seen in the Standard IEC 61508 (which originally introduced this concept) that there is very little difference between SIL 1 and SIL 2 as far as life-cycle activities are concerned. The main difference comes at SIL 3, which demands far greater rigor. There is a considerable body of opinion that SIL 4 safety functions should be avoided (as achieving SIL 4 requires very significant levels of design effort and analysis) and that additional levels of risk reduction need to be introduced such that lower SIL targets are required for each element of the system. In any case, a system with a SIL 4 target would imply a scenario with a high probability of the hazard leading to fatality and only one level of control (i.e. no separate mitigation). It is hard to imagine such a scenario as being acceptable.

These life-cycle activities were introduced in Chapter 17 and are more fully described in *The Safety Critical Systems Handbook.*

Both qualitative and quantitative assessments are therefore required and satisfying the IEC 61508 standard (and most other second-tier guidance) requires the following to be satisfied:

- random hardware failures targets being met (reliability prediction against a quantified target already covered in earlier chapters)
- establishing ALARP (see Chapter 10)
- meeting the safe failure fraction requirements (see below)
- meeting the life-cycle activities requirement (Chapter 17)
- demonstrating adequate functional safety competence as an organisation (Chapter 18).

22.2.2 .2 Safe failure fraction (SFF)

Safe Failure Fraction (SFF) is a term used to describe the proportion of failures that are either 'not hazardous' or 'hazardous but revealed by some auto-test'. In other words it is one minus the proportion of 'unrevealed hazardous failures'. The IEC 61508 standard specifies levels of SFF required to claim conformance to a given SIL target according to the amount of redundancy being employed. There are two tables of rules according to whether an item of equipment or component is simple (with well-defined failure modes, known as Type A) or complex (such as a programable instrument, known as Type B). Table 22.3 shows the requirements.

Table 22.3: Requirements for Safe Failure Fraction

	SIL for Simplex HFT 0*	SIL for (m + 1) HFT 1*	SIL for (m + 2) HFT 2*
Type A SFF			
< 60%	1	2	3
61%–90%	2	3	4
91%–99%	3	4	4
> 99%	3	4	4
Type B SFF			
< 60%	NO**	1	2
61%–90%	1	2	3
91%–99%	2	3	4
> 99%	3	4	4

*Simplex is often referred to as a Hardware Fault Tolerance of zero etc. Simplex implies no redundancy; (m + 1) implies 1 out of 2, 2 out of 3, etc. (m + s 2) implies 1 out of 3, 2 out of 4, etc.
**This configuration is not allowed.

22.2.2 .3 Life-cycle activities

A frequent misunderstanding is to assume that, if the qualitative (life-cycle activities) requirements of a particular SIL target are met then the failure rate (or PFD) of that SIL will somehow magically follow. This is certainly not the case since the different requirements of the IEC 61508 standard address different types of failure. Qualitative requirements address systematic failures and the random hardware failures are a matter for the component failure rates, redundancy, proof-test intervals, which are dealt with quite separately.

22.2.2 .4 Functional safety capability

This is now more frequently referred to as functional safety management. It is necessary to be able to demonstrate not only that the above aspects are met but also that the appropriate procedures and competencey criteria are in place to ensure consistency and thus gurantee that they will continue to be met. This is dealt with more fully in Chapter 18.

22.3 Programable Electronic Systems (PESs)

PESs are now the most common form of control or safety system although hardwired systems are still sometimes favored due to their greater visibility in terms of quantified reliability prediction. There has been controversy since the early 1980s concerning the integrity of programable safety-related systems and, as a result, even now, non-programable controls are still widely used.

For many years there was a general principle that no single software error may lead to a hazardous failure. In practice this meant that where programable control and protection equipment was used a hard-wired or even mechanical/pneumatic protection arrangement was also provided. In this way no software error can cause the hazard without a simultaneous non-programable failure. At one time integrity studies concentrated on establishing the existence of this arrangement.

With the emergence of the SIL principle the use of a simplex software-based safety system has become acknowledged as credible, at the lower SIL levels, provided that it can be demonstrated that the design meets the requirements of the SIL.

There are three basic configurations of system:
- a simplex PES acting alone
- one or more PESs acting in combination with one or more non-programable systems (including safety monitors)
- a number of PESs acting in combination (with or without diversity).

22.4 Current Guidance

There are dozens of 'Second tier' guidance and standards documents in this area. A few of the more relevant documents are briefly described here.

22.4.1 IEC International Standard 61508 (2010): Functional safety of electrical/electronic/programmable electronic safety-related systems: *7 parts*

This is the major (umbrella) document concerning functional safety. It is in seven parts although the normative standard is effectively Parts 1–3.

Part 1 deals with the functional safety competence of an organization and with the matter of setting SIL targets. Part 2 addresses hardware (random hardware failures, safe failure fraction and life-cycle activities). Part 3 addresses software (life-cycle activities). The remaining parts of the standard provide definitions, a bibliography and guidance to the first three parts.

The standard was re-issued in 2010 with a number of updated requirements.

22.4.2 IEC International Standard 61511: Functional safety – Safety instrumented systems for the process industry sector

IEC 61511 is intended as the process industry sector implementation of IEC 61508. It gives application-specific guidance on the use of standard products for the use in 'safety instrumented' systems using the proven-in-use justification. The guidance allows the use of field devices to be selected based on proven-in-use for application up to SIL 3 and for standard off-the-shelf PLCs for applications up to SIL 2.

22.4.3 Institution of Gas Engineers and Managers IGEM/SR/15: programmable equipment in safety-related applications – 5th edition

This is the gas industry 2nd tier guidance to IEC 61508 (2010). It is suitable for oil and gas and process applications. SR/15 describes the approaches to establishing target SILs and a preference for the quantitative approach is stressed. Maximum tolerable risk (i.e. fatality) targets are suggested. More specific design guidance is given for pressure and flow control, gas holder control, burner control and process shutdown systems.

22.4.4 European Standard EN 50126: Railway applications – The specification and demonstration of dependability, reliability, maintainability and safety (RAMS)

EN 50126 is effectively the Europe-wide rail industry 2nd tier general guidance (1999) for IEC 61508. It is often referred to as 'the RAMS standard', as it addresses both reliability and safety issues. EN50126 is intended to cover the railway system in total, while the companion standards, EN 50128 and EN 50129, are more specific. CENELEC describes standard 50126 as being '… intended to provide railway authorities and the railway support industry throughout the European Community with a process which will enable the implementation of a consistent approach to the management of RAMS'.

22.4.5 UK Defence Standard 00-56 (Issue 3.0): Safety Management Requirements for Defence Systems

Def Stan 00-56 now replaces a suite of earlier standards (00-54, 00-55, 00-58). It is less prescriptive and places the onus on the supplier to identify hazards and demonstrate how they are to be mitigated. The structure is:

Part 1: Requirements: this is largely an exhortation to establish safety management, identify hazards and establish a safety case that will reflect risk assessments and the subsequent demonstration of tolerable risks following appropriate risk reduction.

Part 2: Code of Practice: provides more detail on the practices to be adopted to satisfy Part 1. It comprises four volumes:
1.   Interpretation of Part 1: somewhat repetitive, although with more detail such as items for the content of safety cases, aspects of hazard identification, etc.
2.   Risk management: addressing issues such as HAZID, risk classification and SILs (previously covered in 00-56 Issue 2.0).
3.   Software: a successor to 00-55 (below).
4.   Electronic hardware: a successor to 00-54 (below).

22.4.6 RTCA DO-178B/(EUROCAE ED-12B): Software Considerations in Airborne Systems and Equipment Certification

This is a very detailed and thorough standard that is used in civil avionics to provide a basis for certifying software used in aircraft. Drafted by a EUROCAE/RTCA committee, DO-178B was published in 1992 and replaces an earlier version published in 1985. The qualification of software tools, diverse software, formal methods and user-modified software are now included. It defines five levels of software criticality from A (software that can lead to catastrophic failure) to E (no effect). The standard provides guidance that applies to levels A to D.

22.4.7 Documents Related to Machinery

There are three standards relevant to this area. (a) EN ISO 14121 Principles of Risk Assessment, (b) EN 62061 *Functional Safety of E/E/PES*, (c) EN ISO 13849 *Safety Related Parts of Control Systems*. EN ISO 14121 provides guidance on undertaking general risk assessments associated with a machine and, if it is found necessary to provide risk reduction using an active interlock/control mechanism, the evaluation of both the requirements and design of this interlock/control mechanism can be undertaken by using either EN ISO 13849 or EN 62061.

22.4.8 Other Industry Sectors

Other major sectors for which similar guidance documents exist are automotive, nuclear, medical, earthmoving, stage & entertainment and electrical power devices. This is a rapidly changing picture and a book (subject to five-yearly updates) cannot possibly provide an up-to-date picture.

22.4.9 Technis Guidelines, Q124, 2010: Demonstration of product/system compliance with IEC 61508

This 32-page document provides a framework for demonstration/certification of either products or systems (be that by self-demonstration, third-party assessment or certifying body). It is intended for use by experienced functional safety professionals and offers a realistic level of rigor whilst allowing assessors scope for interpretation. It is available from Technis (see end of this book).

22.5 Framework for Certification

It is becoming increasingly necessary to demonstrate (or even certify) conformance to the requirements of IEC 61508. This has been driven by customer demands for certification, coupled with suppliers' aspirations not to be 'left out' of the trend. There are two types of certification.

First: that an organization can demonstrate the generic capability to produce such a product or system (i.e. that it has the necessary procedures and competence in place).

Second: that a specific product or system design meets the requirements outlined in the preceding chapters (i.e. that the above procedures have been implemented).

In the first case it is the raft of procedures and work practices, together with the competence of individuals that is being assessed. This is known as the Functional Safety Capability (FSC) of an organization and is now more commonly referred to as Functional Safety Management (FSM). It is demonstrated by an appropriate quality management system and evidenced by documented audits and examples of the procedures being used.

In the second it is the design and the life-cycle activities of a particular product that are being assessed. This is demonstrated by specifications, design documents, reviews, test specifications and results, failure rate predictions, FMEAs to determine safe failure fraction and so on.

In practice, however, it is not really credible to assess one of the above without evidence of the other. FSM needs to be evidenced by at least one example of a product or project and a product's conformance needs to be evidenced by documentation and life-cycle activities that show overall capability. The options for demonstrating conformance to IEC 61508 safety-integrity requirements are as follows.

22.5.1 Self-Certification

An organization is free to carry out the various assessment tasks (e.g. quantifying the predicted hardware failure rates, assessing safe failure fraction, recording the life-cycle activities) and to produce an assessment report to demonstrate conformance. Clearly the acceptability of such a claim will depend upon the organization's reputation in respect of such work.

22.5.2 Third-Party Assessment

In this case the above activities might be facilitated, or even carried out, by an external body or individual. Again, the acceptability of such work will depend upon the reputation of the external assessor.

22.5.3 Use of a Certifying Body

Additional confidence can be assumed if a certifying body is used and is based on the reputation of the body. In practice, for IEC 61508, there are only two UK-based bodies (at the time of writing) who offer certification. They both have various levels of UKAS accreditation for that certification activity.

Figure 22.3 shows the arrangement at the time of writing.

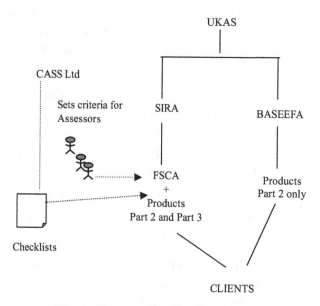

Figure 22.3: Certification framework.

CHAPTER 23

A Case Study: The Datamet Project

This chapter is a revised case study that has been used by the author on reliability courses for over twenty-five years. It is not intended to represent any actual company or product.

The section entitled 'Syndicate Study' suggests a number of areas for thought and discussion. When discussing the contract clauses, two syndicates can assume the two roles of producer and customer, respectively. After separate discussion, the two syndicates can renegotiate the contract under the guidance of a course tutor. This approach has proved both stimulating and effective. It is worth bearing in mind, when criticising the contract clauses, that although the case study is fictional, the clauses were drawn from actual examples.

23.1 Introduction

The communications division of electrosystems Ltd has an annual turnover of £30 million as follows:

	Line Communications	Radio Systems	New Products
UK	£20 million	£4 million	£0.5 million
Export	£2 million	£2 million	£1.5 million

Line communications products include voice and data communications over copper and fiber cable. Radio systems include vhf and uhf voice systems for both onshore and offshore applications. The new products department handles major developments and any new communications projects.

23.2 The Datamet Concept

An overseas inquiry for a remote meteorological telemetry system had been received. It involved a requirement to scan a number of weather-recording instruments and to transmit the information to a terminal station. Each observation would involve an instrument providing an analog signal to a unit, which would poll the instruments once a minute. The information would be sent by frequency-modulated vhf carrier. A number of stations would be associated with a terminal and would transmit on different carrier frequencies.

Reliability, Maintainability and Risk. DOI: 10.1016/B978-0-08-096902-2.00023-4

The inquiry was for 10 such systems, each having 10 remote (unattended) sites. A price in the region of £2 million was thought to be likely and there was believed to be the possibility of additional sales over the next five years, elsewhere, in the order of £6 million.

A project group was formed, in the new products department, and a conceptual design emerged (illustrated in Figure 23.1).

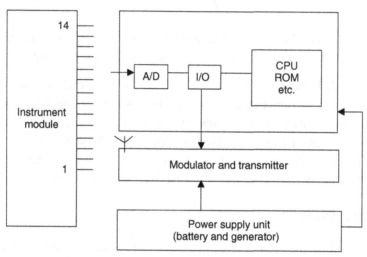

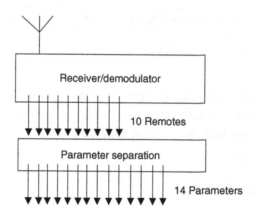

Figure 23.1: Conceptual design

The instrument module provided 14 channels, scanned by a microprocessor-controlled signal processing board. The digital output from this board was used to modulate a vhf carrier. Power was supplied by rechargeable batteries and a solar charger. In the event of charger failure the battery life was seven days.

The potential customer's requirement was that 5 of the 14 weather inputs were classified as major (barometric pressure, wet and dry bulb temperature, wind direction and speed). A 10-year MTBF was required for loss of any major parameter.

Meteorological instruments were available from a specialist supplier. The supplier showed evidence of seven contracts, each for 50 instruments, over a period of five years. Assuming that these were sold evenly across the period, this represents usage of $7 \times 50 \times 5 \times 8760/2 = 7.665$ million hours. Documentation was tabled relating to 15 instruments returned as failed, of which three had clearly been mistreated. The remaining 12 failures involved the various types of transducer and had resulted in grossly inaccurate, or loss of, readings. The pessimistic assumption was made that this reporting represented 50% of actual field failures and hence $2 \times 12 = 24$ were used to assess field reliability. Hence a failure rate of 24/7.665 million = **3.1×10^{-6} per hour** was inferred. In view of the large number of failures, statistical inference at a confidence level (i.e. chi-square) was not appropriate. The Technis FARADIP.THREE database suggests a range of failure rates for pressure/flow/level/temperature instrumentation (given the assumption of a proven history of use) in the range 1–5 pmh. This adds further credibility to the above inference.

An approximate interpretation of the 12 times to failure suggested, from the evidence, that they occurred after the following times (in months) in service: 6, 9, 11, 15, 17, 18, 28, 30, 33, 45, 55, 57. Furthermore there were 40 items that did not fail and which were in service for 60 months. Using the Technis COMPARE package, Weibull analysis (explained in Chapter 6) was carried out to infer a shape parameter of 1.05 with a significance level of 76% that the data did not indicate other than constant failure. On that basis it was assumed that any wearout mechanisms inherent in the instruments would not manifest themselves until after 60 months (5 years).

Failure mode and effect analysis of the proposed units (based on preliminary circuit diagrams) made the following total failure rate assessments:

Switch board	2 pmh
Communications board	2.4 pmh
Power board	0.79 pmh
Battery	0.5 pmh

Assuming, pessimistically, that all failures led to a loss of a major parameter, the predicted failure rate was:

$$[5 \text{ (major instruments)} \times 3.1] + 2 + 2.4 + 0.79 + 0.5 = 21.19 \times 10^{-6} \text{ per hour}$$

which is 5.4 years MTBF, being only 50% of the above target.

Looking at the accuracy of reliability prediction (in Chapter 4, Section 4.4) it can be seen that (using generic data) a predicted MTBF of three times the requirement would be required in order to be 60% sure of meeting the requirement.

The contribution to the failure rate from the instruments was $(5 \times 3.1)/21.19 = 73\%$ and, therefore, it was proposed to suggest a modification to the customer whereby the five major parameters used duplicated (voted) instruments, taking up 10 of the 14 inputs. The remaining four inputs would service less-essential parameters (fog, ground temperature, rainfall, brightness). It was explained that this reduction of five minor input measurements was not an unrealistic proposal in order to offer a more robust design with a better assurance of meeting the MTBF requirement. This was reinforced by the above prediction, which indicated that the customer's original requirement would be unlikely to be met, given the state of the art.

The reliability block diagram in Figure 23.2 models 'loss of a major parameter'. The common cause failure of two identical instruments was assessed using the BETAPLUS model (Chapter 8, Section 8.2.3) and suggested a BETA factor of 10%.

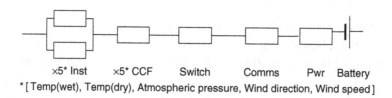

×5* Inst   ×5* CCF   Switch   Comms   Pwr   Battery
* [Temp(wet), Temp(dry), Atmospheric pressure, Wind direction, Wind speed]

Figure 23.2

Assuming a six month (4000 hour) proof-test interval then quantifying the model provides:

$$5 \times (3.1 \times 10^{-6})^2\, 4000 + 5 \times 10\% + 3.1 \times 10^{-6} + 2 \times 10^{-6} + 2.4 \times 10^{-6}$$

$$+ 0.79 \times 10^{-6} + 0.5 \times 10^{-6} = 7.43 \times 10^{-6} \text{ per hour}$$

which is an MTBF of 15 years.

Although only 1.5 times better than the requirement, since a conservative approach had been taken in the failure rate assessment, it was decided to proceed with the design.

23.3 The Contract

The customer stated an intention to place a contract for the 10 systems and tabled a requirements specification together with the draft contract terms shown in Table 23.1.

Table 23.1

(a) Ten years MTBF is required for loss of any major weather parameter (of which there will be five parameters) from each remote station. The supplier will satisfy the customer, by means of a reliability prediction, that the design is capable of meeting this requirement.
(b) The equipment must be capable of operating in a temperature range of 0–50°C with a maximum relative humidity of 80%.
(c) Failure shall consist of the loss of a major parameter or its incorrect measurement.
(d) For two year's operation of the equipment, the contractor will refund the cost of all replacements to the equipment. When a corrective maintenance visit, other than its coinciding with a scheduled preventive maintenance visit, is required the contractor will refund all labor and traveling costs including overtime and incentives at a rate to be agreed.
(e) In the event of a system failure, the maximum repair time to restore the terminal to effective operation shall be one hour. The contractor is required to show that the design is compatible with this target.
(f) In the event of systematic failures, the contractor shall perform all necessary design work and make the necessary modifications to the system.
(g) The contractor is to use components having the most reasonable chance of being available throughout the life of the equipment and is required to state shelf life and number of spares to be carried in the case of any components that might cease to be available.
(h) Interchangeable printed cards may be employed and a positive means of identifying which card is faulty must be provided so that, when the fault occurs, it can be rectified with the minimum effort and skill. The insertion of cards in the wrong position shall be impossible or shall not cause damage to the cards or system.
(i) Maintenance instructions will be provided by the contractor and shall contain all necessary information for the checking and maintenance of the system. These shall be comprehensive and give full operational and functional information. The practice of merely providing a point-to-point component description of the circuits will not, in itself, be adequate.

The contract was signed and detailed development commenced. A 40-year MTBF commitment was placed on the instrument supplier. The cost of a penalty repair visit was estimated to be £2000.

23.4 Detailed Design

Later in the detailed design phase, a more detailed failure mode and effect analysis was carried out for the specific failure mode, 'loss of a major parameter'. It produced a predicted MTBF of 19 years, which provided additional confidence in the design.

At the beginning of the integration test phase a failure reporting form was devised for recording both test and field failures. The customer was approached to the effect that, in view of the penalty clause, full maintenance reporting documentation would be required to accompany all field claims.

It was noticed, during functional test, that a number of failures were due to timing and component tolerance problems rather than to catastrophic component failure.

Late in the development phase the instrument supplier went into liquidation and only 60 of the 100 sets of instruments, necessary for the initial project, could be supplied. Another was found but at 20% greater cost.

23.5 Syndicate Study

First Session

Comment on the development and identify key areas of risk.

Second Session

1. Discuss the contract clauses and construct alternatives either as:
    (i) the Producer
    (ii) the Customer.
2. Set up a role-playing negotiation.

23.6 Hints

Project

Having regard to the project size, as a percentage of turnover, decide if the project risk issues were adequately addressed.

*   Where is the most likely area of risk?
*   Were any activities omitted?
*   Should this have been undertaken as a development contract with separate production contracts to follow?
*   Is MTBF the appropriate parameter for describing the performance (would not the customer have responded to the suggestion of an availability target?).
*   The single source of instruments (a critical component) was not wise. The field reliability data did not apply to the second supplier thus reducing the value of the prediction.
*   What meaning had the 40-year MTBF requirement (on its own) in respect of instrument procurement?
*   The drift-related failures, observed during development, were in addition to the random hardware failures assumed in the FMEA model.
*   The second FMEA offered a 20% improvement – was this significant?

Contract

(a) Are these appropriate parameters?
(b) Is the description of the environment (in the contract) realistic? Rate of change is more important than absolute values.
(c) Is the parameter? 'Incorrect' (See Chapter 19, Section 19.6)
(d1) When does the 'two years' start?
(d2) If the MTBF target is met, what would the penalty visits represent as a percentage of the contract price? What is the risk that the MTBF will be half of that target?
(e) Maximum repair time! Can this be stipulated?
(f) What criteria will allow a failure to be called systematic?

CHAPTER 24

A Case Study: Gas Detection System

This chapter is a case study based on a typical gas detection system. It provides an example of the safety-integrity targeting described in Chapter 22. The topic is fully covered in *The Safety Critical Systems Handbook*, D. J. Smith and K. G. L. Simpson, 3rd edition, ISBN 9780080967813.

Assume that our gas detection system has the primary function of providing an executive action input to some other 'Emergency Shutdown' system in order to protect a chemical process plant. This ESD system will close valves in order to isolate the flammable gases so as to prevent subsequent explosion or fire. It is assumed that successful operation of the gas detection and ESD systems will prevent fatality.

24.1 Safety-Integrity Target

In order to establish a safety-integrity target, it is necessary to establish a maximum tolerable failure rate (or probability of failure on demand) for the safety-related system in question. The table of SIL targets (from IEC 61508) was shown and explained in Chapter 22.

If a maximum tolerable failure rate is to be established then, in turn, a maximum tolerable risk of fatality must first be stated. There are no hard and fast rules but a typical guide, from the Institution of Gas Engineers and Managers Recommendations SR/15, suggests:

Scenario	Maximum Individual Risk of Fatality
voluntary (employee)	10^{-4} pa
involuntary (public)	10^{-5} pa

It might be argued that the operatives in this plant are accepting a voluntary risk and that the 10^{-4} per annum should apply. It might also be argued that there is approximately an order of magnitude of plant-related hazards that could lead to fatality and, thus, the maximum tolerable risk in respect of this gas detection-related hazard should be an order less, namely **10^{-5} per annum**. For the particular application, the following judgements were made:

Person(s) at risk (16 hours per day)	67%
Probability of gas release finding a source of ignition	80%
Probability that subsequent fire/explosion leads to fatality	50%

Reliability, Maintainability and Risk. DOI: 10.1016/B978-0-08-096902-2.00024-6

It follows that the **maximum tolerable failure rate** for this event (given the above **maximum tolerable risk**) is:

$$10^{-5}\,\text{pa}/(0.67 \times 0.8 \times 0.5) = 3.7 \times 10^{-5}\,\text{pa}$$

Further, assume that a fault tree has been constructed to model all the causes of a suitably large gas release which places a demand on this system of mitigation, and that it predicts a demand of 1.5×10^{-3} per annum.

It follows that the **maximum tolerable probability of failure on demand** for the mitigation is:

$$3.7 \times 10^{-5}\,\text{pa}/1.\,5 \times 10^{-3}\,\text{pa} = 2.5 \times 10^{-2}$$

Since neither the gas detection nor the ESD systems must fail, then this target has to be shared between them. Assume that, for other reasons, the probability of failure on demand target for the ESD system has been established as 5×10^{-3}.

It follows that our gas detection system attracts a **maximum tolerable probability of failure on demand (PFD) target of $2.5 \times 10^{-2} - 5 \times 10^{-3} = 2 \times 10^{-2}$ which imposes a SIL 1 target** (see Chapter 22, Table 22.1)

24.2 Random Hardware Failures

One of the five things that IEC 61508 requires to be demonstrated in respect of the SIL target (Chapter 22) is that the hardware reliability meets the target failure rate or PFD for the hazardous failure mode in question. In this case we are required to show a PFD of 2×10^{-2} or better for the failure mode 'fail to detect or respond to a valid gas concentration'.

Each area is served by a group of three pellister hydrocarbon gas detectors each of which is assumed to be able to sense any given release. Any one detector sensing gas will give rise to an alarm. This, with operator response, may well be sufficient to mitigate the hazards. However, the safety function addressed here is as follows. Any two out of three detectors sensing gas will give rise to an executive output, which is the signal to the ESD. It is this latter function that is the SIL 1 targeted safety-function in question.

A simplified but realistic reliability block diagram of the gas detection system is shown in Figure 24.1. Each group of three detectors is spread over three separate input PLC cards in order to maximize the redundancy. The executive action is provided by a single (unvoted) output card with a normally de-energized relay operating to open a closed contact that provides an input signal to the ESD system. The failure data used for the prediction (taken from the TECHNIS FARADIP.THREE data bank) are:

Pellister gas detector	Fail to respond	5×10^{-6} per hr
PLC input card	Fail to respond	0.05×10^{-6} per hr
PLC processor card	Fail to output	0.05×10^{-6} per hr
PLC output card	Fail to release relay	0.05×10^{-6} per hr
Relay coil	Fail to energize	0.05×10^{-6} per hr
Relay contact	Contact s/c	0.045×10^{-6} per hr

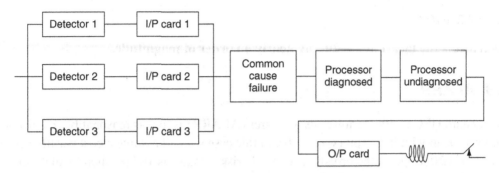

Figure 24.1: Reliability block diagram

Assume that 90% of the dormant failures in the processor module are diagnosed by auto-test and are corrected in one week. Assume, also, that the remainder are revealed by proof test and that the proof-test interval is 8000 hours (i.e. *c.* 12 months).

A common cause BETA factor of 10% is assumed. The block diagram is quantified using the mathematics from Chapters 8 and 9. The probability of failure on demand is:

(a) *Triplicated detector/input card (2 out of 3)*:

$$\lambda^2 T^2 \text{ (from Table 8.4)} = (5 \times 10^{-6} + 0.05 \times 10^{-6})^2 \times 8000^2 = 1.6 \times 10^{-3}$$

(b) *Common cause failure*:

$$10\% \lambda\ T/2 = 10\% \times (5 \times 10^{-6} + 0.05 \times 10^{-6}) \times 4000 = 2 \times 10^{-3}$$

(c) *Processor*:

Diagnosed (by auto-test) failures:

$$90\%\ \lambda\ \text{MDT} = 90\% \times (0.05 \times 10^{-6}) \times 168 = 7.6 \times 10^{-6}$$

Undiagnosed (by auto-test) failures:

$$10\%\ \lambda\ T/2 = 10\% \times 0.05 \times 10^{-6} \times 4000 = 2 \times 10^{-5}$$

(d) *Output card*:

$$\lambda\ T/2 = 0.05 \times 10^{-6} \times 4000 = 2 \times 10^{-4}$$

(e) *Relay coil fails to operate*:

$$\lambda\ T/2 = 0.05 \times 10^{-6} \times 4000 = 2 \times 10^{-4}$$

(f) *Relay contact fails to open*:

$$\lambda\ T/2 = 0.045 \times 10^{-6} \times 4000 = 1.8 \times 10^{-4}$$

Total = 4.2 × 10^{-3}

Which meets the target of 2 × 10^{-2} by nearly an order of magnitude.

24.3 ALARP

The Standard (IEC 61508) requires us to address ALARP (as low as reasonably practicable), which was dealt with in Chapters 3 and 10. In this case we will assume a cost per life saved criterion of £2 000 000. If the broadly acceptable risk is taken as 10^{-6} pa, then by the same token as above we will reduce this by ten to 10^{-7} pa for this hazard.

Now, if the maximum tolerable PFD of 2×10^{-2} corresponds to a risk of 10^{-5} pa, then the 4.2×10^{-3} (predicted) corresponds to a risk of 10^{-5} pa × $4.2 \times 10^{-3}/2 \times 10^{-2} = 2.1 \times 10^{-6}$ pa. This is greater than the broadly acceptable risk and we argue as follows.

Any proposal that would reduce the risk further (to the broadly acceptable level) can be tested, given a two-fatality scenario and a 25-year plant life, by:

$$£2\,000\,000 = (\text{Max cost of proposal})/[(2.1 \times 10^{-6} - 1 \times 10^{-7}) \times 2 \text{ fatalities} \times 25 \text{ years}]$$

Therefore, max. cost of proposal = £200.

Since it is unlikely that any further risk reduction can be achieved within this sum, ALARP might be argued to have been satisfied.

24.4 Architectures

In Chapter 22 a brief explanation of the safe failure fraction was given.

In this case study the SIL 1 target imposes safe failure fraction targets as follows:

The single PLC, being a Type B device:
  Greater than 60%. It is very likely that this could be demonstrated.

The voted pellister gas detector, assuming it can be treated as a Type A device:
  No special requirement.

24.5 Life-Cycle Activities

It will be necessary to demonstrate that the appropriate life-cycle activities have been carried out.

24.6 Functional Safety Capability

It is also necessary to demonstrate not only that the life-cycle activities have been carried out, but also that the company's management system ensures that they will always be carried out. This involves competencies and quality management procedures.

CHAPTER 25

A Case Study: Pressure Control System

This chapter is a case study based on a typical pressure reduction system. It provides an example of the safety-integrity targeting described in Chapter 22. This is Chapter 11 of *The Safety Critical Systems Handbook*, D. J. Smith and K. G. L. Simpson, 3rd edition, ISBN 9780080967813.

This exercise is based on a real scenario. Spaces have been left for the reader to attempt the calculations. The answers are provided in Appendix 8.

25.1 The Unprotected System

Consider a plant supplying gas to offsite via a twin-stream pressure control station. Each stream is regulated by two valves (top of Figure 25.1). Each valve is under the control of its downstream pressure. Each valve is closed by the upstream gas pressure via its pilot valve, J, but only when its pilot valve, K1, is closed. Opening pilot valve K1 relieves the pressure on the diaphragm of valve, V, allowing it to open. Assume that a HAZOP (HAZard and OPerability) study of this system establishes that downstream overpressure, whereby the valves fail to control the downstream pressure, is an event which could lead to one or more fatalities.

Since the risk is offsite, and a two-fatality scenario assumed, a target maximum tolerable risk of 10^{-5} per annum has been proposed.

Assume that a quantified risk assessment has predicted a probability of 20% that failure, involving overpressure, will lead to subsequent pipe rupture and ignition. Furthermore it is predicted that, due to the high population density, fatality is 50% likely.

Assume also that the plant offers approximately 10 risks in total to the same population (e.g. tanker deliveries, other pipelines, site explosion).

It follows that the target failure rate for overpressure of the twin stream sub-system is

$$[10^{-5}/[10\,\text{risks} \times 0.2 \times 0.5] = \mathbf{10^{-5}\ pa}$$

Assume, however, that field experience of a significant number of these twin-stream systems shows that the frequency of overpressure is dominated by the pilots and is $\mathbf{2.5 \times 10^{-3}\,pa}$.

Reliability, Maintainability and Risk. DOI: 10.1016/B978-0-0809-6902-2.00025-8

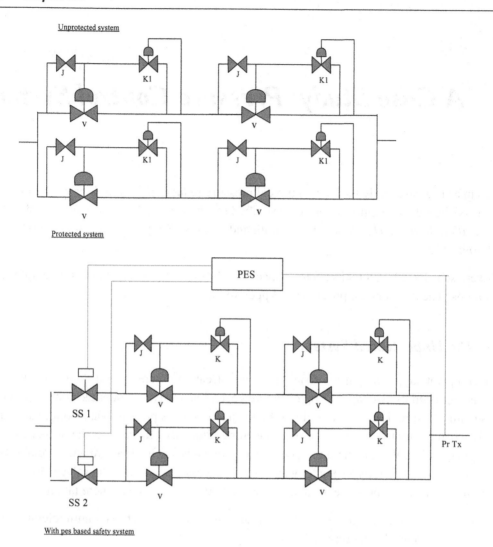

Unprotected system

Protected system

With pes based safety system

Figure 25.1: The system, with and without backup protection

25.2 Protection System

Since 2.5×10^{-3} is greater than 10^{-5}, a design modification is proposed whereby a programable electronic system (PES) closes a valve in each stream, based on an independent measure of the downstream pressure. The valves consist of actuated ball valves (sprung to close). This is illustrated at the bottom of Figure 25.1.

The target unavailability for this 'add-on' safety system is therefore ?..................

Which indicates a SIL of ?............

25.3 Assumptions

The following assumptions are made in order to construct and quantify the reliability model:

(a) Failure rates (symbol λ), for the purpose of this prediction, are assumed to be constant with time. Both early and wearout-related failures are assumed to be removed by burn-in and preventive replacement respectively.

(b) The MTTR (mean time to repair) of a revealed failure is four hours.

(c) The auto-test coverage of the PLC is 90% and occurs at just under five minute intervals. The MDT (mean down time) for failures revealed by this PES auto-test are taken to be the same as the MTTR (mean time to repair) because the MTTR > the auto-test period. The MDT is thus assumed to be four hours. Neither the pressure transmitter nor the valve is assumed to have any self diagnostics.

(d) The manual proof-test is assumed to be 100% effective and to occur annually (*c.* 8000 hours).

(e) One maintenance crew is assumed to be available for each of the three equipment types (PES, instrumentation, pneumatics).

(f) The detailed design assumptions needed for an assessment of the common cause failure BETA factor (see modified proposal) are summarized in Section 25.8.

25.4 Reliability Block Diagram

Figure 25.2 is the reliability block diagram for the add-on safety system. Note that the PES will occur twice in the diagram. This is because the model needs to address those failures revealed by auto-test separately from those revealed by the longer manual proof-test due to their different MDTs.

TO BE FILLED IN BY THE READER (See Appendix 8 for answer)

Figure 25.2: Reliability block diagram.

25.5 Failure Rate Data

The following failure rate data will have been chosen for the protection system components, shown in Figure 25.1. These are the component level failure modes that lead to the hazard under consideration (i.e. downstream overpressure). FARADIP.THREE has been used to obtain the failure rates.

Item	Failure Mode	Failure Rates 10^{-6} per hour	
		Total	Mode
PES	PES low or zero*	5	0.25
Pressure transmitter	Fail low	2	0.5 (25% has been assumed)
Actuated ball valve (sprung to close)	Fail to close	8	0.8**

*This represents any failure of the PES i/p, CPU or o/p causing the low condition.
**10% has been used based on the fact that the most likely failure mode is fail closed.

25.6 Quantifying the Model

The following unavailability calculations address each of the groups (left to right) in Figure 25.2 (see Appendix 8):

(a) Ball valve 1 – unrevealed failures

Unavailability = ...

= ....

(b) Ball valve 2 – unrevealed failures

Unavailability = ...

= ....

(c) PES output 1 failures revealed by auto-test

Unavailability = ...

= ....

(d) PES output 1 failures not revealed by auto-test

Unavailability = ...

= ....

(e) PES output 2 failures revealed by auto-test

Unavailability = ...

= ....

(f) PES output 2 failures not revealed by auto-test

Unavailability = ...

= ....

(g) Pressure Transmitter – unrevealed failures

Unavailability = ...

= ....

The predicted unavailability is obtained from the sum of the unavailabilities in (a) to (e)

= ....?

25.7 Proposed Design and Maintenance Modifications

The proposed system is not acceptable (as can be seen in Appendix 8) and modifications are required.

Before making modification proposals it is helpful to examine the relative contributions to system failure of the various elements in Figure 25.2.

....% from items (a) and (b) ball valve.
....% from items (c) to (f) the PES.
....% from item (g) the pressure transmitter

It was decided to duplicate the pressure transmitter and vote the pair (one out of two). It was also decided to reduce the proof test interval to six months (c4000 hrs).

TO BE FILLED IN BY THE READER (See Appendix 8 for answer) Figure 25.3

25.8 Modeling Common Cause Failure (Pressure Transmitters)

The BETAPLUS method provides a method for assessing the percentage of common cause failures. The scoring for the method was carried out assuming:

- written procedures for system operation and maintenance are evident but not extensive
- there is some training of all staff in CCF awareness
- extensive environmental testing was conducted
- identical (i.e. non-diverse) redundancy
- basic top level FMEA (failure mode analysis) had been carried out

Figure 25.3: Revised reliability block diagram (or fault tree).

- there is some limited field failure data collection
- simple, well proven, pressure transmitters ½ meter apart with cables routed together
- good electrical protection
- annual proof test.

The BETAPLUS software package performs the calculations and was used to calculate a BETA value of 9%.

25.9 Quantifying the Revised Model

The following takes account of the pressure transmitter redundancy, common cause failure and the revised proof test interval. Changed figures are shown in bold in Appendix 8.

Changed figures are shown in bold.

(a) Ball valve SS1 fails open.
   Unavailability = ...
   = ...

(b) Ball valve SS2 fails open.
   Unavailability = ...
   = ...

(c) PES output 1 fails to close valve (undiagnosed failure).
   Unavailability = ...
   = ...

(d) PES output 2 fails to close valve (undiagnosed failure).
   Unavailability = ...
   = ...

(e) PES output 1 fails to close valve (diagnosed failure).
   Unavailability = ...
   = ...

A Case Study: Pressure Control System

(f) PES output 2 fails to close valve (diagnosed failure).
  Unavailability = ...

  = ...

(g) Voted pair of pressure transmitters.
  Unavailability = ...

  = ...

(h) Common cause failure of pressure transmitters.
  Unavailability = ...

  = ...

  The predicted unavailability is obtained from the sum of the unavailabilities in (a) to (h)
  = ...?

25.10 ALARP

Assume that further improvements in CCF can be achieved for a total cost of £1000. Assume, also, that this results in an improvement in unavailability to **4×10^{-4}**. It is necessary to consider, applying the ALARP principle, whether this improvement should be implemented.

The cost per life saved over a 40- year life of the equipment (without cost discounting) is calculated, assuming two fatalities, as follows:

  ...... (see Appendix 8)

25.11 Architectural Constraints

Consider the architectural constraints imposed by IEC 61508 Part 2, outlined in Chapter 22.

Do the pressure transmitters and valves in the proposed system, meet the minimum architectural constraints assuming they are 'TYPE A components'?

Does the PES in the proposed system meet the minimum architectural constraints assuming it is a 'TYPE B component'?

Glossary

A1.1 Terms Related to Failure

A1.1.1 Failure

Termination of the ability of an item to perform its specified function. OR, non-conformance to some defined performance criteria. Failures may be classified by:	Meaningless without perfor-mance spec	
1. Cause –		Chapter 2
Misuse: Caused by operation outside specified stress.		
Primary: Not caused by an earlier failure.		
Secondary: Caused by an earlier failure.		
Wearout: Caused by accelerating failure rate mechanism.		
Design: Caused by an intrinsic weakness.		
Software: Caused by a program error despite no hardware failure		Chapter 17
2. Type –		
Sudden: Not anticipated and no prior degradation.		
Degradation: Parametric drift or gradual reduction in performance.		
Intermittent: Alternating between the failed and operating condition.		
Dormant: A component or unit failure that does not cause system failure but that either hastens it or, in combination with another dormant fault, would cause system failure.		
Random: Failure is equally probable in each successive equal time interval.		
Catastrophic: Sudden and complete.		

A1.1.2 Failure Mode

The outward appearance of a specific failure effect (e.g. open circuit, leak to atmosphere).		Chapter 2

A1.1.3 Failure Mechanism

The physical or chemical process that causes the failure.		Chapter 11

Reliability, Maintainability and Risk. DOI: 10.1016/B978-0-08-096902-2.00026-X

A1.1.4 Failure Rate

The number of failures of an item per unit time. This can be applied to: 1. Observed failure rate: as computed from a sample. 2. Assessed failure rate: as inferred from sample information. 3. Extrapolated failure rate: projected to other stress levels.		Per hour, cycle, operation, etc. Point estimate Involves a confidence level

A1.1.5 Mean Time Between Failures and Mean Time to Fail

The total cumulative functioning time of a population divided by the number of failures. As with failure rate, the same applies to observed, assessed and extrapolated MTBF. MTBF is used for items that involve repair. MTTF is used for items with no repair.		

A1.1.6 Common Cause Failure

The result of an event(s) that, because of dependencies, causes a coincidence of failure states of components in two or more separate channels of a redundant system, leading to the defined system failing to perform its intended function.		Section 8.2

A1.1.7 Common Mode Failure

A subset of *Common Cause Failure* whereby two or more components fail in the same manner.		Section 8.2

A1.2 Reliability Terms
A1.2.1 Reliability

The probability that an item will perform a required function, under stated conditions, for a stated period of time. Since observed reliability is empirical it is defined as the ratio of items that perform their function for the stated period to the total number in the sample.		

A1.2.2 Redundancy

The provision of more than one means of achieving a function. Active: All items remain operating prior to failure. Standby: Replicated items do not operate until needed.		

A1.2.3 Diversity

The same performance of a function by two or more independent and dissimilar means (of particular relevance to software).		Chapter 17

A1.2.4 Failure Mode and Effect Analysis

Determining the outcomes of all known failure modes within an assembly or circuit.		Section 9.3

A1.2.5 Fault Tree Analysis

A graphical method of modeling a system failure using AND and OR logic in tree form.		Section 8.3

A1.2.6 Cause Consequence Analysis (Event Trees)

A graphical method of modeling one or more outcomes of a failure or of an event by means of interconnected YES/NO decision boxes.		Section 8.4

A1.2.7 Reliability Growth

Increase in reliability as a result of continued design modifications resulting from field data feedback.		Section 12.3

A1.2.8 Reliability Centered Maintenance

The application of quantified reliability techniques to optimize discard, times, proof-test intervals and spares levels.		Chapter 16

A1.3 Maintainability Terms
A1.3.1 Maintainability

The probability that a failed item will be restored to operational effectiveness within a given period of time when the repair action is performed in accordance with prescribed procedures.		

A1.3.2 Mean Time to Repair (MTTR)

The mean time to carry out a defined maintenance action.	Usually refers to corrective maintenance	

A1.3.3 Repair Rate

The reciprocal of MTTR.	When used in reliability calculations it is the reciprocal of down time	

A1.3.4 Repair Time

The time during which an item is undergoing diagnosis, repair, checkout and alignment.	Must be carefully defined; may also depend on diagnostics	Chapter 14 and Section 9.2

A1.3.5 Down Time

The time during which an item is not able to perform to specification.	Must be carefully defined	

A1.3.6 Corrective Maintenance

The actions associated with repair time.		

A1.3.7 Preventive Maintenance

The actions, other than corrective maintenance, carried out for the purpose of keeping an item in a specified condition.		

A1.3.8 Least Replaceable Assembly (LRA)

That assembly at which diagnosis ceases and replacement is carried out.	Typically a printed-board assembly	

A1.3.9 Second-Line Maintenance

Maintenance of LRAs that have been removed from the field for repair or for preventive maintenance.		

A1.4 Terms Associated with Software

A1.4.1 Software

All documentation and inputs (for example, tapes, disks) associated with programable devices.		Chapter 17

A1.4.2 Programable Device

Any piece of equipment containing one or more components that provides a computer architecture with memory facilities.		

A1.4.3 High-Level Language

A means of writing program instructions using symbols each of which represents several program steps.		

A1.4.4 Assembler

A program for converting program instructions, written in mnemonics, into binary machine code suitable to operate a programable device.		

A1.4.5 Compiler

A program that, in addition to being an assembler, generates more than one instruction for each statement thereby permitting the use of a high-level language.		

A1.4.6 Diagnostic Software

A program containing self-test algorithms enabling failures to be identified.	Particularly applicable to ATE	

A1.4.7 Simulation

The process of representing a unit or system by some means in order to provide some or all identical inputs, at some interface, for test purposes. A means of prediction.		

A1.4.8 Emulation

A type of simulation whereby the simulator responds to all possible inputs as would the real item and generates all the corresponding outputs.	Identical to the real item from the point of view of a unit under test	

A1.4.9 Load Test

A system test involving simulated inputs in order to prove that the system will function at full load.		

A1.4.10 Functional Test

An empirical test routine designed to exercise an item such that all aspects of the software are brought into use.		

A1.4.11 Software Error

An error in the digital state of a system that may propagate to become a failure.		

A1.4.12 Bit Error Rate

The random incidence of incorrect binary digits.	Expressed 10^{-x}/bit	

A1.4.13 Automatic Test Equipment (ATE)

Equipment for stimulus and measurement controlled by a programed sequence of steps (usually in software).		

A1.4.14 Data Corruption

The introduction of an error by reason of some change to the software already resident in the system. This could arise from electrical interference or from incorrect processing of a portion of the software.		

A1.5 Terms Related to Safety

A1.5.1 Hazard

A scenario whereby there is a potential for human, property or environmental damage.		

A1.5.2 Major Hazard

A general, imprecise, term for large-scale hazards as, for example, in the chemical or nuclear industries.		

A1.5.3 Hazard Analysis

A term that refers to a number of techniques for analyzing the events leading to a hazardous situation.		Chapter 10

A1.5.4 HAZOP

Hazard and Operability Study. A formal analysis of a process or plant by the application of guidewords.		Chapter 10

A1.5.5 LOPA

Levels of Protection Analysis. A systematic way of crediting levels of risk reduction in order to calculate any residual requirement for additional risk reduction.		Chapter 22

A1.5.6 Risk

The likelihood, expressed either as a probability or as a frequency, of a hazard materializing.		Chapters 3 and 10

A1.5.7 Consequence Analysis

Techniques that involve quantifying the outcome of failures in terms of dispersion, radiation, fatality, etc.		

A1.5.8 Safe Failure Fraction

The proportion of failures that are either near-hazardous or are detected hazardous failures.		Chapter 22

A1.5.9 Safety-Integrity

The probability of a system performing specific safety functions in a stated period of time.		

A1.5.10 Safety-Integrity level

One of four discrete target levels for specifying safety-integrity requirements.		

A1.6 General Terms

A1.6.1 Availability (Steady State)

The proportion of time that an item is capable of operating to specification within a large time interval.	Given as: MTBF/ (MTBF + MDT)	

A1.6.2 Unavailability (PFD)

The proportion of time that an item is NOT capable of operating to specification within a large time interval. Since the probability of failure on demand (PFD) is the probability of the item not being 'available', then PFD is the same as unavailability.		

A1.6.3 Burn-In

The operation of items for a specified period of time in order to remove early failures and bring the reliability characteristic into the random failure part of the bathtub curve.		

A1.6.4 Confidence Interval

A range of a given variable within which a random value will lie at a stated confidence (probability).		Chapter 5

A1.6.5 Consumer's Risk

The probability of an unacceptable batch being accepted owing to a favorable sample.		

A1.6.6 Derating

The use of components having a higher strength rating in order to reduce failure rate.		

A1.6.7 Ergonomics

The study of human/machine interfaces in order to minimize human errors due to mental or physical fatigue.		

A1.6.8 Mean

Usually used to indicate the arithmetic mean, which is the sum of a number of values divided by the number thereof.		

A1.6.9 Median

The median is that value such that 50% of the values in question are greater and 50% less than it.		

A1.6.10 PFD

See Unavailability.		

A1.6.11 Producer's Risk

The probability of an acceptable batch being rejected owing to an unfavourable sample.		

A1.6.12 Quality

Conformance to specification.		

A1.6.13 Random

Such that each item has the same probability of being selected as any other.		

A1.6.14 FRACAS

An acronym meaning failure reporting and corrective action system.		

A1.6.15 RAMS

A general term for reliability, availability, maintainability and safety-integrity.		

Percentage Points of the Chi-Square Distribution

α / n	0.9995	0.999	0.995	0.990	0.975	0.95	0.90	0.80	0.70	0.60
1	0.0^6393	0.0^5157	0.0^4393	0.0^3157	0.0^3982	0.0^2393	0.0158	0.0642	0.148	0.275
2	0.0^2100	0.0^2200	0.0100	0.0201	0.0506	0.103	0.211	0.446	0.713	1.02
3	0.0153	0.0243	0.0717	0.115	0.216	0.352	0.584	1.00	1.42	1.87
4	0.0639	0.0908	0.207	0.297	0.484	0.711	1.06	1.65	2.19	2.75
5	0.158	0.210	0.412	0.554	0.831	1.15	1.61	2.34	3.00	3.66
6	0.299	0.381	0.676	0.872	1.24	1.64	2.20	3.07	3.83	4.57
7	0.485	0.598	0.989	1.24	1.69	2.17	2.83	3.82	4.67	5.49
8	0.710	0.857	1.34	1.65	2.18	2.73	3.49	4.59	5.53	6.42
9	0.972	1.15	1.73	2.09	2.70	3.33	4.17	5.38	6.39	7.36
10	1.26	1.48	2.16	2.56	3.25	3.94	4.87	6.18	7.27	8.30
11	1.59	1.83	2.60	3.05	3.82	4.57	5.58	6.99	8.15	9.24
12	1.93	2.21	3.07	3.57	4.40	5.23	6.30	7.81	9.03	10.2
13	2.31	2.62	3.57	4.11	5.01	5.89	7.04	8.63	9.93	11.1
14	2.70	3.04	4.07	4.66	5.63	6.57	7.79	9.47	10.8	12.1
15	3.11	3.48	4.60	5.23	6.26	7.26	8.55	10.3	11.7	13.0
16	3.54	3.94	5.14	5.81	6.91	7.96	9.31	11.2	12.6	14.0
17	3.98	4.42	5.70	6.41	7.56	8.67	10.1	12.0	13.5	14.9
18	4.44	4.90	6.26	7.01	8.23	9.39	10.9	12.9	14.4	15.9
19	4.91	5.41	6.84	7.63	8.91	10.0	11.7	13.7	15.4	16.9
20	5.40	5.92	7.43	8.26	9.59	10.9	12.4	14.6	16.3	17.8
21	5.90	6.45	8.03	8.90	10.3	11.6	13.2	15.4	17.2	18.8
22	6.40	6.98	8.64	9.54	11.0	12.3	14.0	16.3	18.1	19.7
23	6.92	7.53	9.26	10.2	11.7	13.1	14.8	17.2	19.0	20.7
24	7.45	8.08	9.98	10.9	12.4	13.8	15.7	18.1	19.9	21.7
25	7.99	8.65	10.5	11.5	13.1	14.6	16.5	18.9	20.9	22.6
26	8.54	9.22	11.2	12.2	13.8	15.4	17.3	19.8	21.8	23.6
27	9.09	9.80	11.8	12.9	14.6	16.2	18.1	20.7	22.7	24.5
28	9.66	10.4	12.5	13.6	15.3	16.9	18.9	21.6	23.6	25.5
29	10.2	11.0	13.1	14.3	16.0	17.7	19.8	22.5	24.6	26.5
30	10.8	11.6	13.8	15.0	16.8	18.5	20.6	23.4	25.5	27.4
31	11.4	12.2	14.5	15.7	17.5	19.3	21.4	24.3	26.4	28.4
32	12.0	12.8	15.1	16.4	18.3	20.1	22.3	25.1	27.4	29.4
33	12.6	13.4	15.8	17.1	19.0	20.9	23.1	26.0	28.3	30.3
34	13.2	14.1	16.5	17.8	19.8	21.7	24.0	26.9	29.2	31.3
35	13.8	14.7	17.2	18.5	20.6	22.5	24.8	27.8	30.2	32.3
36	14.4	15.3	17.9	19.2	21.3	23.3	25.6	28.7	31.1	33.3
37	15.0	16.0	18.6	20.0	22.1	24.1	26.5	29.6	32.1	34.2
38	15.6	16.6	19.3	20.7	22.9	24.9	27.3	30.5	33.0	35.2
39	16.3	17.3	20.0	21.4	23.7	25.7	28.2	31.4	33.9	36.2
40	16.9	17.9	20.7	22.2	24.4	26.5	29.1	32.3	34.9	37.1
41	17.5	18.6	21.4	22.9	25.2	27.3	29.9	33.3	35.8	38.1
42	18.2	19.2	22.1	23.7	26.0	28.1	30.8	34.2	36.8	39.1
43	18.8	19.9	22.9	24.4	26.8	29.0	31.6	35.1	37.7	40.0
44	19.5	20.6	23.6	25.1	27.6	29.8	32.5	36.0	38.6	41.0
45	20.1	21.3	24.3	25.9	28.4	30.6	33.4	36.9	39.6	42.0

0.50	0.40	0.30	0.20	0.10	0.05	0.025	0.01	0.005	0.001	0.0005	α / n
0.455	0.708	1.07	1.64	2.71	3.84	5.02	6.63	7.88	10.8	12.1	1
1.39	1.83	2.41	3.22	4.61	5.99	7.38	9.21	10.6	13.8	15.2	2
2.37	2.95	3.67	4.64	6.25	7.81	9.35	11.3	12.8	16.3	17.7	3
3.36	4.04	4.88	5.99	7.78	9.49	11.1	13.3	14.9	18.5	20.0	4
4.35	5.13	6.06	7.29	9.24	11.1	12.8	15.1	16.7	20.5	22.1	5
5.35	6.21	7.23	8.56	10.6	12.6	14.4	16.8	18.5	22.5	24.1	6
6.35	7.28	8.38	9.80	12.0	14.1	16.0	18.5	20.3	24.3	26.0	7
7.34	8.35	9.52	11.0	13.4	15.5	17.5	20.1	22.0	26.1	27.9	8
8.34	9.41	10.7	12.2	14.7	16.9	19.0	21.7	23.6	27.9	29.7	9
9.34	10.5	11.8	13.4	16.0	18.3	20.5	23.2	25.2	29.6	31.4	10
10.3	11.5	12.9	14.6	17.3	19.7	21.9	24.7	26.8	31.3	33.1	11
11.3	12.6	14.0	15.8	18.5	21.0	23.3	26.2	28.3	32.9	34.8	12
12.3	13.6	15.1	17.0	19.8	22.4	24.7	27.7	29.8	34.5	36.5	13
13.3	14.7	16.2	18.2	21.1	23.7	26.1	29.1	31.3	36.1	38.1	14
14.3	15.7	17.3	19.3	22.3	25.0	27.5	30.6	32.8	37.7	39.7	15
15.3	16.8	18.4	20.5	23.5	26.3	28.8	32.0	34.3	39.3	41.3	16
16.3	17.8	19.5	21.6	24.8	27.6	30.2	33.4	35.7	40.8	42.9	17
17.3	18.9	20.6	22.8	26.0	28.9	31.5	34.8	37.2	42.3	44.4	18
18.3	19.9	21.7	23.9	27.2	30.1	32.9	36.2	38.6	43.8	46.0	19
19.3	21.0	22.8	25.0	28.4	31.4	34.2	37.6	40.0	45.3	47.5	20
20.3	22.0	23.9	26.2	29.6	32.7	35.5	38.9	41.4	46.8	49.0	21
21.3	23.0	24.9	27.3	30.8	33.9	36.8	40.3	42.8	48.3	50.5	22
22.3	24.1	26.0	28.4	32.0	35.2	38.1	41.6	44.2	49.7	52.0	23
23.3	25.1	27.1	29.6	33.2	36.4	39.4	43.0	45.6	51.2	53.5	24
24.3	26.1	28.2	30.7	34.4	37.7	40.6	44.3	46.9	52.6	54.9	25
25.3	27.2	29.2	31.8	35.6	38.9	41.9	45.6	48.3	54.1	56.4	26
26.3	28.2	30.3	32.9	36.7	40.1	43.2	47.0	49.6	55.5	57.9	27
27.3	29.2	31.4	34.0	37.9	41.3	44.5	48.3	51.0	56.9	59.3	28
28.3	30.3	32.5	35.1	39.1	42.6	45.7	49.6	52.3	58.3	60.7	29
29.3	31.3	33.5	36.3	40.3	43.8	47.0	50.9	53.7	59.7	62.2	30
30.3	32.3	34.6	37.4	41.4	45.0	48.2	52.2	55.0	61.1	63.6	31
31.3	33.4	35.7	38.5	42.6	46.2	49.5	53.5	56.3	62.5	65.0	32
32.2	34.4	36.7	39.6	43.7	47.4	50.7	54.8	57.6	63.9	66.4	33
33.3	35.4	37.8	40.7	44.9	48.6	52.0	56.1	59.0	65.2	67.8	34
34.3	36.5	38.9	41.8	46.1	49.8	53.2	57.3	60.3	66.6	69.2	35
35.3	37.5	39.9	42.9	47.2	51.0	54.4	58.6	61.6	68.0	70.6	36
36.3	38.5	41.0	44.0	48.4	52.2	55.7	59.9	62.9	69.3	72.0	37
37.3	39.6	42.0	45.1	49.5	53.4	56.9	61.2	64.2	70.7	73.4	38
38.3	40.6	43.1	46.2	50.7	54.6	58.1	62.4	65.5	72.1	74.7	39
39.3	41.6	44.2	47.3	51.8	55.8	59.3	63.7	66.8	73.4	76.1	40
40.3	42.7	45.2	48.4	52.9	56.9	60.6	65.0	68.1	74.7	77.5	41
41.3	43.7	46.3	49.5	54.1	58.1	61.8	66.2	69.3	76.1	78.8	42
42.3	44.7	47.3	50.5	55.2	59.3	63.0	67.5	70.6	77.4	80.2	43
43.3	45.7	48.4	51.6	56.4	60.5	64.2	68.7	71.9	78.7	81.5	44
44.3	46.8	49.5	52.7	57.5	61.7	65.4	70.0	73.2	80.1	82.9	45

α / n	0.9995	0.999	0.995	0.990	0.975	0.95	0.90	0.80	0.70	0.60
46	20.8	21.9	25.0	26.7	29.1	31.4	34.2	37.8	40.5	43.0
47	21.5	22.6	25.8	27.4	30.0	32.3	35.1	38.7	41.5	43.9
48	22.1	23.3	26.5	28.2	30.8	33.1	35.9	39.6	42.4	44.9
49	22.8	24.0	27.2	28.9	31.6	33.9	36.8	40.5	43.4	45.9
50	23.5	24.7	28.0	29.7	32.4	34.8	37.7	41.4	44.3	46.9
51	24.1	25.4	28.7	30.5	33.2	35.6	38.6	42.4	45.3	47.8
52	24.8	26.1	29.5	31.2	34.0	36.4	39.4	43.3	46.2	48.8
53	25.5	26.8	30.2	32.0	34.8	37.3	40.3	44.2	47.2	49.8
54	26.2	27.5	31.0	32.8	35.6	38.1	41.2	45.1	48.1	50.8
55	26.9	28.2	31.7	33.6	36.4	39.0	42.1	46.0	49.1	51.7
56	27.6	28.9	32.5	34.3	37.2	39.8	42.9	47.0	50.0	52.7
57	28.2	29.6	33.2	35.1	38.0	40.6	43.8	47.9	51.0	53.7
58	28.9	30.3	34.0	35.9	38.8	41.5	44.7	48.8	51.9	54.7
59	29.6	31.0	34.8	36.7	39.7	42.3	45.6	49.7	52.9	55.6
60	30.3	31.7	35.5	37.5	40.5	43.2	46.5	50.6	53.8	56.6
61	31.0	32.5	36.3	38.3	41.3	44.0	47.3	51.6	54.8	57.6
62	31.7	33.2	37.1	39.1	42.1	44.9	48.2	52.5	55.7	58.6
63	32.5	33.9	37.8	39.9	43.0	45.7	49.1	53.5	56.7	59.6
64	33.2	34.6	38.6	40.6	43.8	46.6	50.0	54.3	57.6	60.5
65	33.9	35.4	39.4	41.4	44.6	47.4	50.9	55.3	58.6	61.5
66	34.6	36.1	40.2	42.2	45.4	48.3	51.8	56.2	59.5	62.5
67	35.3	36.8	40.9	43.0	46.3	49.2	52.7	57.1	60.5	63.5
68	36.0	37.6	41.7	43.8	47.1	50.0	53.5	58.0	61.4	64.4
69	36.7	38.3	42.5	44.6	47.9	50.9	54.4	59.0	62.4	65.4
70	37.5	39.0	43.3	54.4	48.8	51.7	55.3	59.9	63.3	66.4
71	38.2	39.8	44.1	46.2	49.6	52.6	56.2	60.8	64.3	67.4
72	38.9	40.5	44.8	47.1	50.4	53.5	57.1	61.8	65.3	68.4
73	39.6	41.3	45.6	47.9	51.3	54.3	58.0	62.7	66.2	69.3
74	40.4	42.0	40.4	48.7	52.1	55.2	58.9	63.6	67.2	70.3
75	41.1	42.8	47.2	49.5	52.9	56.1	59.8	64.5	68.1	71.3
76	41.8	43.5	48.0	50.3	53.8	56.9	60.7	65.5	69.1	72.3
77	42.6	44.3	48.8	51.1	54.6	57.8	61.6	66.4	70.0	73.2
78	43.3	45.0	49.6	51.9	55.5	58.7	62.5	67.3	71.0	74.2
79	44.1	45.8	50.4	52.7	56.3	59.5	63.4	68.3	72.0	75.2
80	44.8	46.5	51.2	53.5	57.2	60.4	64.3	69.2	72.9	76.2
81	45.5	47.3	52.0	54.4	58.0	61.3	65.2	70.1	73.9	77.2
82	46.3	48.0	52.8	55.2	58.8	62.1	66.1	71.1	74.8	78.1
83	47.0	48.8	53.6	56.0	59.7	63.0	67.0	72.0	75.8	79.1
84	47.8	49.6	54.4	56.8	60.5	63.9	67.9	72.9	76.8	80.1
85	48.5	50.3	55.2	57.6	61.4	64.7	68.8	73.9	77.7	81.1
86	49.3	51.1	56.0	58.5	62.2	65.6	69.7	74.8	78.7	82.1
87	50.0	51.9	56.8	59.3	63.1	66.5	70.6	75.7	79.6	83.0
88	50.8	52.6	57.6	60.1	63.9	67.4	71.5	76.7	80.6	84.0
89	51.5	53.4	58.4	60.9	64.8	68.2	72.4	77.6	81.6	85.0
90	52.3	54.2	59.2	61.8	65.6	69.1	73.3	78.6	82.5	86.0

0.50	0.40	0.30	0.20	0.10	0.05	0.025	0.01	0.005	0.001	0.0005	α / n
45.3	47.8	50.5	53.8	58.6	62.8	66.6	71.2	74.4	81.4	84.2	46
46.3	48.8	51.6	54.9	59.8	64.0	67.8	72.4	75.7	82.7	85.6	47
47.3	49.8	52.6	56.0	60.9	65.2	69.0	73.7	77.0	84.0	86.9	48
48.3	50.9	53.7	57.1	62.0	66.3	70.2	74.9	78.2	85.4	88.2	49
49.3	51.9	54.7	58.2	63.2	67.5	71.4	76.2	79.5	86.7	89.6	50
50.3	52.9	55.8	59.2	64.3	68.7	72.6	77.4	80.7	88.0	90.9	51
51.3	53.9	56.8	60.3	65.4	69.8	73.8	78.6	82.0	89.3	92.2	52
52.3	55.0	57.9	61.4	66.5	71.0	75.0	79.8	83.3	90.6	93.5	53
53.3	56.0	58.9	62.5	67.7	72.2	76.2	81.1	84.5	91.9	94.8	54
54.3	57.0	60.0	63.6	68.8	73.3	77.4	82.3	85.7	93.2	96.2	55
55.3	58.0	61.0	64.7	69.9	74.5	78.6	83.5	87.0	94.5	97.5	56
56.3	59.1	62.1	65.7	71.0	75.6	79.8	84.7	88.2	95.8	98.8	57
57.3	60.1	63.1	66.8	72.2	76.8	80.9	86.0	89.5	97.0	100.1	58
58.3	61.1	64.2	67.9	73.3	77.9	82.1	87.2	90.7	98.3	101.4	59
59.3	62.1	65.2	69.0	74.4	79.1	83.3	88.4	92.0	99.6	102.7	60
60.3	63.2	66.3	70.0	75.5	80.2	84.5	89.6	93.2	100.9	104.0	61
61.3	64.2	67.3	71.1	76.6	81.4	85.7	90.8	94.4	102.2	105.3	62
62.3	65.2	68.4	72.2	77.7	82.5	86.8	92.0	95.6	103.4	106.6	63
63.3	66.2	69.4	73.3	78.9	83.7	88.0	93.2	96.9	104.7	107.9	64
64.3	67.2	70.5	74.4	80.0	84.8	89.2	94.4	98.1	106.0	109.2	65
65.3	68.3	71.5	75.4	81.1	86.0	90.3	95.6	99.3	107.3	110.5	66
66.3	69.3	72.6	76.5	82.2	87.1	91.5	96.8	100.6	108.5	111.7	67
67.3	70.3	73.6	77.6	83.3	88.3	92.7	98.0	101.8	109.8	113.0	68
68.3	71.3	74.6	78.6	84.4	89.4	93.9	99.2	103.0	111.1	114.3	69
69.3	72.4	75.7	79.7	85.5	90.5	95.0	100.4	104.2	112.3	115.6	70
70.3	73.4	76.7	80.8	86.6	91.7	96.2	101.6	105.4	113.6	116.9	71
71.3	74.4	77.8	81.9	87.7	92.8	97.4	102.8	106.6	114.8	118.1	72
72.3	75.4	78.8	82.9	88.8	93.9	98.5	104.0	107.9	116.1	119.4	73
73.3	76.4	79.9	84.0	90.0	95.1	99.7	105.2	109.1	117.3	120.7	74
74.3	77.5	80.9	85.1	91.1	96.2	100.8	106.4	110.3	118.6	121.9	75
75.3	78.5	82.0	86.1	92.2	97.4	102.0	107.6	111.5	119.9	123.2	76
76.3	79.5	83.0	87.2	93.3	98.5	103.2	108.8	112.7	121.1	124.5	77
77.3	80.5	84.0	88.3	94.4	99.6	104.3	110.0	113.9	122.3	125.7	78
78.3	81.5	85.1	89.3	95.5	100.7	105.5	111.1	115.1	123.6	127.0	79
79.3	82.6	86.1	90.4	96.6	101.9	106.6	112.3	116.3	124.3	128.3	80
80.3	83.6	87.2	91.5	97.7	103.0	107.8	113.5	117.5	126.1	129.5	81
81.3	84.6	88.2	92.5	98.8	104.1	108.9	114.7	118.7	127.3	130.8	82
82.3	85.6	89.2	93.6	99.9	105.3	110.1	115.9	119.9	128.6	132.0	83
83.3	86.6	90.3	94.7	101.0	106.4	111.2	117.1	121.1	129.8	133.3	84
84.3	87.7	91.3	95.7	102.1	107.5	112.4	118.2	122.3	131.0	134.5	85
85.3	88.7	92.4	96.8	103.2	108.6	113.5	119.4	123.5	132.3	135.8	86
86.3	89.7	93.4	97.9	104.3	109.8	114.7	120.6	124.7	133.5	137.0	87
87.3	90.7	94.4	98.9	105.4	110.9	115.8	121.8	125.9	134.7	138.3	88
88.3	91.7	95.5	100.0	106.5	112.0	117.0	122.9	127.1	136.0	139.5	89
89.3	92.8	96.5	101.1	107.6	113.1	118.1	124.1	128.3	137.2	140.8	90

n \ α	0.9995	0.999	0.995	0.990	0.975	0.95	0.90	0.80	0.70	0.60
91	53.0	54.9	60.0	62.6	66.5	70.0	74.2	79.5	83.5	87.0
92	53.8	55.7	60.8	63.4	67.4	70.9	75.1	80.4	84.4	88.0
93	54.5	56.5	61.6	64.2	68.2	71.8	76.0	81.4	85.5	88.9
94	55.3	57.2	62.4	65.1	69.1	72.6	76.9	82.3	86.4	89.9
95	56.1	58.0	63.2	65.9	69.9	73.5	77.8	83.2	87.3	90.9
96	56.8	58.8	64.1	66.7	70.8	74.4	78.7	84.2	88.3	91.9
97	57.6	59.6	64.9	67.6	71.6	75.3	79.6	85.1	89.2	92.9
98	58.4	60.4	65.7	68.4	72.5	76.2	80.5	86.1	90.2	93.8
99	59.1	61.1	66.5	69.2	73.4	77.0	81.4	87.0	91.2	94.8
100	59.9	61.9	67.3	70.1	74.2	77.9	82.4	87.9	92.1	95.8

0.50	0.40	0.30	0.20	0.10	0.05	0.025	0.01	0.005	0.001	0.0005	α / n
90.3	93.8	97.6	102.1	108.7	114.3	119.3	125.3	129.5	138.4	142.0	91
91.3	94.8	98.6	103.2	109.8	115.4	120.4	126.5	130.7	139.7	143.3	92
92.3	95.8	99.6	104.2	110.9	116.5	121.6	127.6	131.9	140.9	144.5	93
93.3	96.8	100.7	105.3	111.9	117.6	122.7	128.8	133.1	142.1	145.8	94
94.3	97.9	101.7	106.4	113.0	118.8	123.9	130.0	134.2	143.3	147.0	95
95.3	98.9	102.8	107.4	114.1	119.9	125.0	131.1	135.4	144.6	148.2	96
96.3	99.9	103.8	108.5	115.2	121.0	126.1	132.3	136.6	145.8	149.5	97
97.3	100.9	104.8	109.5	116.3	122.1	127.3	133.5	137.8	147.0	150.7	98
98.3	101.9	105.9	110.6	117.4	123.2	128.4	134.6	139.0	148.2	151.9	99
99.3	102.9	106.9	111.7	118.5	124.3	129.6	135.8	140.2	149.4	153.2	100

Microelectronics Failure Rates

The following table gives rates per million hours showing the highest and lowest values likely to be quoted in databases. The middle column is the geometric mean (Section 4.3). Each group of three columns is labelled for a junction temperature range in degrees Centigrade. The following multipliers apply:

	Multiplier
QUALITY	
Normal commercial procurement	2
Procured to some agreed specification and quality management system	1
100% screening and burn-in	0.4
ENVIRONMENT	
Dormant (little stress)	0.1
Benign (e.g. air-conditioned)	0.5
Fixed ground (no adverse vibration, temperature cycling, etc.)	1
Mobile/portable	4
PACKAGING	
Ceramic	1
Plastic	1 for quality factor 0.4 2 for quality factors 1 or 2

Note: FARADIP.THREE (see Chapter 4) is updated regularly and values will change slightly. These values are from around a year 2000 version.

Logic	<40			40–62		
Bipolar SRAM 64k bits	0.03	0.06	0.13	0.05	0.08	0.13
Bipolar SRAM 256k bits	0.04	0.14	0.50	0.09	0.21	0.50
Bipolar PROM/ROM 256k bits	0.02	0.02	0.03	0.03	0.03	0.03
Bipolar PROM/ROM 16k bits	0.03	0.03	0.04	0.03	0.04	0.06
MOS SRAM 16k bits	0.02	0.02	0.03	0.02	0.03	0.05
MOS SRAM 4m bits	0.08	0.19	0.44	0.20	0.30	0.44
MOS DRAM 64k bits	0.02	0.02	0.02	0.02	0.02	0.03
MOS DRAM 16m bits	0.05	0.11	0.23	0.09	0.14	0.23
MOS EPROM 16k bits	0.03	0.05	0.07	0.04	0.05	0.07
MOS EPROM 8m bits	0.06	0.13	0.30	0.07	0.14	0.30

Reliability, Maintainability and Risk. DOI: 10.1016/B978-0-08-096902-2.00028-3

Logic	62–87			>87		
Bipolar SRAM 64k bits	0.13	0.14	0.15	0.13	0.25	0.48
Bipolar SRAM 256k bits	0.30	0.39	0.50	0.50	0.70	0.96
Bipolar PROM/ROM 256k bits	0.03	0.05	0.08	0.03	0.03	0.03
Bipolar PROM/ROM 16k bits	0.03	0.07	0.15	0.03	0.12	0.47
MOS SRAM 16k bits	0.02	0.05	0.13	0.02	0.09	0.38
MOS SRAM 4m bits	0.44	0.59	0.80	0.44	1.09	2.70
MOS DRAM 64k bits	0.02	0.03	0.05	0.02	0.05	0.13
MOS DRAM 16m bits	0.23	0.25	0.28	0.23	0.46	0.92
MOS EPROM 16k bits	0.04	0.05	0.07	0.04	0.05	0.07
MOS EPROM 8m bits	0.14	0.20	0.30	0.30	0.33	0.36

Logic	<40			40–62		
Linear Bipolar 50 tr	0.01	0.01	0.02	0.01	0.02	0.03
Linear MOS 50 tr	0.02	0.03	0.04	0.03	0.03	0.04
Logic Bipolar 50 gate	0.01	0.01	0.02	0.01	0.01	0.02
Logic Bipolar 500 gate	0.01	0.01	0.02	0.01	0.02	0.03
Logic MOS 50 gate	0.01	0.01	0.02	0.01	0.02	0.03
Logic MOS 500 gate	0.01	0.02	0.03	0.01	0.02	0.05
MicroProc Bipolar 8 bits	0.01	0.03	0.07	0.01	0.04	0.14
MicroProc Bipolar 16 bits	0.01	0.03	0.08	0.01	0.05	0.23
MicroProc Bipolar 32 bits	0.01	0.03	0.11	0.01	0.06	0.40
MicroProc MOS 8 bits	0.02	0.04	0.10	0.02	0.05	0.14
MicroProc MOS 16 bits	0.02	0.06	0.18	0.02	0.08	0.30
MicroProc MOS 32 bits	0.02	0.08	0.32	0.02	0.10	0.55
ASIC/PLA/FPGA Bip'lr 1k gate	0.05	0.06	0.07	0.05	0.07	0.12
ASIC/PLA/FPGA MOS 1k gate	0.05	0.05	0.06	0.05	0.05	0.06
GaAs/MMIC 100 element	0.06	0.06	0.07	0.06	0.06	0.07

Logic	62–87			>87		
Linear Bipolar 50 tr	0.01	0.03	0.10	0.01	0.06	0.34
Linear MOS 50 tr	0.05	0.07	0.10	0.03	0.10	0.34
Logic Bipolar 50 gate	0.01	0.02	0.04	0.01	0.03	0.10
Logic Bipolar 500 gate	0.01	0.02	0.06	0.01	0.04	0.18
Logic MOS 50 gate	0.02	0.03	0.04	0.02	0.03	0.06
Logic MOS 500 gate	0.02	0.03	0.06	0.02	0.04	0.10
MicroProc Bipolar 8 bits	0.01	0.07	0.54	0.01	0.14	2.00
MicroProc Bipolar 16 bits	0.01	0.10	1.00	0.01	0.20	4.00
MicroProc Bipolar 32 bits	0.01	0.14	2.00	0.01	0.28	7.70
MicroProc MOS 8 bits	0.02	0.07	0.26	0.02	0.10	0.50
MicroProc MOS 16 bits	0.02	0.10	0.50	0.02	0.14	1.00
MicroProc MOS 32 bits	0.02	0.14	1.00	0.02	0.20	2.00
ASIC/PLA/FPGA Bip'lr 1k gate	0.05	0.14	0.40	0.05	0.26	1.40
ASIC/PLA/FPGA MOS 1k gate	0.05	0.05	0.06	0.05	0.06	0.07
GaAs/MMIC 100 element	0.06	0.06	0.07	0.06	0.06	0.07

General Failure Rates

This appendix, which is an extract from an **early** version of FARADIP.THREE, provides some failure rates. The multiplying factors for quality and environment, together with an explanation of the columns, are given in Appendix 3. **Up-to-date versions of FARADIP. THREE will contain more items and revised values.**

Item	Failure Rate in Failures per Million Hours		
Accelerometer	10		30
Air compressor	70		250
Air supply (instrument)	5	6	10
Alarm bell	2		10
Alarm circuit			
– Simple	4		
– Panel	45		
Alarm siren	1	6	20
Alternator	1		9
Analyzer			
– CO_2	100		500
– Conductivity	500	1 500	2 000
– Dewpoint	100		200
– Geiger	15		
– Hydrogen	400		100
– Oxygen	50	60	200
– pH	650		
– Scintillation	20		
– Bourdon/Geiger	5		
– H_2S	100		200
Antenna	1		5
Attenuator	0.01		
Battery			
– Lead-acid	0.5	1	3
– Ni-Cd/Ag-Zn	0.2	1	3
– Lead-acid (vehicle)			
per million miles	30		
– Dry primary	1		30
Battery charger			
– Simple rectifier	2		
– Stabilized/float	10		
– Motor generator	100		
Battery lead	3		

Reliability, Maintainability and Risk. DOI: 10.1016/B978-0-08-096902-2.00029-5

(*Cont.*)

Item	Failure Rate in Failures per Million Hours			
Bearings				
– Ball, light	0.1	1	10	
– Ball, heavy	2		20	
– Roller	0.3		5	
– Sleeve	0.5		5	
– Jewel	0.4			
– Brush	0.5			
– Bush	0.05		0.4	
Bellows, simple expandable	2	5	10	
Belts	4		50	
Busbars				
– 11 kV	0.02		**0.2**	
– 3.3 kV	0.05		2	
– 415 V	0.6		2	
Cable (power) per km				
– Overhead <600 V	0.5			
600–15 kV	5		15	
>33 kV	3		7	
– Underground <600 V	2			
600–15 kV	2			
– Subsea	2.5			
Capacitors				
– Paper	0.001	**0.15**		
– Plastic	0.001	**0.01**	0.05	
– Mica	0.002	0.03	**0.1**	
– Glass	0.002			
– Ceramic	0.0005	0.1		
– Tant. sol.	0.005	0.1		
– Tant. non-sol.	0.001	**0.01**	0.1	
– Alumin. (gen.)	0.3			
– Variable	0.005	**0.1**	2	
Card reader	**150**		4 000	
Circuit breaker				
– <600 V or A	0.5		1.5	
– >3 kV	0.5		2	
– >100 kV	3		10	
Clutch				
– Friction	0.5		3	
– Magnetic	2.5		**6**	
Compressor				
– Centrifugal, turbine driven	150			
– Reciprocating, turbine driven	500			
– Electric motor driven	100		300	
Computer				
– Mainframe	4 000		8 000	
– Mini	100	200	500	
– Micro (CPU)	30		100	
– PLC	20		50	

(Cont.)

Item	Failure Rate in Failures per Million Hours			
Connections				
– Hand solder	0.0002		0.003	
– Flow solder	0.0003		0.001	
– Weld	0.002			
– Wrapped	0.00003		0.001	
– Crimped	0.0003		0.007	
– Power cable	0.05		0.4	
– Plate th. hl.	0.0003			
Connectors				
– Coaxial	0.02		0.2	
– PCB	0.0003		0.1	
– Pin	0.001		0.1	
– r.f.	0.05			
– Pneumatic	1			
– DIL	0.001			
Counter (mech.)	0.2	2		
Crystal, quartz	0.02	0.1	0.2	
Detectors				
– Gas, pellistor	3		8	
– Smoke, ionization	2		6	
– Ultra-violet	5		15	
– Rate of rise (temp.)	3		9	
– Temperature level	0.2	2	8	
– Fire, wire/rod	10			
Diesel engine	**300**	6000		
Diesel generator	125		4000 (0.97 start)	
Diodes				
– Si, high power	**0.1**	0.2		
– Si, low power	0.01	**0.04**	0.1	
– Zener	0.005	0.03	**0.1**	
– Varactor	0.06		0.3	
– SCR (Thyristor)	0.01		0.5	
Disk memory	100	500	2000	
Electricity supply	100			
Electropneumatic converter (I/P)	2		4	
Fan	2		50	
Fiber optics				
– Connector	0.1			
– Cable/km	0.1			
– LED	0.2		0.5	
– Laser	0.3		0.5	
– Si avalanche photodiode	0.2			
– Pin avalanche photodiode	0.02			
– Optocoupler	0.02		0.1	
Filter				
– Blocked	0.5	1	10	
– Leak	0.5	1	10	

(Cont.)

Item	Failure Rate in Failures per Million Hours			
Fire sprinkler (spurious)	0.05	**0.1**	0.5	0.02 probability of non-operation
Fire water pump system	150	200	800	
Flow instruments				
– Transmitter	1	**5**	20	
– Controller	25		50	
– DP sensor	80		200	
– Switch	4		40	
– Rotary meter	5	15		
Fuse	0.02		0.5	(Mobile 2-20)
Gaskets	0.05	**0.4**	3	
Gear				
– per mesh	0.05	**0.5**	1	
– Assembly	10		50	Proportional to size
Generator				
– a.c.	3		30	
– d.c.	1		10	
– Turbine set	10	**200**	800	
– Motor set	30		70	
– Diesel set	125		4000	(Standby 8-200)
Hydraulic equipment				
– Accumulator/damper	**20**	200		
– Actuator	15			
– Piston	1			
– Motor	5			
Inductor (l.f., r.f.)	0.2		0.5	
Joints				
– Pipe	0.5			
– O ring	0.2		0.5	
Lamps				
– Filament	0.05	**1**	10	
– Neon	0.1	0.2	1	
LCD				
– per character	0.05			
– per device	2.5			
LED				
– Indicator	0.06		0.3	
– Numeral (per char.)	0.01		0.1	
Level instruments				
– Switch	2	**5**	20	
– Controller	4		20	
– Transmitter	10		20	
– Indicator	1		10	
Lines (communications)				
– Speech channel, land	100		250	
– Coaxial/km	1.5			
– Subsea/km	2.4			
Load cell	100		400	
Loudspeaker	10			
Magnetic tape unit, incl. drive	200		500	
Meter (moving coil)	1		5	

(Cont.)

Item	Failure Rate in Failures per Million Hours			
Microwave equipment				
– Fixed element	0.01			
– Tuned element	0.1			
– Detector/mixer	0.2			
– Waveguide, fixed	1			
– Waveguide, flexible	2.5			
Motor (Electrical)				
– a.c.	1	**5**	20	
– d.c.	5	**15**		
– Starter	4		10	
Optodevices				See Fiber optics
Photoelectric cell	15			
Pneumatic equipment				
– Connector	1.5			
– Controller	1		2	Open or short
– Controller	10		20	Degraded
– I/P converter	2		10	
– Pressure relay	20			
Power supply				
– d.c./d.c. converter	2	5	20	
– a.c./d.c. stabilized	5	20	100	If possible carry out FMEA
Pressure instruments				
– Switch	1	**5**	40	
– Sensor	2		10	
– Indicator	1	**5**	10	
– Controller	1	**10**	30	1 catastrophic, 20 degraded
– Transmitter (P/I) (I/P)	5		20	
Printed Circuit Boards				
– Single sided	0.02			
– Double (plated through)	0.01		0.3	
– Multilayer	0.07		0.1	
Printer (Line)	300		1 000	
Pumps				
– Centrifugal	10	50	100	
– Boiler	100		700	
– Fire water – diesel	200		3 000	
– Electrical	200		500	
– Fuel	3		180	
– Oil lubrication	6		70	
– Vacuum	10		25	
Pushbutton	0.1	0.5	10	
Rectifier (power)	3		5	
Relays				
– Armature general	0.2		0.4	
– Crystal can	0.15			
– Heavy duty	2		5	

(Cont.)

Item	Failure Rate in Failures per Million Hours			
Relays				
– Polarized	0.8			
– Reed	0.002	**0.2**	2	
– BT	0.02		0.07	
– Contactor	1		6	
– Power	1		16	
– Thermal	0.5		10	
– Time delay	0.5	**2**	10	
– Latching	0.02		1.5	
Resistors				
– Carbon comp	0.001		0.006	
– Carbon film	0.001		0.05	
– Metal oxide	0.001	**0.004**	0.05	
– Wire wound	0.001	**0.005**	0.5	
– Networks	0.05		0.1	
– Variable WW	0.02	**0.05**	0.5	
– Variable comp.	0.5		1.5	
Solenoid	0.4	**1**	4	
Stepper motor	0.5		5	
Surge arresters				
– >100 kV	0.5		1.5	
– Low power	0.003		0.02	
Switches (per contact)				
– Micro	0.1		1	
– Toggle	0.03		1	
– DIL	0.03	0.5	1.8	
– Key (low power)	0.003		2	
(high power)	5		10	
– Pushbutton	**0.2**	1	10	
– Rotary	0.05		0.5	
– Thermal delay	0.5		3	
Synchros and resolvers	**3**		15	
Temperature instruments				
– Sensor	0.2		10	
– Switch	**3**		20	
– Pyrometer	250		1 000	
– Transmitter	10			
– Controller	20		40	
Thermionic tubes				
– Diode	5	20	70	
– Triode and Pentode	20	30	100	
– Thyratron	50			
Thermocouple/thermostat	1	**10**	20	
Timer (electromech.)	2	15	40	
Transformers				
– Signal	0.005	0.2	**0.3**	
– Mains	0.03	0.4	**3**	
– ≥415 V	**0.4**	1	7	

(Cont.)

Item	Failure Rate in Failures per Million Hours			
Transistors				
– Si npn low power	0.01	**0.05**	0.2	
– Si npn high power	0.1		0.4	
– Si FET low power	0.05			
– Si FET high power	0.1			
Turbine, steam	30	40		
TV receiver	2.3			1984 figure
Valves (mechanical, hydraulic, pneumatic, gas (not high temp. nor corrosive substances))				
– Ball	0.2	3	10	
– Butterfly	1	**20**	30	
– Diaphragm (single)	2.6	**10**	20	
– Gate	1	10	30	
– Needle	1.5	20		
– Non-return	1		20	
– Plug	1		18	
– Relief	2		8	
– Globe	0.2		2	
– Solenoid	1		8	De-energize to trip
– Solenoid	8	20		Energize to trip
Valve diaphragm	1	5		
VDU	10	200	500	

Failure Mode Percentages

Just as the failure rates in the preceding tables must vary according to a large number of parameters, then so must the relative percentages of the different failure modes. However, the following figures will provide the reader with some general information that may be of assistance in carrying out a failure mode analysis where no more-specific data are available. The total item failure rate may be multiplied by the appropriate failure mode percentage in order to estimate the mode failure rate.

Item	Mode	Percentage
Battery	Catastrophic open	10
	Catastrophic short	20
	Leak	20
	Low output	50
Bearing	Binding	40
	Worn	60
Capacitor		
– Electrolytic	Open circuit	20
	Short circuit	80
– Mica, ceramic, glass,		
paper	Open circuit	1
	Short circuit	99
– Plastic	Open circuit	50
	Short circuit	50
Circuit breaker	Arcing and damage	10
	Fail to close	5
	Fail to open	40
	Spurious open	45
Clutch (mechanical)	Bind	55
	Slip	45
Connection (solder)	Break	50
	Dry	40
	No solder	10
Connector	High resistance	10
	Intermittent	20
	Open circuit	60
	Short	10
Diesel engine	Air and fuel	23

(Cont.)

Reliability, Maintainability and Risk. DOI: 10.1016/B978-0-08-096902-2.00031-3

Item	Mode	Percentage
	Blocks and heads	7
	Elec., start, battery	1
	Lube and cooling	23
	Misc. and seals	16
	Moving mech. parts	30
Diode (junction)	High reverse	60
	Open	25
	Short	15
Diode (zener)	Open	50
	Short	50
Fuse	Fails to open	15
	Opens	10
	Slow to open	75
Gear	Binding	80
	No transmission	20
Generator	Drift or intermittent	80
	Loss of output	20
Inductor	Open	75
	Short	25
Lamp	Open	100
Meter (moving coil)	Drift	30
	No reading	70
Microelectronics (digital)	o/p high or low	80
	Loss of function	20
Microelectronics (linear)	Output frozen	20
	Drift	20
	o/p max	30
	o/p min (zero)	30
Motor	Failed	
	– Brush 15 – Commutator 10 – Lube 15 – Rotor 10 – Stator 15 }	65
	Performance (degraded)	
	– Brush 15 – Commutator 5 – Lube 15 }	35
Pump	Leak	50
	No transmission	50
Relay	Coil	10
	Contact	90
Relay, contact	Fail to operate	90
	Fail to release	10
Resistor		
– comp.	Open	50
	Drift	50
– Film	Open	50
	Drift	50

(Cont.)

Item	Mode			Percentage
– Var.	Open			40
	Intermittent			60
– Wire	Open			90
	Short			10
SCR	Open			2
	Short			98
Switch				
– Micro	High Resistance			60
	No Function			10
	Open			30
– Pushbutton	Open			80
	Short			20
Transformer	Open			
	– Primary	50 }		60
	– Secondary	10		
	Short			
	– Primary	30 }		40
	– Secondary			
Transistor	High leakage			20
	Low gain			20
	Open circuit			30
	Short circuit			30
Valve (mechanical)	Blocking			5
	External leak			15
	Passing (internal)			60
	Sticking			20
Valve actuator	Fail			10
	Spurious			90

Note: Can be spurious open or spurious close, fail open or fail close, depending on the hydraulic logic.

Human Error Probabilities

The following is an overview of the range of human error rates. It must be emphasized that these are broad guidelines. In any particular situation the human-response reliability will be governed by a number of shaping factors which were explained in Chapter 9. They include:

Environmental factors	– Physical
	– Organizational
	– Personal
Intrinsic error	– Selection of Individuals
	– Training
	– Experience
Stress factors	– Personal
	– Circumstantial

The following examples are specific assessments for various tasks which were part of site safety studies arried out by the author. They have been obtained from a combination and comparison of:

- Anecdotal site data
- HEART assessments
- TESEO assessments

It must be stressed that these are examples only and involve specific site scenarios and therefore specific shaping factors. They should not be used other than as a guide. In general they apply to well managed scenarios involving good training and documentation in reasonably well controlled environments. Less benign factors would easily increase the error rates by an order of magnitude.

Task	Assessed Error Probability
Fail to respond to a normal temp/pressure/level alarm	0.02
Close or open manual valve in error	0.02
Re-open a valve at the wrong time	0.03
Open wrong or too many valves	0.04
Tanker driver drives-off despite SO_2 delivery incomplete	0.0005
Fail to react to jet fire on slug catcher	0.2
Supervisor does not notice a subordinate technician's error (e.g. failure to close a valve)	0.5

Reliability, Maintainability and Risk. DOI: 10.1016/B978-0-08-096902-2.00031-3

(Cont.)

Task	Assessed Error Probability
Fail to carry out 2 yearly maintenance as planned to the vessel	0.006
Fail to continue to hold open the oil delivery arm valve by deliberate inhibit	0.2
Fail to dip tank prior to new distillation run	0.2
Fail to select goggles prior to entering laser laboratory (regular visitor)	0.01
Pass rail signal at red	0.0002

The following table presents a broader range of human error probabilities in order to provide some perspective.

	Error probability (per task)		
	Read/ reason	Physical operation	Everyday yardstick
Simplest Possible Task			
Read single alphanumeric wrongly	0.0002		
Read 5-letter word with good resolution wrongly	0.0003		
Select wrong switch (with mimic diagram)		0.0005	
Fail to notice major cross-roads			0.0005
Routine Simple Task			
Read a checklist or digital display wrongly	0.001		
Set switch (multiposition) wrongly		0.001	
Check for wrong indicator in an array	0.003		
Wrongly carry out visual inspection for a defined criterion (e.g. leak)	0.003		
Fail to correctly replace PCB		0.004	
Select wrong switch among similar		0.005	
Read analogue indicator wrongly	0.005		
Read 10-digit number wrongly	0.006		
Leave light on			0.003
Routine Task with Care Needed			
Mate a connector wrongly		0.01	
Fail to reset valve after some related task		0.01	
Record information or read graph wrongly	0.01		
Let milk boil over			0.01
Type or punch character wrongly		0.01	
Do simple arithmetic wrongly	0.01–0.03		
Wrong selection – vending machine			0.02
Wrongly replace a detailed part		0.02	
Do simple algebra wrongly	0.02		
Read 5-letter word with poor resolution wrongly	0.03		
Put 10 digits into calculator wrongly	0.05		
Dial 10 digits wrongly	0.06		

	Error probability (per task)		
	Read/ reason	**Physical operation**	**Everyday yardstick**
Complicated Non-Routine Task			
Fail to notice adverse indicator when reaching for wrong switch or item		0.1	
Fail to recognize incorrect status in roving inspection	0.1		
New workshift – fail to check hardware, unless specified	0.1		
General (high stress)	0.25		
Fail to notice wrong position of valves	0.5		
Fail to act correctly after 1 min in emergency situation	0.9		

This final table presents an overall view of the range of human error probabilities:

Activity	Error Probability
Activity performed under extreme stress (e.g. a major incident in a plant or refinery)	0.99
Skilled task or task performed under some moderate stress factors	0.1
Average scenario as shown in the first part of this Appendix	0.01
Highly practiced straightforward task, well documented, well motivated etc.	0.001
Approximately the minimum error probability to be assumed (e.g. signal passed at red – see above)	0.0001

Fatality Rates

The following are approximate fatality rates for the UK (summarized from numerous sources) for a number of occupational, voluntary, involuntary and travel risks. They are expressed as rates, which for small values may be taken as probabilities. Some values are expressed on the basis of **annual** and others on the basis of **exposed** hours. A rate per year expresses the probability of an individual becoming a fatality in one year, given a normal exposure to the risk in question. However, for activities with a limited amount of exposure time a more realistic comparison is achieved by expressing the fatality rate per exposed hour. A FAFR (Fatal Accident Frequency Rate) is expressed on the basis of the number of expected fatalities per 100 million exposed hours.

	Per year	FAFR	Other
Travel			
Air (scheduled)	2×10^{-6}	120	1×10^{-7} per landing
–			5×10^{-5} per lifetime
			2×10^{-10} per km
– Train		3–5	1×10^{-9} per km
– Bus		4	5×10^{-10} per km
– Car	5×10^{-5}	50–60	c. 3 500 per year
			4×10^{-10} per km
– Canoe		400	
– Gliding		3 000	
– Motorcycle	2×10^{-2}	800	10^{-7} per km
– Water (general)	2×10^{-6}		9×10^{-9} per km
Occupation			
– British industry		2–4 (USA 7)	c. 800 per year (UK)
– Chemical industry	5×10^{-5}	4	
– Construction	1×10^{-4}		
– Construction erectors		10–70	
– Mining (coal)	1×10^{-4}	10 (USA 30)	
– Nuclear	4×10^{-5}		
– Railway shunting	2×10^{-4}	45	
– Boxing		20 000	
– Steeplejack		300	
– Boilers (100% exposure)	3×10^{-5}	0.3	
– Agriculture	7×10^{-5}	10 (USA 3)	
– Mechanical, manufacturing		8	
– Oil and gas extraction	1×10^{-3}		

Reliability, Maintainability and Risk. DOI: 10.1016/B978-0-08-096902-2.00032-5
Copyright © 2011 by Elsevier Ltd

	Per year	FAFR	Other
– Furniture		3	
– Clothing/textiles	2×10^{-5}	0.2	
– Electrical engineering	1×10^{-5}		
– Shipping	9×10^{-4}	8	c. 250 per year
Voluntary			
– Smoking (20 per day)	500×10^{-5}		
– Drinking (3 pints per day)	8×10^{-5}		
– Football	4×10^{-5}		
– Car racing	120×10^{-5}		
– Rock climbing	14×10^{-5}	4000	4×10^{-5} per hour
– The pill	2×10^{-5}		
– Horse riding		28000	
– Swimming		1300	
Involuntary			
– Earthquake, UK	2×10^{-8}		
– Earthquake, California	2×10^{-6}		
– Lightning (in UK)	1×10^{-7}		
– Skylab strike	5×10^{-12}		
– Pressure vessels	5×10^{-8}		
– Nuclear (1 km)	1×10^{-7}		
– Run over	6×10^{-5}		
– Falling aircraft	2×10^{-8}		
– Venomous Bite	2×10^{-7}		
– Petrol/chemical transport	2×10^{-8}		1 in 670 million miles
– Leukemia	8×10^{-5}		
– Influenza	2×10^{-4}		
– Meteorite	6×10^{-11}		
– Firearms/explosive	1×10^{-6}		
– Homicide	1×10^{-5}		
– Drowning	1×10^{-5}		
– Fire	2×10^{-5}		
– Poison	1.5×10^{-5}		
– Suicide	8×10^{-5}		
– Falls	1×10^{-4}		
– Staying at home		1–4	
– Electrocution	1.2×10^{-6}		
– Cancer	25×10^{-4}		
– All accidents	3×10^{-4}		
– Natural disasters (general)	2×10^{-6}		
– All causes*	1×10^{-2}		

*See *A Healthier Mortality* ISBN 0-952 5072-1-8.

Answers to Exercises

Chapter 2

(a)	(b)
1. 114	1.1
2. 0.99	0.42 (0.12*)
3. 10^{-5}	10^{-3}
4. 2.2×10^{-3}	0.18 (0.22*)
5. Negligible	Negligible
6. Unavailability × 2	Unavailability × 2

*Beware the approximation. λt is large (i.e. 0.876). Furthermore, if the 10 000 hr per year approximation were used, then λt = 1.

Chapter 5

1. Accumulated time $T = 50 \times 100 = 5000$ hr. Since the test was time truncated $n = 2(k + 1)$. Therefore,

   (a) $n = 6$, $T = 5000$, $\alpha = 0.4$. From Appendix 2, $\chi^2 = 6.21$

   $$\text{MTBF}_{60\%} = \frac{2T}{\chi^2} = \frac{10000}{6.21} = 1610 \text{ hr}$$

   (b) $n = 2$, $T = 5000$, $\alpha = 0.4$. From Appendix 2, $\chi^2 = 1.83$

   $$\text{MTBF}_{60\%} = \frac{2T}{\chi^2} = \frac{10000}{1.83} = 5464 \text{ hr}$$

2. If $k = 0$ then $n = 2$ and since confidence level = 90%, $\alpha = 0.1$
   Therefore $\chi^2 = 4.61$

   $$\text{MTBF}_{90\%} = 5000 = \frac{2T}{\chi^2} = \frac{2T}{4.61}$$

   Therefore $T = \dfrac{5000 \times 4.61}{2} = 11\ 525$ hr

   Since there are 50 devices the duration of the test is $\dfrac{11525}{50} = 231$ hr.

3. From Figure 5.7. If $c = 0$ and $P_{0-c} = 0.85$ ($\alpha = 0.15$) then $m = 0.17$

Therefore $T = m\theta = 0.17 \times 1000 = 170$ hrs

If MTBF is 500 hrs then $m = T/\theta = 170/500 = 0.34$ which shows $\beta = 70\%$

If $c = 5$ then $m = 3.6$ at $P_{0-c} = 0.85$

Therefore $T = m\theta = 3.6 \times 1000 = 3600$ hrs

If MTBF is 500 hrs then $m = T/\theta = 3600/500 = 7.2$ which shows $\beta = 28\%$

NB: Do not confuse α meaning (1 – confidence level) with α as producer's risk.

Chapter 6

1. From the example $R(t) = \left[\left(\frac{-t}{1110}\right)^{1.5}\right]$

If $R(t) = 0.95$ Then $\left(\frac{-t}{1110}\right)^{1.5} = 0.051$

Therefore $1.5 \log (t/1110) = \log 0.051$

Therefore $\log (t/1110) = 1.984$

Therefore $t/1110 = 0.138$

Therefore $t = 153$ hrs

2. Using the table of median ranks, sample size 10, as given in Chapter 6, plot the data and verify that a straight line is obtained.

Note that $\beta = 2$ and that $\eta = 13\,000$ hrs

Therefore

$$R(t) = exp\left[\left(\frac{-t}{13\,000}\right)^{2}\right]$$

and

MTBF $= 0.886 \times 13\,000 = 11\,500$ hrs

Chapter 7

1. $R(t) = e^{-\lambda t}[2e^{-\lambda t} - e^{-2\lambda t}]$

$= 2e^{-2\lambda t} - e^{-3\lambda t}$

$$MTBF = \int_0^\infty R(t)\ dt = \frac{1}{\lambda} - \frac{1}{3\lambda} = \frac{2}{3\lambda}$$

NB: Not a constant failure rate system despite λ being constant.

2. This is a conditional redundancy problem. Consider the reliability of the system if (a) B does fail and (b) B does not fail. The following two block diagrams describe the equivalent systems for these two possibilities.

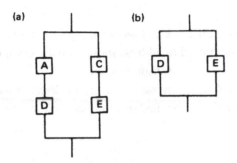

Using Bayes theorem the reliability is given as:

Reliability of diagram (a) × probability that B fails (i.e. $1 - R_b$)

    PLUS

Reliability of diagram (b) × probability that B does not fail (i.e. R_b)

Therefore System Reliability

$$= [R_a R_d + R_c R_e - R_a R_d R_c R_e] (1 - R_b) + [R_d + R_e - R_d R_e] R_b$$

Chapter 9

1(a) *Loss of supply* – Both streams have to fail, i.e. the streams are in parallel, hence the reliability block diagram is

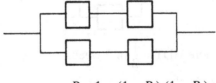

$$R = 1 - (1 - R_s) (1 - R_s)$$

where R_s is the reliability of each stream from Section 7.3

$$R = 1 - (1 - 0.885) (1 - 0.885) = \underline{0.9868}$$

1(b) *Overpressure* – occurs if *either* stream fails open, hence the streams are in series from a reliability point of view, and the block diagram is:

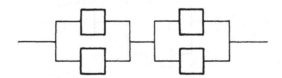

$R = R_s^2$ where R_s is the reliability of each stream from Section 7.4.2

$R = (0.999)^2 = \underline{0.998}$

Notes

The twin stream will *reduce* the risk of loss of supply, but *increase* the risk of over-pressure. The same principles can be used to address more realistic complex systems with non-return valves, slam shut pressure transducers, etc.

R will be increased if loss of supply in one stream can be detected and repaired while the other stream supplies. The down time of a failed stream is then relevant to the calculation and different.

2 (a)

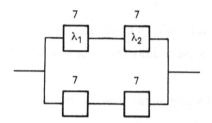

λ (Stream) $= \lambda_1 + \lambda_2 = 14 \times 10^{-6}$ per hr

Thus:

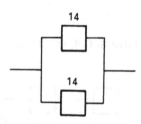

Failure Rate $\approx 2\lambda^2 MDT$ where $MDT = \frac{1}{2}$ of 2 weeks

$\qquad = 2(14 \times 10^{-6})^2 \times 168$

$\qquad = 0.0659 \times 10^{-6}$

$MTBF = 1/\lambda = 1733$ years

2(b)

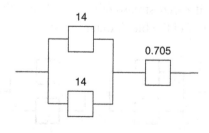

Failure Rate $= 0.0659 \times 10^{-6} + 0.705 \times 10^{-6}$

$$= 0.7709 \times 10^{-6}$$

MTBF $= 1/\lambda = 148$ years.

3. The overall Unavailability is 0.01. Calculating the Unavailability for each cutset:

MOTOR 0.0084	(76%)
PUMP 0.00144	(13%)
UV DETECTOR 0.00084	(8%)
PANEL 0.00024	(2%)
PSU and STANDBY 0.00024	(2%)

Note that the ranking, and the percentage contributions, are not the same as for failure rate.

Chapter 12

1.

Cumulative hours	Failures	Anticipated	Deviation	CUSUM
3 000	1	1	0	0
6 000	2	1	1	1
9 000	2	1	1	2
12 000	1	1	0	2
15 000	2	1	1	3
18 000	0	1	-1	2
21 000	1	1	0	2
24 000	2	1	1	3
27 000	1	1	0	3
30 000	1	1	0	3
33 000	2	1	1	4
36 000	2	1	1	5
39 000	0	1	-1	4
42 000	1	1	0	4
45 000	0	1	-1	3
48 000	0	1	-1	2
51 000	0	1	-1	1
54 000	0	1	-1	0
57 000	0	1	-1	-1
60 000	0	1	-1	-2

2.

$T_1 = 50 \times 8760 \times 0.25 = 109\,500$

$\theta_1 = 109\,500/20 = 5475$ hrs

$T_2 = 109\,500 + 100 \times 8760 \times 0.25 = 328\,500$

$\theta_2 = 328\,500/35 = 9386$ hrs

$\theta_2/\theta_1 = (T_2/T_1)^\alpha$ Therefore $1.714 = 3^\alpha$ Therefore $\alpha = 0.5$

$\theta = k\,T^\alpha$ So $5475 = k \times 331$ Therefore $k = 16.5$ (Note the use of the 2nd data pair to orient the new slope – see Figure 12.6).

For MTBF to be $12\,000$, $T^{0.5} = 12\,000/16.5$ so $T = 528\,900$ hours.

Which is another $200\,400$ hours.

Which will take $c.$ 2000 hours with the number on trial.

If $\alpha = 0.6$, k changes as follows:

$k\,(328\,500)^{0.6} = 9386$ Therefore $k = 4.6$

Now MTBF is $12\,000$ at $T^{0.6} = 12\,000/4.6$ so $T = 491\,800$ hours.

Which is another $163\,300$ hours.

Which will take $c.$ 1600 hours with the number on trial.

Chapter 25

25.2: Protection System

The target Unavailability for this "add-on" safety system is therefore

10^{-5} pa/2.5×10^{-3} pa = $\mathbf{4 \times 10^{-3}}$ which indicates **SIL 2**.

25.4: Reliability Block Diagram

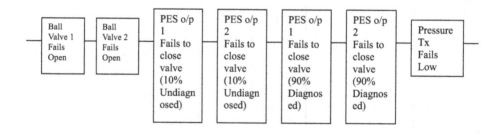

25.6: Quantifying the Model

(a) Ball valve SS1 fails open.

     Unavailability = λ MDT = $0.8 \times 10^{-6} \times 4000$

                  = 3.2×10^{-3}

(b) Ball valve SS2 fails open.

     Unavailability = λ MDT = $0.8 \times 10^{-6} \times 4000$

                  = 3.2×10^{-3}

(c) PES output 1 fails to close valve (Undiagnosed Failure).

Unavailability = 10% λ MDT = $0.025 \times 10^{-6} \times 4000$

$= 1 \times 10^{-4}$

(d) PES output 2 fails to close valve (Undiagnosed Failure).

Unavailability = 10 % λ MDT = $0.025 \times 10^{-6} \times 4000$

$= 1 \times 10^{-4}$

(e) PES output 1 fails to close valve (Diagnosed Failure).

Unavailability = 90% λ MDT = $0.225 \times 10^{-6} \times 4$

$= 9 \times 10^{-7}$

(f) PES output 2 fails to close valve (Diagnosed Failure).

Unavailability = 90% λ MDT = $0.225 \times 10^{-6} \times 4$

$= 9 \times 10^{-7}$

(g) Pressure Transmitter fails low.

Unavailability = λ MDT = $0.5 \times 10^{-6} \times 4000$

$= 2 \times 10^{-3}$

The predicted Unavailability is obtained from the sum of the unavailabilities in (a) to (g)

$= \mathbf{8.6 \times 10^{-3}}$. (Note: the target was 4 10^{-3}.)

This is higher than the unavailability target. We chose to calculate an unavailability target (rather than to simply fall within the SIL range) and thus it is NOT met.

74% from items a) and b) the valves.

23% from item g) the pressure transmitter.

Negligible from items c)-f) the PES.

25.7 Revised diagrams:

Reliability Block Diagram

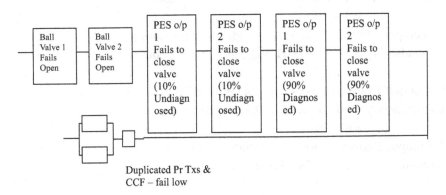

Duplicated Pr Txs &
CCF – fail low

Equivalent Fault Tree

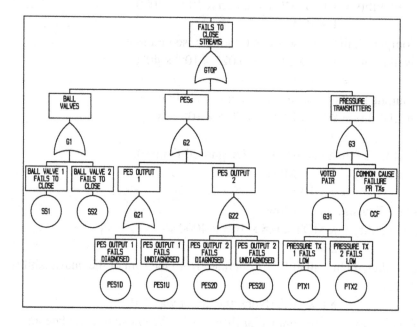

25.9 Quantifying the revised Model

Changed figures are shown in bold.

(a) Ball valve SS1 fails open.

Unavailability = λ MDT = 0.8 × 10^{-6} × **2000**

$= \mathbf{1.6 \times 10^{-3}}$

(b) Ball valve SS2 fails open.

Unavailability = λ MDT = 0.8 × 10^{-6} × **2000**

$= \mathbf{1.6 \times 10^{-3}}$

(c) PES output 1 fails to close valve (Undiagnosed Failure).

Unavailability = 10% λ MDT = 0.025 × 10^{-6} × **2000**

$= \mathbf{5 \times 10^{-5}}$

(d) PES output 2 fails to close valve (Undiagnosed Failure).

Unavailability = 10 % λ MDT = 0.025 × 10^{-6} × **2000**

$= \mathbf{5 \times 10^{-5}}$

(e) PES output 1 fails to close valve (Diagnosed Failure).

Unavailability = 90% λ MDT = 0.225 × 10^{-6} × 4

$= 9 \times 10^{-7}$

(f) PES output 2 fails to close valve (Diagnosed Failure).

Unavailability = 90% λ MDT = 0.225 × 10^{-6} × 4

$= 9 \times 10^{-7}$

(g) Voted pair of pressure transmitters.

Unavailability $= \lambda^2 T^2/3 = [0.5 \times 10^{-6}]^2 \times 4000^2/3$

$= 1.3 \times 10^{-6}$

(h) Common cause failure of pressure transmitters.

Unavailability $= 9\% \lambda \, MDT = 0.09 \times 0.05 \times 10^{-6} \times 2000$

$= 9 \times 10^{-5}$

The predicted Unavailability is obtained from the sum of the unavailabilities in (a) to (h) $= \mathbf{3.3 \times 10^{-3}}$ which meets the target.

25.10 ALARP

Assume that further improvements, involving CCF and a further reduction in proof test interval, could be achieved for a total cost of £1000. Assume, also, that this results in an improvement in unavailability, of the safety-related system, from $\mathbf{3.3 \times 10^{-3}}$ to the PFD associated with the Broadly Acceptable limit of $\mathbf{4 \times 10^{-4}}$. It is necessary to consider, applying the ALARP principle, whether this improvement should be implemented.

If the target unavailability of $\mathbf{4 \times 10^{-3}}$ represents a maximum tolerable risk of $\mathbf{10^{-5}}$ **pa** then it follows that $\mathbf{3.3 \times 10^{-3}}$ represents a risk of $10^{-5} \times 3.3/4 = \mathbf{8.3 \times 10^{-6}}$ **pa**. If $\mathbf{10^{-6}}$ pa is taken as the boundary of the negligible risk then the proposal remains within the tolerable range and thus subject to ALARP.

Assuming a two-fatality scenario, the cost per life saved over a 40 year life of the equipment (without cost discounting) is calculated as follows:

$\mathbf{3.3 \times 10^{-3}}$ represents a risk of 8.3×10^{-6}

$\mathbf{4 \times 10^{-4}}$ represents a risk of 10^{-6}

Cost per life saved $= £1000 / (40 \times 2 \text{ lives} \times [8.3 - 1] \, 10^{-6})$

$= \mathbf{£1\,700\,000}$

On this basis, if the cost per life saved criterion were £1,000,000, then justification for the further improvement would be considered marginal as the benefit is just below (but close to) the criteria. On the other hand it would be justified if the criterion were £2,000,000.

25.11 Architectural Constraints

(a) PES

The safe failure fraction for the PESs is given by 90% diagnosis of 5% of the failures, which cause the failure mode in question, PLUS the 95% which are "fail safe"

Thus $(90\% \times 5\%) + 95\% = 99.5\%$

Consulting the tables in Chapter 22 then:

If the simplex PES is regarded as Type B then SIL 2 can be considered if this design has >90% safe failure fraction.

(b) Pressure transmitters

The safe failure fraction for the transmitters is given by the 75% which are 'fail safe'

If they are regarded as Type A then SIL 2 can be considered since they are voted and require less than 60% safe failure fraction.

Incidentally, in the original proposal, the simplex pressure transmitter would not have met the architectural constraints.

(c) Ball valves

The safe failure fraction for the valves is given by the 90% which are 'fail safe'

If they are are regarded as Type A then SIL 2 can be considered since they require more than 60% safe failure fraction.

Bibliography

This is by no means exhaustive and, for a book that is updated approximately every five years, can never be up-to-date. However, it provides a basic starting point that, together with web searches, can lead the reader in the right direction.

Books

Carter, A. D. S., *Mechanical Reliability*, 2nd edn, Macmillan, London (1986).

Collins, J. A., *Failure of Materials in Mechanical Design*, Wiley, New York (1981).

Fullwood, R. F., *Probabilistic Safety Assessment in the Chemical and Nuclear Industries*, Butterworth-Heinemann, Oxford (1999) ISBN 0 7506 7208 0.

Goldman, A. S. and Slattery, T. B., *Maintainability – A Major Element of System Effectiveness*, Wiley, New York (1964).

Jensen, F. and Petersen, N. E., *Burn In*, Wiley, New York (1982).

Kapur, K. C. and Lamberson, L. R., *Reliability in Engineering Design*, Wiley, New York (1977).

Kivensen, G., *Durability and Reliability in Engineering Design*, Pitman, London (1972).

Moubray, J., *Reliability-centred Maintenance*, Butterworth-Heinemann, Oxford (1997) ISBN 0 7506 3358 1.

Myers, G. J., *Software Reliability, Principles and Practice*, Wiley, New York (1976).

O'Connor, P. D. T. O., *Practical Reliability Engineering*, 3rd edn, Wiley, Chichester (1991).

Shooman, M., *Software Engineering – Reliability, Design, Management*, McGraw-Hill, New York (1984).

Smith, C. O., *Introduction to Reliability in Design*, McGraw-Hill, New York (1976).

Smith, D. J., *Statistics Workshop*, Technis, Tonbridge (1991) ISBN 0 9516562 0 1.

Smith, D. J. and Simpson, K. G. L. *The Safety Critical Systems Handbook*, 3rd edn – ISBN 978 0080967813.

Snedecor, G. W. and Cochran, W. G., *Statistical Methods*, Iowa State University Press, Ames (1967).

Whittingham, R. B., *The Blame Machine*, Oxford (2004) Elsevier ISBN 07506 5510-0.

Other publications

Human Reliability Assessors Guide (SRDA-R11), UKAEA, Thomson House, Risley, Cheshire WA3 6AT (1995), ISBN 085 3564 205.

Nomenclature for Hazard and Risk Assessment in the Process Industries, IChemE, 165-171 Railway Terrace, Rugby, CV21 3HQ (1985 reissued 1992), ISBN 0852951841.

Tolerability of Risk for Nuclear Power Stations, UK Health and Safety Executive, ISBN 0118863681.

HSE. *Reducing Risks, Protecting People, HSE's decision-making process*, HSE Books (2001).

HAZOP – A Guide to Hazard and Operability Studies, Chemical Industries Association, Alembic House, 93 Albert Embankment, London SE1 7TU (1977).

A Guide to the Control of Industrial Major Accident Hazards Regulations, HMSO, London (1984).

UPM3.1: A Pragmatic Approach to Dependent Failures Assessment for Standard Systems, UKAEA, ISBN 0853564337.

SINTEF, Reliability Prediction Method for Safety Instrumented Systems STF50 A06031.

Technis Guidelines Q124, Demonstration of Product/System Compliance with IEC 61508 ISBN 0951 6562 79.

Standards and guidelines

BS 2011 *Basic environmental testing procedures.*

BS 4200 *Guide on the reliability of electronic equipment and parts used therein.*

BS 4778 *Glossary of terms used in quality assurance.* (Section 3.2: 1991 is Reliability).

BS 5760 *Reliability of systems, equipment and components.*

BS 6651: *1990 Code of practice for protection of structures against lightning.*

UK DEF STD 00-40 *Reliability and maintainability.*

UK DEF STD 00-41 *MOD Practices and procedures in reliability and maintainability.*

UK DEF STD 00-55 *The procurement of safety critical software in defense equipment.*

UK DEF STD 00-56 *Hazard analysis and safety classification of the computer and programmable electronic system elements of defense equipment.*

UK DEF STD 00-58 *A guideline for HAZOP studies on systems which include programmable electronic systems.*

UK DEF STD 07-55 *Environmental testing.*

US Military Handbook 217E (Notice 1) *Reliability Prediction of Electronic Equipment,* 1990.

US Military Handbook 338 *Electronic Reliability Design Handbook.*

US Military Standard 470 *Maintainability program for systems and equipment.*

US Military Standard 471A *Maintainability/Verification/Demonstration/Evaluation,* 1973.

US Military Handbook 472 *Maintainability Prediction,* 1966.

US Military Standard 721B *Definitions of Effectiveness Terms for Reliability.*

US Military Standard 756 *Reliability Prediction.*

US Military Standard 781C *Reliability Design Qualification and Production Acceptance Tests.*

US Military Standard 785A *Reliability Programe for Systems and Equipment Development and Production,* 1969.

US Military Standard 810 *Environmental Test Methods.*

US Military Standard 883 *Test Methods and Procedures for Microelectronic Devices.*

US Military Standard 1629A *Procedures for Performing a Failure Mode, Effects and Criticality Analysis.*

US Military Standard 52779 (AD) *Software Quality Assurance Requirements.*

IGEM Publication SR15, *Programmable Equipment in Safety Related Applications,* 5th edn (2010).

IGEM Publication SR24, *Risk Assessment Techniques* (1998) ISBN 0717700933.

IEE, *Competency Guidelines for Safety Related Systems Practitioners* (1999), ISBN 085296787X.

IEC Publication 271, *Preliminary List of Basic Terms and Definitions for the Reliability of Electronic Equipment and Components (or parts) used therein.*

IEC 61508 (2010) *Functional Safety: Safety Related Systems* – 7 Parts.

IEC International Standard 61511: *Functional Safety – Safety Instrumented Systems for the Process Industry Sector.*

IEEE Standard 500, *Reliability Data for Pumps, Drivers, Valve Actuators and Valves* (1994), Library of Congress 83-082816.

European Standard BSEN 51026: *Railway Applications – The Specification and Demonstration of Dependability, Reliability, Maintainability and Safety (RAMS).*

RTCA DO–178B/(EUROCAE ED–12B) – *Software Considerations in Airborne Systems and Equipment Certification.*

Scoring Criteria for BETAPLUS Common Cause Model

A10.1 Checklist and Scoring for Equipment Containing Programable Electronics

Score between 0 and 100% of the indicated maximum values.

1. Separation/segregation	A Max. Score	B Max. Score
Are all signal cables separated at all positions?	15	52
Are the programable channels on separate printed circuit boards?	85	55
OR are the programable channels in separate racks?	90	60
OR in separate rooms or buildings?	95	65
MAXIMUM SCORE	110	117

2. Diversity	A Max. Score	B Max. Score
Do the channels employ diverse technologies; 1 electronic + 1 mechanical/pneumatic?	100	25
OR 1 electronic or CPU + 1 relay based?	90	25
OR 1 CPU + 1 electronic hardwired?	70	25
OR do identical channels employ enhanced voting? i.e. 'M out of N' where N > M + 1.	40	25
Were the diverse channels developed from separate requirements from separate people with no communication between them?	20	–
Were the two design specifications separately audited against known hazards by separate people and were separate test methods andmaintenance applied by separate people?	12	25
MAXIMUM SCORE	132	50

Reliability, Maintainability and Risk. DOI: 10.1016/B978-0-08-096902-2.00035-0

3. Complexity/design/application/maturity/ experience	A Max. Score	B Max. Score
Does cross-connection between CPUs preclude the exchange of any information other than the diagnostics?	30	–
Is there > 5 years experience of the equipment in the particular environment?	–	10
Is the equipment simple < 5 PCBs per channel;	–	20
OR < 100 lines of code		
OR < 5 ladder logic rungs		
OR < 50 I/O and < 5 safety functions?		
Are I/O protected from over-voltage and over-current and rated > 2:1?	30	–
MAXIMUM SCORE	60	30

4. Assessment/analysis and feedback of data	A Max. Score	B Max. Score
Has a combination of detailed FMEA, fault tree analysis and design review established potential CCFs in the electronics?	–	140
Is there documentary evidence that field failures are fully analyzed with feedback to design?	–	70
MAXIMUM SCORE	–	210

5. Procedures/human interface	A Max. Score	B Max. Score
Is there a written system of work on site to ensure that failures are investigated and checked in other channels? (including degraded items that have not yet failed)	30	20
Is maintenance of diverse/redundant channels staggered at such an interval as to ensure that any proof-tests and cross-checks operate satisfactorily between the maintenance?	60	–
Do written maintenance procedures ensure that redundant separations, as, for example, signal cables, are separated from each other and from power cables and must not be re-routed?	15	25
Are modifications forbidden without full design analysis of CCF?	–	20
Is diverse equipment maintained by different staff?	15	20
MAXIMUM SCORE	120	85

6. Competence/training/safety culture	A Max. Score	B Max. Score
Have designers been trained to understand CCF?	–	100
Have installers been trained to understand CCF?	–	50
Have maintainers been trained to understand CCF?	–	60
MAXIMUM SCORE	–	210

7. Environmental control	A Max. Score	B Max. Score
Is there limited personnel access?	40	50
Is there appropriate environmental control? (e.g. temperature, humidity)	40	50
MAXIMUM SCORE	80	100

8. Environmental testing	A Max. Score	B Max. Score
Has full EMC immunity or equivalent mechanical testing been conducted on prototypes and production units (using recognized standards)?	–	316
MAXIMUM SCORE	–	316

	A Max score	B Max score
TOTAL MAXIMUM SCORE	502	1118

A10.2 Checklist and Scoring for Non-Programable Equipment

Only the first three categories have different questions as follows:

1. Separation/segregation	A Max. Score	B Max. Score
Are the sensors or actuators physically separated and at least 1 metre apart?	15	52
If the sensor/actuator has some intermediate electronics or pneumatics, are the channels on separate PCBs and screened?	65	35
OR if the sensor/actuator has some intermediate electronics or pneumatics, are the channels indoors in separate racks or rooms?	95	65
MAXIMUM SCORE	110	117

2. Diversity	A Max. Score	B Max. Score
Do the redundant units employ different technologies? e.g. 1 electronic or programmable + 1 mechanical/pneumatic.	100	25
OR 1 electronic, 1 relay based?	90	25
OR 1 PE, 1 electronic hardwired?	70	25
OR do the devices employ 'M out of N' voting where; N > M+?	40	25
Were separate test methods and maintenance applied by separate people?	32	52
MAXIMUM SCORE	132	50

3. Complexity/design/application/maturity/ experience	A Max. Score	B Max. Score
Does cross-connection preclude the exchange of any information other than the diagnostics?	30	–
Is there > 5 years experience of the equipment in the particular environment?	–	10
Is the equipment simple e.g. non-programable type sensor or single actuator field device?	–	20
Are devices protected from over-voltage and over-current and rated > 2:1 or mechanical equivalent?	30	–
MAXIMUM SCORE	60	30

4. ASSESSMENT/ANALYSIS and FEEDBACK OF DATA
As for programable electronics (see above).

5. PROCEDURES/HUMAN INTERFACE
As for programable electronics (see above).

6. COMPETENCE/TRAINING/SAFETY CULTURE
As for programable electronics (see above).

7. ENVIRONMENTAL CONTROL
As for programable electronics (see above).

8. ENVIRONMENTAL TESTING
As for programable electronics (see above).

	A Max. Score	B Max. Score
TOTAL MAXIMUM RAW SCORE (Both programable and non-programable lists)	502	1118

The diagnostic interval is shown for each of the two (programable and non-programable) assessment lists. The (**C**) values have been chosen to cover the range one to three in order to construct a model that caters for the known range of BETA values.

For Programable Electronics

Diagnostic Coverage	Interval < 1 min	Interval 1–5 mins	Interval 5–10 mins	Interval > 10 mins
98%	3	2.5	2	1
90%	2.5	2	1.5	1
60%	2	1.5	1	1

For Sensors and Actuators

Diagnostic Coverage	Interval < 2 hrs	Interval 2 hrs–2 days	Interval 2 days–1 week	Interval > 1 week
98%	3	2.5	2	1
90%	2.5	2	1.5	1
60%	2	1.5	1	1

A score of C > 1 may only be proposed if the resulting action, initiated by the diagnostics, has the effect of preventing or invalidating the effect of the subsequent CCF failure. For example, in some process industry equipment, even though the first of the CCF failures was diagnosed before the subsequent failure, there would nevertheless be insufficient time to take action to maintain the process. The subsequent (second) CCF failure would thus occur before effective action could be taken. Therefore, in such a case, the diagnostics would not help in defending against CCF and a C > 1 score cannot be proposed in the assessment.

The result is then modified according to the level of 'M out of N' redundancy as shown in Chapter 8.

AVAILABLE IN SOFTWARE FORM, AS BETAPLUS, FROM THE AUTHOR:
See end of book.

Example of HAZOP

Sour gas consisting mainly of methane (CH$_4$) but with 2% hydrogen sulfide (H$_2$S) is routed to an amine absorber section for sweetening. The absorber uses a 25:75 diethanolamine (amine)/water solution to remove the H$_2$S in the absorber tower. Sweet gas is removed from the tower top and routed to fuel gas. Rich amine is pressurized from the tower bottom under level control and then routed to an amine regeneration unit on another plot. Regenerated amine is returned to the amine absorber section and stored in a low-pressure buffer storage tank.

A11.1 Equipment Details

Absorber tower operating pressure = 20 bar gauge.

The buffer storage tank is designed for low pressure, with weak seam roof and additional relief provided by a hinged manhole cover.

A11.2 HAZOP Worksheets

The HAZOP worksheets with this example will demonstrate the HAZOP method for just one node, i.e. the line from the buffer storage tank to the absorber tower.

Nodes that could have been studied in more detail are:

- amine buffer tank
- line to absorber tower from amine buffer tank
- sour gas line to absorber tower
- absorber tower
- sweet gas line out of absorber tower
- rich amine line out of absorber tower.

A11.3 Potential Consequences

The importance of the consequences identified for a process deviation, and how these are used to judge the adequacy of safeguards, cannot be over emphasized. In this example, the consequences of reverse flow include:

Reliability, Maintainability and Risk. DOI: 10.1016/B978-0-08-096902-2.00036-2

- possible tank damage
- release of a flammable gas near a congested unit which could lead to an explosion
- release of a highly toxic gas.

The latter two consequences alone are deemed sufficient for the matter to be referred back for more consideration. If only the first consequence applied, tank damage could be deemed acceptable if the incident were unlikely, no hazardous substance involved and no personnel would be present. In the common case of a pump tripping and a non-return valve failing, even this may not be deemed acceptable to the HAZOP team if excessive costs from lost production followed from tank damage.

Considerable judgement is called for by the team in making this decision. It is essential that the team be drawn from personnel with sufficient practical knowledge of the process under study.

Although the main action in this example is to consider fitting a slam-shut valve, it could be that an alarm and manual isolation is acceptable. This decision cannot, however, be made without full consideration of the unit manning levels, what duties the operator has that could cause distraction from responding to an alarm, whether the operator's training will be sufficient to understand the implications of that alarm, and how far the control panel is from the nearest manual isolation valve.

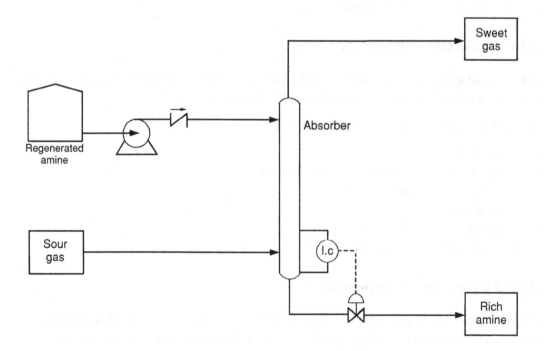

Figure A11.1: Amine absorber section

Worksheet

Company : Any Town Gas Producers
Facility : Amine Absorber Section
Session : 1 25-07-96
Node : 1 Line from amine tank via pump to absorber tower
Parameter : Flow

Deviation	Causes	Consequences	Safeguards	Recommendations	By
No flow	Amine buffer tank empty	Damage to pump	Level indication	Consider a low level alarm	
		Loss of fresh amine to absorber tower giving H_2S in the sweet gas line	Ditto	Ditto	
	Line frozen	Ditto	Ditto	Check freezing point of water/amine mixture	
	Valve in line shut	Possible damage to line as pump dead heads, i.e. runs against closed discharge line	Operator training	Check line for maximum pump pressure	
More flow	None (fixed by maximum pump discharge)				
Less flow	Line partially plugged or valve partially closed	Possible damage to line as pump dead heads grind against closed discharge line	None	Check freezing point of water/amine mixture and check pipe spec against pump dead head pressure	
Reverse flow	Pump trips	Back flow of 20 bar gas to amine tank	Non-return valve (which may not be reliable in amine service)	In view of the potential consequence of the release and its likelihood, undertake a full study of the hazards involved, and safeguards appropriate to these hazards proposed (possibly installing a chopper valve to cut in and prevent back flow)	

(Cont.)

Deviation	Causes	Consequences	Safeguards	Recommendations	By
		Resulting in: 1. Possible rupture of tank 2. Major H_2S release to plant causing potential toxic cloud and possible vapor cloud explosion if cloud reaches congested part of the plant	Tank weak seam None	As above As above	
High temperature	Failure of cooling on the amine regeneration unit resulting in hot amine in amine tank	Possibility of poor absorber tower efficiency	Temperature alarm on amine regeneration unit		
Low temperature	Cold conditions	Possible freezing of line	None at present – but see action under 'No flow' to investigate freezing point		
High pressure	Pump dead head	Possibility of overpressure of pipe	None – but see action under 'No flow' to check pipe spec		
	Reverse flow from absorber tower	Ditto	None	In previous action to check pipe spec against pump dead head pressure also include checking spec against operating pressure in absorber tower	
Low pressure	None identified	Not seen as a problem	Line good for vacuum conditions	None	

HAZID Checklist

1. Acceleration/shock	Change in velocity, impact energy of vehicles, components or fluids	1. Structural deformation 2. Breakdown by impact 3. Displacement of parts or piping 4. Seating or unseating valves or electrical contacts 5. Loss of fluid pressure head (cavitation) 6. Pressure surges in fluid systems 7. Disruption of metering equipment
2. Chemical energy	Chemical disassociation or replacement of fuels, oxidizers, explosives, organic materials or components	1. Fire 2. Explosion 3. Non-explosive exothermic reaction 4. Material degradation 5. Toxic gas production 6. Corrosion fraction production 7. Swelling of organic compounds
3. Contamination	Producing or introducing contaminants to surfaces, orifices, filters, etc.	1. Clogging or blocking of components 2. Friction between moving surfaces 3. Deterioration of fluids 4. Degradation of performance sensors or operating components 5. Erosion of lines or components 6. Fracture of lines or components by fast-moving large particles
4. Electrical energy	System or component potential energy release or failure. Includes shock both thermal and static	1. Electrocution 2. Involuntary personnel reaction 3. Personnel burns 4. Ignition of combustibles 5. Equipment burnout 6. Inadvertent activation of equipment or ordinance devices 7. Necessary equipment unavailable for functions or caution and warning 8. Release on holding devices

(*Cont.*)

Reliability, Maintainability and Risk. DOI: 10.1016/B978-0-08-096902-2.00037-4

5. Human capability	Human factors including perception, dexterity, life support and error	1. Personal injury due to: • restricted routes • hazardous location • inadequate visual/audible warnings
	PROBABILITY	2. Equipment damage by improper operation due to: • inaccessible control location • inadequate control/display identification
6. Human hazards	Conditions that could cause skin abrasions, cuts, bruises, etc.	1. Personal injury due to: • sharp edges/corners • dangerous heights • unguarded floor/wall openings
7. Interface/interaction	Compatibility between systems/subsystems/ facilities/software	1. Incompatible materials reaction 2. Interfacing reactions 3. Unintended operations caused/prevented by software
8. Kinetic energy	System/component linear or rotary motion	1. Linear impact 2. Disintegration of rotating components
9. Material deformation	Degradation of material by corrosion, ageing, embrittlement, oxidation, etc.	1. Change in physical or chemical properties 2. Structural failure 3. Delamination of layered material 4. Electrical short circuiting
10. Mechanical energy	System/component potential energy such as compressed springs	1. Personal injury or equipment damage from energy release
11. Natural environment	Conditions including lightning, wind, projectiles, thermal, pressure, gravity, humidity, etc.	1. Structural damage from wind 2. Electrical discharge 3. Dimension changes from solar heating
12. Pressure	System/component potential energy, including high/low or changing pressure	1. Blast/fragmentation from container overpressure rupture 2. Line/hose whipping 3. Container implosion/explosion 4. System leaks 5. Heating/cooling by rapid changes 6. Aeroembolism, bends, choking or shock

(Cont.)

13. Radiation	Conditions including electromagnetic, ionizing, thermal or ultraviolet radiation	1. Electronic equipment interference 2. Human tissue damage 3. Charring of organic materials 4. Decomposition of chlorinated hydrocarbons into toxic gases 5. Ozone or nitrogen oxide generation
14. Thermal	System/component potential energy, including high, low or changing temperature	1. Ignition of combustibles 2. Ignition of other reactions 3. Distortion of parts 4. Expansion/contraction of solids or fluids 5. Liquid compound stratification 6. Personal injury
15. Toxicants	Adverse human effects of inhalants or ingests	1. Respiratory system damage 2. Blood system damage 3. Body organ damage 4. Skin irritation or damage 5. Nervous system effects
16. Vibration/sound	System/component produced energy	1. Material failure 2. Personal fatigue or injury 3. Pressure/shock-wave effects 4. Loosening of parts 5. Chattering of valves or contacts 6. Contamination interface

Markov Analysis of Redundant Systems

In Chapter 8, it was explained that the traditional use of Markov analysis in respect of redundant systems is fundamentally flawed in that repair/down times are not a random process and depend upon earlier events than the one immediately preceding the failure in question. The repair of the second failure is dependent on the progress of the first in that a single repair crew will continue to deal with that first failure.

However, up to the point of the second failure occurring the Markov assumption, given that there are multiple repair crews, holds good. It is, therefore, reproduced in this Appendix for interest since it is so frequently referred to in the literature.

The Markov method for calculating the MTTF of a system with repair is to consider the 'states' in which the system can exist. Figure A13.1 shows a system with two identical units each having failure rate λ and repair rate (reciprocal of mean down time) μ. The system can be in each of three possible states.

State (0)	Both units operating
State (1)	One unit operating, the other having failed
State (2)	Both units failed

It is important to remember one rule with Markov analysis, namely, that the probabilities of changing state are dependent only on the state itself. In other words, the probability of failure or of repair is not dependent on the past history of the system – not so in real life.

Let $P_i(t)$ be the probability that the system is in state (i) at time t and assume that the initial state is (0).

Therefore

$$P_0(0) = 1 \text{ and } P_1(0) = P_2(0) = 0$$

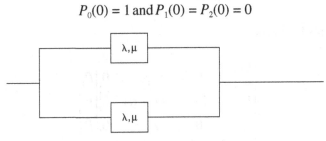

Figure A13.1.

Reliability, Maintainability and Risk. DOI: 10.1016/B978-0-08-096902-2.00038-6

Therefore

$$P_0(t) + P_1(t) + P_2(t) = 1$$

We shall now calculate the probability of the system being in each of the three states at time $t + \Delta t$. The system will be in state (0) at time $t + \Delta t$ if:

1. The system was in state (0) at time t and no failure occurred in either unit during the interval Δt, or
2. The system was in state (1) at time t, no further failure occurred during Δt, and the failed unit was repaired during Δt.

The probability of only one failure occurring in one unit during that interval is simply $\lambda \Delta t$ (valid if Δt is small, which it is). Consequently $(1 - \lambda \Delta t)$ is the probability that no failure will occur in one unit during the interval. The probability that both units will be failure free during the interval is, therefore,

$$(1 - \lambda \Delta t)(1 - \lambda \Delta t)1 - 2\lambda \Delta t$$

The probability that one failed unit will be repaired within Δt is $\mu \Delta t$, provided that Δt is very small. This leads to the equation:

$$P_0(t + \Delta t) = [P_0(t) \times (1 - 2\lambda \Delta t)] + [P_1(t) \times (1 - \lambda \Delta t) \times \mu \Delta t]$$

Similarly, for states 1 and 2:

$$P_1(t + \Delta t) = [P_0(t) \times 2\lambda \Delta t] + [P_1(t) \times (1 - \lambda \Delta t) \times (1 - \mu \Delta t)]$$

$$P_2(t + \Delta t) = [P_1(t) \times \lambda \Delta t] + P_2(t)$$

Now the limit as Δt 0 of $[P_i(t + \Delta t) - P_i(t)]/\Delta t$ is $\dot{P}_i(t)$ and so the above yield:

$$\dot{P}_0(t) = -2\lambda P_0(t) + \mu P_1(t)$$

$$\dot{P}_1(t) = 2\lambda P_0(t) - (\lambda + \mu)P_1(t)$$

$$\dot{P}_2(t) = P_1(t)\lambda$$

In matrix notation this becomes:

$$\begin{vmatrix} \dot{P}_0 \\ \dot{P}_1 \\ \dot{P}_2 \end{vmatrix} = \begin{vmatrix} -2\lambda & \mu & 0 \\ 2\lambda & -(\lambda + \mu) & 0 \\ 0 & \lambda & 0 \end{vmatrix} \begin{vmatrix} P_0 \\ P_1 \\ P_2 \end{vmatrix}$$

The elements of this matrix can also be obtained by means of a Transition Diagram. Since only one event can take place during a small interval, Δt, the transitions between states involving only one repair or one failure are considered. Consequently, the transitions (with transition rates) are:

by failure of either unit

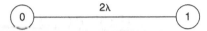

by failure of the remaining active unit,

by repair of the failed unit of state 1.

The transition diagram is:

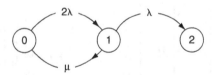

Finally closed loops are drawn at states 0 and 1 to account for the probability of not changing state. The rates are easily calculated as minus the algebraic sum of the rates associated with the lines leaving that state. Hence:

A (3×3) matrix, $(a_{i,j})$, can now be constructed, where $i = 1, 2, 3; j = 1, 2, 3; a_{i,j}$ is the character on the flow line pointing from state j to state i. If no flow line exists the corresponding matrix element is zero. We therefore find the same matrix as before.

The MTTF is defined as

$$\theta_s = \int_0^\infty R(t)\, dt$$

$$= \int_0^\infty [P_0(t) + P_1(t)]\, dt$$

$$= \int_0^\infty P_0(t)\, dt + \int_0^\infty P_1(t)\, dt$$

$$= T_0 + T_1$$

The values of T_0 and T_1 can be found by solving the following:

$$\int_0^\infty \begin{vmatrix} \dot{P_0}(t) \\ \dot{P_1}(t) \\ \dot{P_2}(t) \end{vmatrix} dt = \int_0^\infty \begin{vmatrix} -2\lambda & \mu & 0 \\ 2\lambda & -(\lambda+\mu) & 0 \\ 0 & \lambda & 0 \end{vmatrix} \begin{vmatrix} P_0 \\ P_1 \\ P_2 \end{vmatrix} dt$$

Since the (3×3) matrix is constant we may write

$$\int_0^\infty \begin{vmatrix} \dot{P_0}(t) \\ \dot{P_1}(t) \\ \dot{P_2}(t) \end{vmatrix} dt = \begin{vmatrix} -2\lambda & \mu & 0 \\ 2\lambda & -(\lambda+\mu) & 0 \\ 0 & \lambda & 0 \end{vmatrix} \int_0^\infty \begin{vmatrix} P_0 \\ P_1 \\ P_2 \end{vmatrix} dt$$

or

$$\begin{vmatrix} \int_0^\infty \dot{P_0}(t)\, dt \\ \int_0^\infty \dot{P_1}(t)\, dt \\ \int_0^\infty \dot{P_2}(t)\, dt \end{vmatrix} = \begin{vmatrix} -2\lambda & \mu & 0 \\ 2\lambda & -(\lambda+\mu) & 0 \\ 0 & \lambda & 0 \end{vmatrix} \begin{vmatrix} \int_0^\infty P_0(t)\, dt \\ \int_0^\infty P_1(t)\, dt \\ \int_0^\infty P_2(t)\, dt \end{vmatrix}$$

or

$$
\begin{vmatrix} P_0(\infty) - P_0(0) \\ P_1(\infty) - P_1(0) \\ P_2(\infty) - P_2(0) \end{vmatrix} = \begin{vmatrix} -2\lambda & \mu & 0 \\ 2\lambda & -(\lambda+\mu) & 0 \\ 0 & \lambda & 0 \end{vmatrix} \begin{vmatrix} T_0 \\ T_1 \\ T_2 \end{vmatrix}
$$

Taking account of

$$
P_0(0) = 1; P_1(0) = P_2(0) = 0
$$

$$
P_0(\infty) = P_1(\infty) = 0; P_2(\infty) = 1
$$

we may reduce the equation to

$$
\begin{vmatrix} -1 \\ 0 \\ 1 \end{vmatrix} = \begin{vmatrix} -2\lambda & \mu & 0 \\ 2\lambda & -(\lambda+\mu) & 0 \\ 0 & \lambda & 0 \end{vmatrix} \begin{vmatrix} T_0 \\ T_1 \\ T_2 \end{vmatrix}
$$

or

$$
-1 = -2\lambda T_0 + \mu T_1
$$

$$
0 = 2\lambda T_0 - (\lambda + \mu)T_1
$$

$$
1 = \lambda T_1
$$

Solving this set of equations

$$
T_0 = \frac{\lambda + \mu}{2\lambda^2} \text{ and } T_1 \frac{1}{\lambda}
$$

so that

$$
\theta_S = T_0 + T_1 = \frac{1}{\lambda} + \frac{\lambda + \mu}{2\lambda^2} = \frac{3\lambda + \mu}{2\lambda^2}
$$

that is,

$$
\theta_S = \frac{3\lambda + \mu}{2\lambda^2}
$$

Hence $\lambda_S = \dfrac{2\lambda^2}{3\lambda + \mu} = 2\lambda^2$ MDT (see Chapter 8, Section 8.1.2, Table 8.1).

Index

 is a straightforward and fast Fault Tree Analysis package which can be assimilated in one hour. It is used on TECHNIS Reliability Workshops and caters for the majority of fault tree sizes encountered in practice.

It provides:

* A SIMPLE AND LOGICAL TREE INPUT METHOD WHICH ENABLES TREE LOGIC TO BE VERIFIED

* CLEAR OUTPUTS PROVIDING CUT SET RANKING BY UNAVAILABILITY AND BY FAILURE RATE (SIX MEASURES OF CUTSET AND EVENT IMPORTANCE)

* CLEAR GRAPHICS WHICH CAN BE EASILY IMPORTED INTO A WORD PROCESSING PACKAGE (GATE SUPRESSION FACILITY)

* THE TECHNIS BETA+ METHOD FOR TAKING ACCOUNT OF COMMON CAUSE FAILURE

* FACILITIES FOR ONE SHOT PROBABILITY SUCH AS HUMAN ERROR

AVAILABLE FROM

TECHNIS

26 ORCHARD DRIVE,
TONBRIDGE,
KENT TN10 4LG
01732 352532
technis.djs@virgin.net

£675

BETAPLUS

BETAPLUS is a Common Cause Failure (partial β model) assessment technique developed by David J Smith (Technis). It develops the partial BETA model beyond the currently available theories to include:

* RECOGNISING THAT PROOF TEST AND AUTO-TEST INTERVALS INFLUENCE THE VALUE OF BETA

* A POSITIVE CALIBRATION OF THE MODEL USING COMMON MODE/CAUSE FAILURE DATA, ANALYSED BY THE AUTHOR

* POSITIVE SCORING OF BETA RELATED SUB-FACTORS RATHER THAN SUBJECTIVE ASSESSMENT OF A VARIABLE FOR EACH GROUP OF FACTORS

* THE FACILITY OF TESTING THE EFFECT, ON BETA, OF ALTERNATIVE DESIGN PROPOSALS AND MODIFICATIONS

* THE OPPORTUNITY TO ADD SCORING CRITERIA AND (VIA TECHNIS) TO RE-CALIBRATE AGAINST NEW COMMON MODE/CAUSE FAILURE DATA

* THE OPPORTUNITY (VIA TECHNIS) TO ALTER THE WEIGHTINGS OF EACH GROUP OF FACTORS

AVAILABLE FROM:

26 ORCHARD DRIVE,
TONBRIDGE,
KENT TN10 4LG

01732 352532   technis.djs@virgin.net

£125

Printed in the United States
By Bookmasters